ECOLOGICAL
SIMULATION
PRIMER

BIOLOGICAL RESOURCE MANAGEMENT

A Series of Primers on the Conservation and Exploitation of Natural and Cultivated Ecosystems

Wayne M. Getz, Series Editor
University of California, Berkeley

Adaptive Management of Renewable Resources, by Carl Walters
Building Models for Wildlife Management, by Anthony Starfield and A. L. Bleloch
Mathematical Programming for Economic Analysis in Agriculture, by Peter B. R. Hazell and Roger D. Norton
Range Economics, by John P. Workman
Forest Management and Economics: A Primer in Quantitative Methods, by Joseph Buongiorna and J. Keith Gilless
Ecological Simulation Primer, by Gordon L. Swartzman and Stephen P. Kaluzny

ECOLOGICAL SIMULATION PRIMER

GORDON L. SWARTZMAN

STEPHEN P. KALUZNY

MACMILLAN PUBLISHING COMPANY
NEW YORK

Collier Macmillan Publishers
LONDON

To Peg and Susan
and
In memory of our mothers
Sylvia and Barbara

Macmillan Publishing Company
866 Third Avenue, New York, NY 10022

Collier Macmillan Canada, Inc.

Printed in the United States of America

printing number   year
1 2 3 4 5 6 7 8 9 10   7 8 9 0 1 2 3 4 5 6

Library of Congress Cataloging-in-Publication Data

Swartzman, G. L. (Gordon L.)
  Ecological simulation primer.

  (Biological resource management)
  Bibliography: p.
  Includes index.
  1. Ecology—Mathematical models. 2. Ecology—
Data processing. I. Kaluzny, S. P. II. Title.
III. Series.
QH541.15.M3S89   1987   574.5′0724   87–5802
ISBN 0–02–947960–6

Contents

Preface

Our *Ecological Simulation Primer* has been written in response to what we perceive as a void—the lack of a text devoted to ecological simulation modeling that covers the basics of model development, techniques for model analysis, and methods for model evaluation as well as giving many examples over a wide range of ecosystems and processes.

The text is designed for a senior or graduate level course to be taught over two quarters. The breadth of scale and the detail in many of the methods as well as the mathematical rigor of the book will make it useful as a reference for scientists already doing ecological simulation. The book is organized to be tutorial and self-contained so that, if necessary, it could be used for individual study. However, the best milieu for this book is group study. In the class taught by the senior author students have worked in teams of three or four. This is good preparation for working in interdisciplinary teams, which is commonly done in ecosystem simulation work. Often, ecologists studying different aspects of a problem work with mathematically oriented persons who generally assume computer implementation and analysis duties. This approach leads to a cross fertilization of ideas and meets a frequent need for definition and explanation at a basic level. Almost all team members have benefited greatly from the interaction.

Two issues that were the most difficult to address were what is the proper mathematical level for this text and what notation scheme should we use. Regarding mathematical background we felt it important that a mathematics background through at least calculus and basic statistics would be mandatory. To try to cover this material would make the book too weighty and would not serve the major audience to whom the text is directed. On the other hand, jumping into matrices and eigenanalysis without providing some background review seemed to put the book beyond the reach of a majority of the students we have encountered in this course. To allow more mathematically advanced students to proceed without interruption we provided reviews of these subjects in appendices. The same applied to some of the more advanced statistical techniques discussed, such as quantile–quantile plots. We develop the derivations of some of the techniques used, such as the Euler technique, to give a grounding in the basis of simulation and insight into the sources of error inherent in the technique. Some computer programming background seemed essential. We did not think that this necessarily be in Fortran—our language of choice for the book. To best allow multiple computer backgrounds we gave many examples in pseudocode, a structured, flow oriented form of text that parallels the actual code while not being computer language specific. We also provided the code for the major models presented so that the specifics of coding ecological simulation models could be easily mastered.

The notation was a story we will return to in the text when we give a justification for our eventual choice. This turned out to be a mnemonic notation, which we have found very useful for past model communication. We have had more discussion about this notation than any other aspect of the book and actually changed it once during our writing. The basic problems are that no standard notation exists for these kinds of models and that we are covering a wide variety of ecological systems. We chose a notation scheme that blended the desirable model feature of identifying types of variables by a specific letter and the ecological feature of having different kinds of variables (e.g., photosynthesis, weight, light) denoted by specific mnemonics (e.g., *lt* for light, *pht* for photosynthesis, *wt* for weight). The notation scheme is built up of primitives that identify the process and further specify that process (or variable) through other, adjective-like, primitives.

Writing the exercises proved a special challenge to us. We wanted them to develop concepts and skills while being based on examples from the literature or at least to be realistic. This involved extra work to find proper examples and assure that the solutions are tractable. Real life problems often do not have neat solutions. Some exercises replicate methods shown in the text with a slight modification. Most exercises, however, are used to introduce new concepts, especially those anticipating topics in later chapters but in a simplified framework. Several of the exercises, especially those in the later chapters, can be research projects. Many require computer work, either using programs we have provided or software packages to accomplish some (usually statistical) procedure. We have provided answers to those exercises that have short answers.

Ecological simulation is in its infancy. It is a forward looking tool in that it represents the dynamics of ecosystems as a collection of hypotheses. As such, it is ahead of its time for ecology, where experiments are designed to keep most environmental conditions constant to be able to understand one aspect of a process. With simulation the interactions between processes are important. We have not arrived in ecology at the type of experiments that focus on interactions and allow many environmental variables to vary in unison as they do in real systems. The dependence of ecological simulation on this one-step-at-a-time experimental evidence has hampered its predictive ability.

Despite its limitations ecological simulation is the only tool for examining dynamic interactions between parts of ecosystems. It is also an excellent aid for studying individual processes. Ecological simulation models are too complex and "realistic" to generalize to multiple sites let alone multiple ecosystems. We believe that the tools developed in doing ecological simulation will be applicable when their time of real need comes. Let up hope that environmental problems can command the research emphasis they deserve and that ecological simulation plays an important part in helping us to understand how to achieve some semblance of balance that we, in our environmental depredations, have gone a long way toward disrupting.

We would like the tools presented here, that are taught in step by step fashion, to prove useful in the future quest as well as in the present training of mathematical ecologists. A course in simulation and in use of computers are future prerequisites of ecological research. Ecological problems are intriguing and will, we hope, come to be recognized as of ultimate importance to the survival of mankind. We hope that this book can attract good minds into the field by providing intellectual stimulation and an appreciation of the great potential of this burgeoning field.

The writing of this book was originated through encouragement by Lev Ginzburg, although it was spawned through a long incubation period of teaching. We wish to thank above all Wayne Getz, who kept us to the task as series editor. He and Dick Park provided extremely thoughtful and thorough critical reviews. We appreciate the many colleagues and friends who took time to review selected chapters. These include Virginia Dale, Simon Levin, Kenneth Rose, Patrick Sullivan, Chuck Fowler, Steve Running, Susan Kaluzny, Jeanne Simpson, Daniel McKenzie, and Frieda Taub. Most illustrations were drawn by Nancy Williams-Nelson who carefully worked long hours to meet our deadlines. We appreciate David Ford, who supported our effort with encouragement and clerical resources. Our students in QS450 did all the exercises in the first half of this book, found mistakes in the text, some of which we would have been embarassed to appear, and helped sort the wheat from the chaff in the exercises.

Writing a book of this sort is time consuming. We appreciate all the encouragement and support of our wives Peg and Susan who had to put up with seeing less of us than they would have liked through our time of book writing. Finally, we want to acknowledge the contributions of the many researchers in simulation modeling. Without their clear communication of their results much of this book could not have been written.

Gordon Swartzman and Stephen Kaluzny
Seattle, Washington

Chapter 1

Simulation Modeling Overview

1.1 Introduction

Ecological simulation is part of the life blood of our growing use of technology to help in the understanding and management of ecosystems. A scientist studying medusae feeding on fish larvae, a judge in litigation involving the effect of power plants on striped bass populations in the Hudson River, a toxicologist in an Environmental Protection Agency laboratory, and a program planner in charge of land use decisions for the U.S. Forest Service have all been involved with understanding, using, or interpreting the results of ecological simulation models. These and many others throughout the country and the world today have experienced the effects that computers have had in aiding the study of ecological processes and of ecosystems. Simulation models are the only tool presently available for translating a collection of hypotheses for ecological processes into a representation of how the whole ecosystem functions. Furthermore, they depict ecosystem function by changes over time or space (or both) in measurable quantities, thereby allowing some test of the set of process hypotheses at the ecosystem level.

The first simulation models for ecosystems were developed in the mid-1940s by Gordon Riley for marine ecosystem primary and secondary production. Oddly, these models were developed before the advent of the digital computer and required laborious calculations. Present day ecosystem simulation requires use of a digital computer and simulation is now inseparable from the digital computer. As such, much emphasis is placed on proper and effective use of this primary tool for simulation modeling.

This text is designed to introduce readers to the applications and future potential of simulation models in ecological study. Upon completing this text you should be able to design and run a simulation model and be able to read, comprehend, and evaluate modeling articles in the literature.

Mathematical modeling is a tool that can be looked at as analogous to translation of languages. Like translation there are rules for how to go from one

1

medium (for ecological models these are data and experiments) to another (a mathematical framework). Like translation, there is a technical side analogous to the vocabulary, syntax, and grammar of the language. Like translation there is also a creative or artistic side in how to express the material to give it the flavor of the original and to read well in translation. This is less easily taught and comes both with experience and through example. Based on this analogy we speak of the art and science of simulation modeling.

Although simulation modeling is built on a collection of tools and methods that can be taught in a discursive or lecture style, the practice of building simulation models requires experience gained only with hands-on interaction with the computer and with the data that form the backbone of all models. Also, an understanding of ecological modeling demands familiarity with the wide variety of applications that must be built on a case study basis. The believability of model formulations rests on the data and experiments used to generate the model.

We have designed this text using a spiral approach originally developed to teach computer languages, where the tools are introduced using simplified examples to develop the basic working skill and then the applications are expanded through examples of increasing complexity. Each section is introduced with an overview, followed by more specific development of the tool or example at hand. Computer code is included for illustrative purposes and to ease the pain of getting a model running.

1.2   What are simulation models?

Simulation models, the models we will primarily be considering in this book, consist of a collection of hypotheses, in equation form, for how the major elements of the model (*state variables*) change over time. These hypotheses are usually categorized into several processes controlling the rates of change of the state variables. Such processes include photosynthesis and respiration for plants, consumption, respiration, "natural," and predation mortality for animals, and evaporation, transpiration, translocation, and uptake for water flow in plants, to give a few examples. The models of primary interest here are *mechanistic* in that we include, to the best of our ability, mathematical descriptions of the mechanisms that control the various model processes. This is opposed to *empirical* models where a relationship between the process and controlling variables is established without considering the underlying mechanisms.

The choice of processes to consider depends on the major variables in the model and on the degree of detail desired to represent the changes in these variables. These depend in turn on model objectives and on the ecosystem under study. The process rates are combined into a system of *differential equations*. These equations give the rate of change of the state variables as a function of their own condition and of other state variables and environmental conditions, called *driving variables* in modeling jargon. These differential equations are generally nonlinear and are usually too complex to be solved analytically, thus necessitating numerical solution techniques. These techniques solve a continuous time problem by a discrete *time step* approximation. Because the number of computations are enormous, computers are a necessary component of these models. In this book numerical solution techniques are included, along with computer code for implementing them.

The simplest type of model assumption is that the rate of flow of material between state variables or compartments depends only on the value of the variable from which the flow is coming (the *donor* compartment). Such a model is called a first order ordinary differential equation. This distinguishes it from higher order differential equations with second, third, and higher derivatives, and from partial differential equations where derivatives are taken with respect to more than one variable in the same equation, usually time and space. The simple model is also called a donor controlled model because rates of flow depend only on the variable initiating the flow, the donor. Use of many simulation modeling tools can be illustrated with these simple donor controlled models and, therefore, we begin with them for our examples (Chaps. 2 and 3). By the time we get to the more realistic and more complicated nonlinear models you should feel confident of the tools and be ready to focus on some of the more creative and artistic aspects of modeling.

1.3 Preliminary ecosystem description

Preliminary to any ecosystem modeling effort must come a description of the problem and the ecosystem to be modeled. Because the important features of this description are usually subsumed in the resultant model the verbal description is rarely presented. It is, however, important in model formulation. We give an example of a description of the Point Barrow, Alaska ecosystem (Miller et al., 1975).

The Point Barrow ecosystem is simple relative to many other ecosystems. There are about 100 vascular plant species dominated by only 10 species. The major herbivore is the brown lemming whose diet consists largely of three plant species. Herbivorous insects are absent. Predators are abundant and are dominated by the snowy owl and the pomarine jaeger.

The dominant abiotic feature is a 40-cm to over 300-m permafrost. This impedes water infiltration, causing the thawed soils in summer to be water saturated in most areas. The permafrost causes the major topographical features of this tundra area including polygons of ridges and troughs, mass erosion and soil upheaval by ice lensing making plant recolonization very slow. The permafrost also impedes root growth and lemming burrowing.

Lakes and old lake beds play a major role in this ecosystem. Plant succession is connected with lake succession. The lakes elongate in a direction perpendicular to the prevailing wind, then coalesce and spread. Eventually drainage canals form that empty the lakes and leave low flat meadows to be recolonized by vegetation.

Lemming densities are highly variable from year to year and a number of hypotheses have been developed to explain this variability. One important hypothesis emphasizes the interaction between the lemmings and their food supply. Heavy grazing by lemmings decreases the insulating capacity of the vegetation leading to deeper thaw in summer. This in turn releases more nutrients for summer plant growth. Improved herbage quality then leads to even greater lemming growth, followed by a crash as food supplies are depleted. Plant recovery reduces nutrients and makes lemming forage quality poor. These destabilizing factors could lead to wide swings in lemming populations.

The Point Barrow ecosystem description suggests an emphasis in the modeling on lemming population dynamics and on vegetation–soil processes interaction. This was the direction taken by the U.S. International Biological Program's Tundra Biome project investigators, who conducted the Point Barrow study. The physical dimensions or boundaries of the ecosystem depended in this case on the region over which data collection occurred. The region has a significant amount of topographical variability, including many small ponds, all of which were lumped into a single, representative ecosystem. The lemming population and relevant plant species were treated as single populations having average growth, feeding (for the lemmings), and mortality rates. The model time scale, 3 to 5 years, was suggested by the question at hand; the cycling of the lemming population at Point Barrow.

1.4   Model applications

Table 1.1 gives an overview of several types of simulation models. The table is organized by type of ecosystem (biome) and also by the kind of process modeled. For example, in the grassland biome there are models of ecological succession of grasslands after abandonment by agriculture, models of natural prairies, models of grazer energetics, and models of radionuclide flow in grazers. Each type of model is classified according to how many trophic levels are treated. These include *abiotic* (nonbiological) elements of the model and *primary producers* (plants), *primary consumers* (grazers or herbivores), *secondary consumers* (carnivores), *tertiary consumers* (carnivore eaters and carnivores themselves), and *decomposers* (e.g., bacteria). The flows column in Table 1.1 shows the major elements defining the units of the model variables. For example, if the major flows are in weight units (e.g., grams), then the model is expected to describe the change in weight over time of the major (state) variables. The management applications column shows what major questions are addressed by the model. The time step/run time column gives the time units used in the model and the length of the time period the model runs. These give a good indication of the natural time scale over which the processes considered are operating and for how long a time the system must be sampled to address the major questions.

As you can see from Table 1.1 many models have been built to examine natural resource management questions. These include such questions as how to best manage ecosystems (e.g., managing of a big game animal harvest) and the effect of human development (e.g., environmental impact) on the biota in the ecosystem. Other types of management questions involve the search for management alternatives that will help to maintain stability in the ecosystem. An example of this is developing models to examine the effect of various pest control scenarios in a forest, such as pesticide spraying versus control by tree thinning. Because humans have become involved in the dual role of managing and protecting ecosystems we must understand complex ecological relationships such as the host–parasite relationship. Our own activities in the past have neglected these interactions too long and have, therefore, produced many crisis situations. A good example is in forestry where the policy of preventing forest fires has created more crowded tree conditions. These make the trees more susceptible to insect epidemics and more prone to rapidly spreading and often uncontrollable fires than would occur in natural unprotected stands.

Models have also been built to increase our understanding of a particular ecological process such as photosynthesis. Models of this sort are frequently detailed in their representation of the process at hand. Scientists hope, by representing all the details known about the physiological aspects of photosynthesis, to have the model represent this process over a wide range of environmental conditions. This enables them to conduct experiments with the model that are too expensive or difficult to conduct in the real system. Also, they hope the model will reveal interactions between the physiological aspects that are not apparent from studying them separately. Other models of this sort have been built to work in conjunction with controlled animal feeding experiments or with miniecosystems in jars (called *microcosms*) holding small organisms to enable better understanding of how the processes controlling the dynamics in these small systems work. Here the models serve as a tool for bridging the gap between experiments and the real system. Models can also be used to help explain a given phenomenon, such as the population cycles of rodents in the tundra. Here the implications of various process hypotheses to population cycling are explored using *simulation experiments*.

Modeling is often used for educational purposes (you will be seeing some of those as examples here). Other applications are to help guide a field research effort by pinpointing where further data and information are needed, to suggest hypotheses about which factors most strongly affect behavior in an ecosystem, or about how the ecosystem might change with the change in some environmental factor (such as increasing temperature from a power plant—thermal loading) or biotic factor (such as introducing a new fish species into a lake), and to organize available data about an ecosystem or process into a whole systems framework.

The time step and time scale of each model suggests the appropriate sampling interval to check the models and the appropriate time scale for experiments on model processes used to construct the model. Models with long time scales depend on long historical records for corroboration. Models with short time steps, such as some photosynthesis models, require almost continuous recording of the variables of interest (e.g., light, CO_2 uptake, temperature) to get information to build the model.

Table 1.1 Overview of types of ecological simulation models with application areas

Biome	Type of System or Model	Trophic Levels	Flows	Management Applications	Time Step/ Run Time
Grassland	Succession	1	Biomass	Effect of grazing	Season/ 10–30 yr
	Ecosystem	4	Biomass, water, population, nutrients	Effect of grazing, irrigation, fertilization	Day/1–5 yr
	Animal energetics	1–2	Weight	Animal production. Water placement in arid lands	Day/yr
	Chemical	1	Chemical concentration	Toxicant fate, nutrient fate	Day/yr
Forest	Ecosystem/stand	1	Water, biomass	Defoliation, toxicant effects	Day/yr
	Succession	1	Population	Reseeding	Yr/100 yr
	Photosynthesis	1	Carbon		Minute/days
	Forest insect	1–2	Population	Pest control	Day–month/yr
	Watershed hydrology	1	Water	Watershed management. Clearcut effect. Snow accumulation	Day/yr
	Nutrient	1	Nutrient concentration	Toxicant fate	Day/yr
Desert	Grazer energetics	1	Population weight	Game management	Day–week/yr

Aquatic	Fishery	2–4	Population, weight	Multispecies fish management	Day/yr
	Phytoplankton and nutrient	1/2	Biomass. Nutrient concentration	Effect of eutrophication. Thermal loading	Day/yr
Aquatic	Whole ecosystem	4–5	Biomass, Weight. Numbers	Effect of dam changing water level. Species introduction	Day/yr
	Feeding process	1	Numbers, weight		Minute–day/month–yr
	Stream periphyton	2	Biomass	Importance of detritus, Macrophytes	Day/yr
	Nutrient cycling	1	Nutrient concentration	Eutrophication. Pesticide or PCB bioaccumulation	Day/yr
	Fish growth	1	Weight	Aquaculture	Day/yr
	Marine plankton	2	Biomass	Thermal loading	Hour–day/1–10 yr
	Salt marsh	1	Biomass	Impoundment	Day/yr
Tundra	Rodent population dynamics	1–2	Numbers. Weight	Find cause for cycling	Month/100 yr
	Lichen–herbivore interaction	2	Biomass. Numbers. Weight	Management of game harvest	Month/10 yr
Tropical	Marsh	1	Biomass	Fire effects	Day/yr
	Nutrient cycling	1	Concentration	Radionuclide cycling	Day/yr

Models exist at many degrees of detail and complexity. The best model is not necessarily the one with the most complexity or detail. In fact, simplicity and parsimony of both variables and process descriptions are desirable because with fewer processes it is easier to understand how specific processes influence model behavior and because less data are required to estimate the parameters of the process equations.

1.5   Computer use in simulation

Simulation modeling, like all other branches of science, depends on the use of specific tools. The simulation aspect of these models requires the use of computers, usually digital computers. The bookkeeping of simulation is greatly facilitated by computer languages such as Fortran, Pascal, or C, which help us to organize our equations into manageable blocks and to efficiently perform the manipulations necessary to these models. In this text we use Fortran 77 because it is the most commonly used language for scientific computing.

In addition to the general computer programming languages several simulation languages are applicable to ecological simulation. These come under a variety of labels such as DYNAMO, SIMCOMP, GPSS, and Simula. These simulation languages usually include an event scheduler for scheduling all blocks of computations as well as convenient input–output for the model and also a flexible interactive sampling scheme (for sampling model output much as a natural ecosystem is sampled).

1.6   Text overview

This text assumes that the student has completed at least an introductory course in computer programming and at least introductory calculus and basic statistics including analysis of variance and linear regression. These are reviewed briefly as needed in appendices. Some suggested review texts are given in the annotated bibliography at the end of this chapter. Although the student need not have formal ecological training, a willingness to refer to supporting ecological papers will be invaluable for someone not in the ecological discipline and many supporting references with brief annotations are given at the end of each chapter.

We begin (Chaps. 2 and 3) with the simplest form of simulation models—linear differential equation models—using matrices to present the tools and techniques including direct solution and numerical approximation methods. Examples are interspersed with sections on computer coding methods.

The following chapters (4–6) on nonlinear simulation models introduce the steps in model development for "realistic" ecological models and a notation scheme and methods for model documentation and coding. Examples span the wide breadth of ecosystems and processes represented by simulation models. Coding methods are expanded to include the more complicated process hypotheses of nonlinear models. Two full, though simple, example ecological models are developed and used to consider how to run nonlinear models and to use available data effectively. Stochastic methods, which allow real world variability to enter models, are treated in Chap. 7 and model evaluation and analysis methods are discussed in Chap. 8.

Chapter 9 considers applications of models in ecosystem management using case studies for explication. Finally, in Chap. 10 we consider applications involving spatial aspects of ecosystems and optimization models. Our treatment of these topics is introductory and cursory and gives only a flavor of the ongoing work in these burgeoning fields.

The book is organized to be a self-contained unit with chapters oriented toward building skills through examples of increasing complexity. This style of presentation parallels the development of ecological modeling as a discipline, which began with linear, donor controlled models and advanced through simple nonlinear models into the complex collections of hypotheses and tools for analyzing and evaluating model behavior available today. Examples designed to introduce refinements and provoke thought are developed in exercise sections at the end of the chapters. These exercises are preferably done working in small groups, so that the interactive skills necessary for successful ecological simulation modeling can be developed. These groups, which have proven so effective in one of the author's (Swartzman) ecological simulation course, both for promoting cooperation and for letting the people of differing skills learn from each other, reminds us of the story of Nesreddin Hodja, a Muslim scholar of the mid-thirteenth century in Jerusalem who, it is told, was asked to speak in front of a Christian congregation. Fearful that his beliefs would be seen as heretical to the Christians, Nesreddin asked "Does anyone know what I am to talk about today?" "No, the congregation members answered." "Well, in that case there is no use in my speaking," he replied. Asked to return again, but no more certain of a warm reception, Nesreddin asked the same question. This time he was assured that everyone knew what he was going to talk about. "Well, since you all know, there is no use my saying it." When the third week came along and he had been invited back again, he was indeed in a dither. But his wisdom did not desert him. He rejoindered with the same question. This time the congregation conferred and replied, "Well, Nesreddin, some of us know and some of us don't know." "In that case, let those who know tell those who don't know." In this way Nesreddin saved his skin. And the moral of the story is that interactive groups work because those who know can tell those who do not know. And if no one knows and you cannot find out from the text you may be helped by the additional readings at the end of each chapter.

1.7 Annotated Bibliography

Arya, J. C., and R. W. Lardner: *Mathematics for the Biological Sciences,* Prentice Hall, Englewood Cliffs, N.J., 1979.
  Could be used as a background mathematics review book for this text.
Clow, D. J., and N. S. Urquhart: *Mathematics in Biology: Calculus and Related Topics,* W. W. Norton & Co., New York, 1974.
  Gives a review of many mathematical topics needed to understand parts of this text

including integration and differentiation, probability and probability distributions, series and limits, and transformations.

DeWit, C. T., and J. Goudriaan: *Simulation of Ecological Processes,* Centre for Agricultural Publishing and Documentation, Wageningen, Netherlands, 1974.
This book emphasizes applications of ecological modeling to agricultural problems. Code and problems are included. The code is written in the simulation language DYNAMO, which makes it limited in scope.

Gold, H. J.: *Mathematical Modeling of Biological Systems—An Introductory Guidebook,* Wiley, New York, 1977.
An introductory book on simple mathematical models in biology.

Grant, W. E.: *Systems Analysis and Simulation in Wildlife and Fisheries Science,* Wiley, New York, 1986.
A modeling textbook oriented mostly toward population models, but it also treats some processes. This book emphasizes the steps in model development and evaluation with applications in fishery and wildlife management. Code in BASIC is included for several of the models and exercises are given with answers in the back.

Hall, C. A. S., and J. W. Day, Jr.: Systems and models: terms and basic principles, in *Ecosystem Modeling in Theory and Practice: An Introduction with Case Studies,* ed. C. A. S. Hall and J. W. Day Jr., Wiley, New York, 1977a, pp. 5–36.
This article emphasizes the terminology and philosophy of ecological simulation.

Innis, G. S.: A spiral approach to ecosystem simulation, I, in *Systems Analysis of Ecosystems,* ed. G. S. Innis and R. V. O'Neill, Statistical Ecology, 9, International Cooperative Publishing House, Fairland, Maryland, 1979, pp. 211–386.
This three-chapter introduction to ecological simulation emphasizes simple ecological applications and includes computer code to help do the problems, of which there are many.

Jeffers, J. N. R.: *An Introduction to Systems Analysis: With Ecological Applications,* University Park Press, Baltimore, 1978.
The emphasis in this book is on the use of statistical techniques for analysis of ecological data involving one or two species.

Patten, B. C.: A primer for ecological modeling and simulation with analog and digital computers, in *Systems Analysis and Simulation in Ecology, vol. 1,* ed. B. C. Patten, Academic Press, New York, 1971, pp. 3–121.
Written at a time when simulation ecology was young this chapter gives an introduction to the basics of ecological simulation as it was then. At that time analog computers were still commonly used for simulation and these are emphasized.

Shoemaker, C. A.: Mathematical construction of ecological models, in *Ecosystem Modeling in Theory and Practice: An Introduction with Case Studies,* ed. C. A. S. Hall and J. W. Day Jr., Wiley, New York, 1977, pp. 75–114.
This article reviews the principles of modeling. It focuses on simple examples to bring out the modeling process and applications.

Shugart, H. H., and R. V. O'Neill, eds.: *Systems Ecology,* Benchmark Papers in Ecology, 9, Dowden, Hutchinson and Ross, Stroudsburg, Penn., 1979.
This volume presents selected landmark journal articles in systems ecology.

Starfield, A. M., and A. L. Bleloch: *Building Models for Conservation and Wildlife Management,* Macmillan, New York, 1986.
This book provides a modeling perspective on wildlife management. The approach is largely intuitive, with very little required in the way of mathematical background. Many examples are given and the book is easy reading.

Straskraba, M., and A. H. Gnauck: *Freshwater Ecosystems Modelling and Simulation,* Developments in Environmental Modelling, 8, Elsevier, New York, 1985.

Gives quite a thorough review of modeling approaches to freshwater ecosystems with many details about individual processes and tables reviewing relevant models. Documentation is given for several models. The best part of the book is the breadth and depth of the review and the neat pictures at the beginning of each chapter guiding the reader through that chapter. The weaknesses are the notation, which is not standard and is often not defined, the writing style, which is laborious, and the organization, which is not clear. The book is much better as a review of than an introduction to freshwater ecosystem modeling.

Wiegert, R. G.: Simulation models of ecosystems. *Annu. Rev. Ecol. Syst.,* 6: 311–338, 1975.

A good broad-brush review of the scope of models and model building. Emphasizes the modeling process, terminology, and philosophy and reviews application areas.

Chapter 2

Linear Models

2.1 Introduction

In this chapter we get right down to the business of looking at ecological models. We start with the simplest model that is found in the ecological modeling literature—the linear model. Although linear simulation models are not commonly used to represent ecosystems, they have been commonly used in the past. For some applications, namely succession, radionuclide transfer, and global carbon transfer models, the linear differential equation represents the system surprisingly well. We chose to begin with these simpler models primarily because they serve as a good mechanism for introducing many simulation tools—primarily numerical simulation methods. Also, we can solve linear differential equations exactly, thereby providing a means for evaluating the accuracy of the numerical approximation techniques for this case.

After an introduction to the mathematical representation of linear differential equation models and their natural pictorial means of presentation, the compartment diagram, we proceed to derive methods for solving these models both numerically and exactly. Difference equations are also presented and illustrated with the Leslie matrix age structured population model and with the Markov chain—a probability model. Reviews of matrix techniques used to represent and solve these equations are given in Appendices 1 and 2. Examples from the literature are used to tie down the concepts presented and to illustrate the solution techniques.

2.2 Compartment models

Compartment models represent the *rates of change* of material in compartments as the sum of all *flow rates* into the compartment minus the sum of all the flow rates out of the compartment. Each compartment is a state variable and is denoted by x_i. Envision two compartments x_1 and x_2. We denote the flow rate from 1 to 2 by f_{21} and the flow rate from 2 to 1 by f_{12}. These flows are depicted

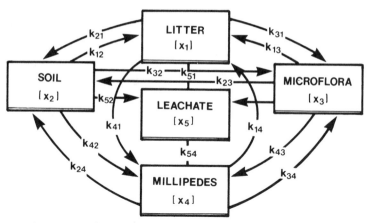

Fig. 2.1.   Compartment diagram for cesium-134 flow through a litter–soil–leachate microcosm. Arrows indicate the possible routes of cesium transfer through the ecosystem. The k_{ij} are the flow rate constants from compartment j to compartment i. From "Systems Analysis of Radiocesium" by B. C. Patten and M. Witkamp, *Ecology*, 1966, *50*, 22–32. Copyright © 1966 by the Ecological Society of America. Reprinted by permission.

(Fig. 2.1) by directed arrows from compartments 2 to 1 (f_{12}) and 1 to 2 (f_{21}). Although this notation appears a reverse of the natural order of things there is a good reason for it, which will become apparent when we write the compartment model in matrix form. The compartment from which the flow is coming is called the *donor* compartment and the one to which the flow goes is called the *recipient* compartment. The system in Fig. 2.1 has all flows beginning and ending within the system. This is termed a *closed system* because material in the system at the start cannot leave the system and no new material is introduced into the system from outside. Most ecosystems do not have this property because they derive energy or material from outside the system and release them outside as well. Such systems are termed *open systems*.

2.3   Donor controlled models

We represent a general differential equation model mathematically as a set of equations, one for each state variable or compartment. Each equation represents the rate of change of a compartment as the sum of all rates of flow into the compartment minus the sum of all flows out. Thus, for each state variable x_i:

$$\dot{x}_i = \frac{dx_i}{dt} = \sum_{j=1}^{n} f_{ij} - \sum_{j=1}^{n} f_{ji} \quad \text{for } i = 1, 2, \ldots, n \text{ and } j \neq i$$

where f_{ij} is the flow from compartment j to i and where j are the states that *communicate with* i (that is have flows to or from compartment i) and n is the total number of variables. Notice that the *dot* notation (e.g., $\dot{x}_i$) is used to denote differentiation with respect to time. The simplest assumption for f_{ij} possible is the linearity assumption $f_{ij} = k_{ij}x_j$. Here the k_{ij}'s are constants called the flow rates per unit of variable x_j and are in units of $time^{-1}$. Unit time can be any time base convenient to the problem at hand (e.g., seconds for gas exchange or chemical processes, days for growth processes, weeks for decomposition, and years for succession). This is called a *linear donor controlled* model because flow rates depend on the magnitude of the donor compartment in a linear fashion. Here:

$$\dot{x}_i = \frac{dx_i}{dt} = \sum_{j \neq i}^{n} k_{ij}x_j - \sum_{j \neq i}^{n} k_{ji}x_i \qquad i = 1, 2, \ldots, n$$

These are a system of linear differential equations and can be written in matrix notation (see Appendix 1 for a review of matrices):

$$\dot{\mathbf{x}}(t) = \mathbf{A}\mathbf{x}(t)$$

where $\mathbf{x}(t)$ and $\dot{\mathbf{x}}(t)$ are vectors representing the state variables and their respective rates of change at any time $t > 0$ ($t = 0$ is the starting or initial time). Here $\mathbf{x}(t)$ is a vector of state variables $x_i(t)$:

$$\mathbf{x}(t) = \begin{bmatrix} x_1(t) \\ x_2(t) \\ \cdot \\ \cdot \\ \cdot \\ x_n(t) \end{bmatrix}$$

and $\mathbf{A}$ is a matrix of rate coefficients k_{ij}, that is:

$$\mathbf{A} = \begin{bmatrix} k_{11} & k_{12} & k_{13} & \cdots & k_{1n} \\ k_{21} & k_{22} & k_{23} & \cdots & k_{2n} \\ \cdot & \cdot & \cdot & & \cdot \\ \cdot & \cdot & \cdot & & \cdot \\ \cdot & \cdot & \cdot & & \cdot \\ k_{n1} & k_{n2} & k_{n3} & \cdots & k_{nn} \end{bmatrix}$$

To have conservation of material, the flow rates out of any compartment must equal the sum of the flow rates from that compartment into each of the others it flows into. Thus, the diagonal matrix element k_{jj} is the negative of the sum of all flow rates out of compartment j. (i.e., $k_{jj} = -\sum_{i \neq j} k_{ij}$). This assures that the coefficients in each column sum to zero.

In general, the flow rates f_{ij} in ecological models will not be linear coefficients, but functions of the state variables x_i. Sometimes these functions can be quite complex. An example is the hyperbolic function used for predator x_j's feeding rate on prey x_i represented as a flow from i to j f_{ji}:

$$f_{ji} = \left[\frac{x_i}{x_i + k_{ij}} \right] x_j$$

where k_{ij} is a *parameter* that controls the shape of the function.

A related model is the *recipient controlled* model in which the flow rate between two compartments depends only on the amount of material in the compartment to which the flow is going, the recipient compartment. The particulars of this model are given in Ex. 2.3.

Donor controlled models have been frequently used to represent ecological systems (Patten and Witkamp, 1967; Bledsoe and VanDyne, 1971; Hett, 1971; O'Neill and Burke, 1971; Patten, 1972; Hett and O'Neill, 1974). Examples include models for grassland and forest land succession after either clearing, clearcut, or abandonment of agricultural land, radionuclide and pesticide flow through an ecosystem, and carbon flow in relatively simple ecosystems such as a salt marsh or a spring-fed pond. The donor controlled assumption is a crude approximation to ecological reality, yet it has been able to fit surprisingly well to data in some applications.

2.4 Compartment diagrams

Figure 2.1 is an example of a compartment diagram showing the compartments (state variables) in boxes and the flows as directed arrows between compartments. Diagrams of this sort arose from diagrams of annual budgets of material in ecosystems where the material was sampled annually (or in some cases seasonally) in different compartments, for example, producers, primary consumers, carnivores, and detritus in a study of Silver Springs in Florida (Odum, 1957), or *Spartina,* insects, spiders, algae, detritus, crabs and worms, and mud crabs in a study of salt marsh ecosystem (Teal, 1962). Flows into the system from outside, for example, addition of radiocesium into litter are denoted by arrows directed toward the recipient compartment but with no originating compartment (Fig. 2.2A). Flows representing losses from the system are denoted by directed arrows away from the compartment sustaining the loss but not directed to any other compartment (Fig. 2.2B).

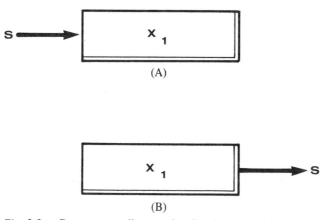

(A)

(B)

Fig. 2.2.   Compartment diagram showing A. a material source and B. a material sink as directed arrows to and from the model x_1 compartment, respectively.

2.4.1   A radionuclide flow model

A good example of a linear donor controlled model examines radiocesium flow through a laboratory controlled leaf litter–soil–millipede–arthropod system presented by Patten and Witkamp (1967). The experiment is illustrated by the compartment diagram in Fig. 2.1. Oak leaf litter of a known radiocesium (Cs-134) dose was defined as the value of the initial litter compartment x_1 at time $t = 0$ (the initial condition). Eight milliliters of distilled water were dropped onto the litter every 2 or 3 days to simulate rainfall. After leaching through the litter, the leachate was collected and sampled for radiocesium dose. The radioactive tracer was followed over a 30-day period as it flowed through various compartments. The terrarium microcosm (a simplified representation of an ecosystem maintained in the laboratory) was studied with different experiments of increasing biotic complexity including: (1) leaf litter and leachate (x_5); (2) leaf litter and soil (x_2); (3) leaf litter, microflora (x_3), and leachate; (4) leaf litter, soil, microflora, and leachate; (5) leaf litter, microflora, millipedes (x_4), and leachate; and (6) all five compartments together. This is a closed system because no material leaves the system.

2.4.2   Old field succession model

A compartment model for old field succession in Oklahoma by Bledsoe and VanDyne (1971) (Fig. 2.3) is somewhat different in that what is flowing in this system is the fraction of the total area in a particular successional stage after abandoment of farmland. The model was based on a study of 30 sites categorized as being in one of five successional stages. The stages are identified with dominant or co-dominant grass species. *Helianthus annuus* (sunflower) and *Digitaria sanguinalis* identify the first stage, *Aristida oligantha* (three-awn grass) the second stage, *Aristada basiramea* (red three-awn) the third stage, *Eragrostis secundiflora* (lovegrass) the fourth stage, and *Andropogon scoparius* (little bluestem) and *Bouteloua curtipendula* (sideoats grama) the final, climax stage.

The flow rate matrix of k_{ij}'s for this model is:

$$\mathbf{A} = \begin{bmatrix} -1 & 0 & 0 & 0 & 0 \\ 0.8 & -1 & 0 & 0 & 0 \\ 0.2 & 1 & -0.7 & 0 & 0 \\ 0 & 0 & 0.7 & -0.8 & 0 \\ 0 & 0 & 0 & 0.8 & 0 \end{bmatrix}$$

Notice that the flow rate out of two of the compartments is -1. This does not mean that there is 100 percent transfer from that compartment in one time step. It is possible to have flow rates greater than 1.0 (or more negative than -1.0) in donor controlled differential equations and still maintain positive values in a compartment. This is because we are considering a rate and not an amount in differential equations. Having a large rate of loss from a compartment causes the compartment to drop toward zero rapidly (unless it is being rapidly

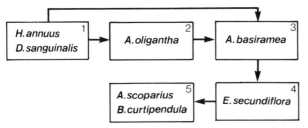

Fig. 2.3.   Compartment diagram for the change of area dominated by various species on abandoned cropland in central Oklahoma.

replenished from some other compartment). The amount going out of the compartment continues to decrease. For example, the differential equation $\dot{x}(t) = -3x(t)$ with an initial value $x(0)$ of 1.0, has the solution $x(t) = e^{-3t}$. At $t = 0$ the rate of loss is -3, at $t = 1$ the value of x is e^{-3} (about 0.05), and at $t = 2$ x is e^{-6} (about 0.0025); small but still positive.

2.5   Linear compartment model solution techniques

To evaluate and study these models we must first consider techniques for solving differential equations. Two types of solutions will be considered: namely approximate numerical solution algorithms and exact solution using *eigenanalysis*. Remember, we are dealing with systems of linear differential equations so that the solutions are not as simple as the solution for a single linear differential equation. For flow matrices **A** larger than 3×3 the computer is desirable for either of these solutions, although in the second case the solution may be written in closed form. For large matrices the exact solution techniques are much less cost effective than the approximate numerical techniques. As we move to nonlinear models in later chapters exact solutions will not be possible and we will rely solely on the numerical approximation techniques introduced here.

2.5.1   Euler technique

There are many techniques available for obtaining approximate numerical solutions to differential equations. The most basic of these is the Euler technique developed by the Swiss mathematician Leonhard Euler in the 1700s. The Euler technique may be used to numerically integrate any system of first order ordinary differential equations represented by:

$$\frac{d\mathbf{x}(t)}{dt} = \dot{\mathbf{x}}(t) = \mathbf{f}(\mathbf{x},t) \tag{2.1}$$

where $\mathbf{x}(t)$ is a vector of state variables, $\dot{\mathbf{x}}(t)$ is a vector of their time derivatives, and $\mathbf{f}(\mathbf{x},t)$ is a vector of functions of the state variables at time t.

The basis for the Euler technique is the Taylor series expansion of the function

$\mathbf{x}(t)$ in powers of time increment Δ and derivatives of $\mathbf{x}$ with respect to t. This is:

$$\mathbf{x}(t + \Delta) = \mathbf{x}(t) + \Delta \frac{d\mathbf{x}}{dt} + \frac{\Delta^2}{2!} \frac{d^2\mathbf{x}}{dt^2} + \cdots + \frac{\Delta^n}{n!} \frac{d^n\mathbf{x}}{dt^n} + \cdots \quad (2.2)$$

Since, from Eq. 2.1 it follows that $\dfrac{d^{n+1}x}{dt^{n+1}} = \dfrac{d^n\mathbf{f}(\mathbf{x},t)}{dt^n}$, Eq. 2.2 becomes:

$$\mathbf{x}(t + \Delta) = \mathbf{x}(t) + \Delta\mathbf{f}(\mathbf{x},t) + \frac{\Delta^2}{2!} \frac{d\mathbf{f}(\mathbf{x},t)}{dt} + \cdots + \frac{\Delta^n}{n!} \frac{d^{n-1}\mathbf{f}(\mathbf{x},t)}{dt^{n-1}} + \cdots$$

The Taylor series expansion gives an approximation to $\mathbf{x}(t + \Delta)$ that approaches the exact value as more terms are included. The Euler technique ignores all but the first two terms of the Taylor expansion and, therefore, is considered a first order approximate solution (because only the terms up to Δ are considered).

$$\mathbf{x}(t + \Delta) = \mathbf{x}(t) + \Delta\mathbf{f}(\mathbf{x},t) \quad (2.3)$$

Notice that the Euler technique requires only a numerical integration time step Δ, the expression $\mathbf{f}(\mathbf{x},t)$, and an initial value of $\mathbf{x}$ at some point in time. For a one dimensional problem the differential equation gives a rule for how the slope of the solution function changes with changing t. For any value of x and t the differential equation tells how rapidly x is changing with respect to t and this specifies a slope at that point. Points of equal slope can be connected to form a slope contour much like a contour map. For example, take the equation $\dot{x}(t) = -x(t)$. Loci of equal slope here are straight lines parallel to the t axis, because the slope of the curve is equal to minus the value of x and all lines parallel to the t axis have the same value of x. The exact solution to the differential equation depends on the slope contours as well as on the *initial conditions*. Any solution will follow a curve such that its rate of change equals the negative of its value at any point. Figure 2.4 shows a family of such curves for our

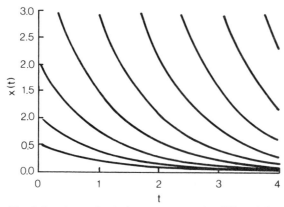

Fig. 2.4.   A set of solution contours to the differential equation $\dot{x} = -x$. Solutions are of the form $x(t) = x(0)^{-t}$, where $x(0)$ is the boundary or initial value of x.

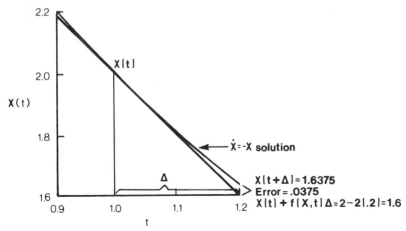

Fig. 2.5.   Diagram showing the Euler approximation solution for the differential equation $\dot{x} = -x$ with $x(1) = 2.0$ and a solution time step Δ of 0.2. The Euler approximation is shown over one time step.

example. Because the slope becomes more negative with increasing values of x, all solution curves slope downward, with the slope becoming less negative (more flat) as time increases. The general solution in this case is $x(t) = x(0)e^{-t}$, with each curve having a different value of $x(0)$.

The Euler technique uses slopes at intervals of length Δ on the time axis to generate an approximate solution to a system of differential equations as in Eq. 2.1. The Euler solution algorithm proceeds as follows: Starting with $\mathbf{x}(0)$ (or a known value at some other time, the initial conditions) choose a step size Δ and compute $\mathbf{f}(\mathbf{x}(0),0)$ from Eq. 2.1 to give $\mathbf{x}(\Delta)$ using Eq. 2.3. Then compute $\mathbf{f}(\mathbf{x}(\Delta),\Delta)$ from Eq. 2.1 and use Eq. 2.3 to compute $\mathbf{x}(2\Delta)$ and so forth until the entire trajectory of $\mathbf{x}(t)$ is generated.

The source of error in the Euler technique comes in neglecting the higher order derivatives of $\mathbf{x}$ in the Taylor expansion. How large is the error? It depends on the function $\mathbf{f}$ and the time increment Δ. Figure 2.5 illustrates the error for our single dimensional example problem $\dot{x} = -x$. Here x is given at time $t = 1$ ($x(1) = 2.0$) and the exact solution is $x(t) = 2e^{1-t}$. The Euler technique projects $x(t)$ to $x(t) + \Delta f(x,t)$, leaving an error ε. This error continues to influence the future solution because the slope at $x(t) + \Delta f(x,t)$ is not the same as that at the true solution $x(t + \Delta)$, which means that the Euler approximation will use an erroneous slope at the next time step (because $f(x,t + \Delta) \neq f((x,t) + \Delta f(x,t),t + \Delta)$). In our example we have $\dot{x}(1) = -2$. The Euler technique, for $\Delta = 0.2$, approximates the solution to be at the point (1.2, 1.6), whereas the exact solution is (1.2, $2e^{-0.2}$), which is (1.2, 1.6375). Clearly the error can be reduced by using a smaller Δ, although this is at the expense of requiring more computations to get the solution. Because the Euler technique requires many computations when Δ is made small enough to give accurate solutions (small enough needs to be determined by testing the solution with different values of Δ), a computer is required.

For linear differential equations $\dot{\mathbf{x}} = \mathbf{Ax}$ the Euler approximate solution may be written:

$$\mathbf{x}(t + \Delta) = \mathbf{x}(t) + \Delta\mathbf{Ax}(t) = [\mathbf{I} + \Delta\mathbf{A}]\mathbf{x}(t) \qquad (2.4)$$

where $\mathbf{I}$ is the identity matrix having 1's on the diagonal and 0's elsewhere. The solution after a general number of time steps n is

$$\mathbf{x}(n\Delta) = [\mathbf{I} + \Delta\mathbf{A}]^n\mathbf{x}(0)$$

2.5.2 Runge–Kutta technique

An alternative numerical approximation to systems of differential equations is due to the German mathematicians Runge and Kutta. It has the advantage over the Euler technique of being significantly more accurate for a given step size Δ, this being done at the expense of more computations per time step. There are several versions of the Runge–Kutta algorithm, each identified as being of a certain order. The order tells the accuracy; a second order Runge–Kutta being accurate to terms of order Δ^2 and a fourth order being accurate to terms of order Δ^4 (as opposed to the Euler, which is accurate to terms of order Δ and is, therefore, a first order algorithm). Being accurate to order Δ^2 means that any terms involving Δ^3 or higher powers of Δ are ignored and are a potential source of error in the approximation. We will derive the equations for the second order Runge–Kutta, but will use the fourth order Runge–Kutta equations in future applications owing to its improved accuracy. (We are not deriving it here because it is complicated and thus difficult to explain.)

All Runge–Kutta methods approximate $\mathbf{x}(t + \Delta)$ by $\mathbf{x}(t)$ plus Δ times some vector function $\mathbf{\Phi}(\mathbf{x},t,\Delta)$. That is:

$$\mathbf{x}(t + \Delta) = \mathbf{x}(t) + \Delta\mathbf{\Phi}(\mathbf{x},t,\Delta) \qquad (2.5)$$

Second order Runge–Kutta technique

The second order Runge–Kutta algorithm supposes $\mathbf{\Phi}$ to equal:

$$\mathbf{\Phi}(\mathbf{x},t,\Delta) = a\mathbf{f}(\mathbf{x},t) + b\mathbf{f}(\mathbf{x} + \alpha\Delta\mathbf{f}(\mathbf{x},t), t + \alpha\Delta) \qquad (2.6)$$

This is equivalent to saying that instead of, as in the Euler technique, moving ahead distance Δ at slope $\mathbf{f}(\mathbf{x},t)$ to estimate $\mathbf{x}(t + \Delta)$ we move at slope $\mathbf{\Phi}$, which is a weighted average of the slope $\mathbf{f}(\mathbf{x},t)$ and the slope obtained from the differential equation at a distance $\alpha\Delta$ from $\mathbf{x}(t)$ moving at slope $\mathbf{f}(\mathbf{x},t)$. Our procedure for setting proper values for the parameters a and b is to equate Eq. 2.6 for $\mathbf{\Phi}$ with the equivalent $\mathbf{\Phi}$ obtained from the first three terms of the Taylor series expansion of $\mathbf{x}(t + \Delta)$ around $\mathbf{x}(t)$ (the second order term is included, making accuracy to the order of Δ^2). To make this comparison we first expand the slope $\mathbf{f}$ of $\mathbf{x}(t) + \alpha\Delta\mathbf{f}(\mathbf{x},t)$ around $\mathbf{f}(\mathbf{x},t)$, the slope at $\mathbf{x}(t)$, in powers of $\alpha\Delta$ using the Taylor series expansion. In the Euler technique we used the Taylor series expansion to approximate $\mathbf{x}(t + \Delta)$ in powers of Δ around $\mathbf{x}(t)$. The

Taylor series expansion can, however, also be used with any function as, for example, $\mathbf{f}(\mathbf{x},t)$. For notational convenience we call the slope of the function at $\mathbf{x}(t) + \alpha\Delta\mathbf{f}(\mathbf{x},t)$ $\mathbf{f}(\mathbf{x},t + \alpha\Delta)$. The Taylor expansion of $\mathbf{f}(\mathbf{x},t + \alpha\Delta)$ in powers of $\alpha\Delta$ around $\mathbf{f}(\mathbf{x},t)$ is:

$$\mathbf{f}(\mathbf{x},t + \alpha\Delta) = \mathbf{f}(\mathbf{x},t) + \alpha\Delta\frac{d\mathbf{f}(\mathbf{x},t)}{dt} + \frac{(\alpha\Delta)^2}{2!}\frac{d^2\mathbf{f}(\mathbf{x},t)}{dt^2} + \cdots \quad (2.7)$$

The above equation is actually a Taylor series expansion in both variables, $\mathbf{x}$ and t, and thus has partial derivative terms in it. The terms shown in Eq. 2.7 are total derivative terms that are obtained by substituting some of the partial derivatives of $\mathbf{f}$ with the original differential Eq. 2.1. These steps are omitted here and are tricky (see the annotated bibliography for references to explain this further). Neglecting terms in $(\alpha\Delta)^2$ and higher and substituting Eq. 2.7 into Eq. 2.6 we get:

$$\Phi(\mathbf{x},t,\Delta) = (a + b)\mathbf{f}(\mathbf{x},t) + b\alpha\Delta\frac{d\mathbf{f}(\mathbf{x},t)}{dt} \quad (2.8)$$

The three parameters a, b, and α in Eq. 2.8, are constrained. That is, their values must be chosen such as to satisfy the Taylor series expansion of $\mathbf{x}(t + \Delta)$ around $\mathbf{x}(t)$ in powers of Δ (Eq. 2.2) as far as the second order term in the expansion. The Taylor expansion up to the second order term is

$$\mathbf{x}(t + \Delta) = \mathbf{x}(t) + \Delta\mathbf{f}(\mathbf{x},t) + \frac{\Delta^2}{2!}\frac{d\mathbf{f}(\mathbf{x},t)}{dt} \quad (2.9)$$

From Eq. 2.9 $\Phi(\mathbf{x},t,\Delta)$ is:

$$\Phi(\mathbf{x},t,\Delta) = \mathbf{f}(\mathbf{x},t) + \frac{\Delta}{2}\frac{d\mathbf{f}(\mathbf{x},t)}{dt} \quad (2.10)$$

Equating terms in Eqs. 2.8 and 2.10 gives $a + b = 1$ and $b\alpha = \frac{1}{2}$. We have some latitude in choosing values of a, b, and α because we have two equations in three unknowns. In general α is chosen equal to 1, which gives $a = b = \frac{1}{2}$. Substituting these values into Eq. 2.6 has the second order Runge–Kutta approximating $\mathbf{x}(t + \Delta)$ by moving a distance Δ with slope

$$\Phi(\mathbf{x},t,\Delta) = \tfrac{1}{2}\mathbf{f}(\mathbf{x},t) + \tfrac{1}{2}\mathbf{f}(\mathbf{x} + \Delta\mathbf{f}(\mathbf{x},t), t + \Delta).$$

The second order Runge–Kutta technique is illustrated in Fig. 2.6 with an example using the *logistic equation*:

$$\dot{x}(t) = 6x(t) - .007(x(t))^2$$

starting with the value $x(0) = 1.0$ and with a time step $\Delta = 0.5$ (much larger than would be chosen in most computer implementations of the method). The exact solution to the equation is given by the solid curve, which, after one time step Δ, has a value at point A of 19.65. The slope at the point $(0.0, 1.0)$ is evaluated and the dashed line going through that point having that slope reaches

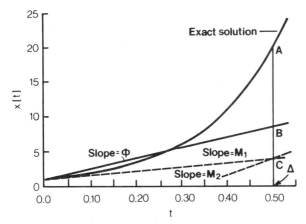

Fig. 2.6.   Diagram showing the second order Runge–Kutta numerical approximation over one solution time step $\Delta = 0.5$ for the logistic differential equation $\dot{x} = 6x - 0.007x^2$. The slope Φ is halfway between the two slopes M_1 and M_2 computed in this method.

$t = 0.5$ at point C, having a value there of 4.00. This would be the approximate solution after one time step from the Euler technique. The solution curve to the logistic equation through this point (0.5, 4.00) has a slope given by the second, shorter, dashed line. The average of these two dashed line slopes is Φ, which is the final slope used by the second order Runge–Kutta approximation. The line beginning at (0.0, 1.0) having this slope is shown with a solid line and reaches $t = 0.5$ at point B, having a value of 8.47. In the general multivariable problem, the slopes Φ are the average of the slopes at $\mathbf{x}(t)$, and the slopes at $\mathbf{x}(t) + \Delta \mathbf{f}(\mathbf{x},t)$.

Fourth order Runge–Kutta technique

The fourth order Runge–Kutta technique requires the slope to be evaluated four times for each iteration but is much more accurate than the second order Runge–Kutta technique, to the order of Δ^4. These four slope computations are combined into a single slope vector Φ given by:

$$\Phi(\mathbf{x},t,\Delta) = \frac{\mathbf{M}_1 + 2\mathbf{M}_2 + 2\mathbf{M}_3 + \mathbf{M}_4}{6}$$

where the four slope vectors $\mathbf{M}_1$–$\mathbf{M}_4$ are given by:

$$\mathbf{M}_1 = \mathbf{f}(\mathbf{x},t)$$

$$\mathbf{M}_2 = \mathbf{f}\left(\mathbf{x} + \frac{\Delta}{2}\,\mathbf{f}(\mathbf{x},t),\ t + \frac{\Delta}{2}\right)$$

$$\mathbf{M}_3 = \mathbf{f}\left(\mathbf{x} + \frac{\Delta}{2}\,\mathbf{M}_2,\ t + \frac{\Delta}{2}\right)$$

$$\mathbf{M}_4 = \mathbf{f}(\mathbf{x} + \Delta\mathbf{M}_3,\ t + \Delta)$$

(2.11)

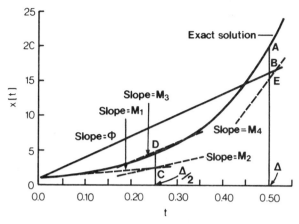

Fig. 2.7.   Diagram showing the fourth order Runge–Kutta numerical approximation over one solution time step $\Delta = 0.5$ for the logistic differential equation $\dot{x} = 6x - 0.007x^2$. The slope Φ is the weighted average of four slopes M_1–M_4 calculated in this algorithm.

These slopes used in the fourth order Runge–Kutta technique are illustrated for the logistic equation in Fig. 2.7. Here M_2 and M_3 are estimated by going a half time step $\Delta/2$ at slopes $f(x,t)$ and M_2 respectively. M_4 is the slope at the point A distance Δ from t moving with slope M_3 from $(t,x(t))$, the starting point. With $\Delta = 0.5$ and $x(0) = 1.0$, we show the four slopes M_1–M_4. The line with the initial slope (that at $(0.0, 1.0)$) intersects $t = 0.25$ at point C, where slope M_2 is evaluated from the original differential equation (remember that the differential equation gives the slope at any point). The line having slope M_2 and going through $(0,1)$ intersects $t = 0.25$ at point D where the slope M_3 is obtained. A line having slope M_3 and going through $(0.0, 1.0)$ intersects $t = 0.5$ at point E, where slope M_4 is calculated. The weighted average slope Φ is shown as a solid line, and the fourth order Runge–Kutta solution after one time step is at point B. This is the point $(0.5, 16.13)$, compared with the true solution of $(0.5, 19.65)$. In general, the Runge–Kutta technique, used with a much smaller time step than that illustrated here, gives significantly more accurate solutions than the Euler technique. Because of the increased number of computations per iteration, however, the Runge–Kutta technique might not always be cost effective compared with the Euler technique. An Euler technique with one-fourth the time step would have about the same number of computations as the fourth order Runge–Kutta technique. In most technical applications the Runge–Kutta is the preferred method.

2.6   Eigenanalysis: exact solutions to linear differential equations

The exact solution to systems of first order linear differential equations, like those generated by the radiocesium and grassland succession models, may be obtained by finding the *latent roots* or *eigenvalues* of the rate matrix **A** as will be shown later. The condition for finding these eigenvalues, denoted by λ_j, is that the determinant of the matrix $(\mathbf{A} - \lambda\mathbf{I})$ is equal to 0. This relationship (i.e., $|\mathbf{A} - \lambda\mathbf{I}| = 0$) is called the *characteristic equation* and results in a polynomial in λ of the same order n as matrix $\mathbf{A}$ (see Appendix 2 for an introduction to eigenanalysis). Each root of the polynomial (there are n roots in total) is an eigenvalue λ_j and each eigenvalue is involved in the solution to the set of differential equations, which is

$$\mathbf{x}(t) = \sum_j c_j \mathbf{v}_j e^{\lambda_j t} \tag{2.12}$$

if all the eigenvalues are distinct. The c_j are constants that adjust the solution to the initial conditions. The $\mathbf{v}_j$ are *eigenvectors* that satisfy the equation:

$$\lambda_j \mathbf{v}_j = \mathbf{A}\mathbf{v}_j \qquad j = 1, \ldots, n \tag{2.13}$$

These are vectors that, when premultiplied by the rate matrix $\mathbf{A}$, give a resultant vector that is a scalar multiple of the original vector. Stated in another way the linear transformation $\mathbf{A}\mathbf{v}_j$ results in no translation or rotation of the eigenvector $\mathbf{v}_j$ but only an elongation or shortening of it. We use Eq. 2.13 to solve for the $\mathbf{v}_j$, there being one vector equation for each λ_j. Each set of linear equations (Eq. 2.13) results in a nonunique $\mathbf{v}_j$ because there is always one redundant equation. This means that the $\mathbf{v}_j$'s are not unique. Any arbitrary value may be chosen for one element of each $\mathbf{v}_j$. In the solution (Eq. 2.12) the initial conditions tie down the specific coefficients c_j that fit the solution. The c_j's are chosen so that $\sum_j c_j \mathbf{v}_j = \mathbf{x}(0)$. Whatever eigenvectors $\mathbf{v}_j$ result from solutions to Eq. 2.13, the c_j's adjust them to the initial or boundary condition $\sum_j c_j \mathbf{v}_j = \mathbf{x}(0)$. Thus, the initial conditions remove the redundancy in the $\mathbf{v}_j$ inherent in Eq. 2.13.

We have presented the solution only for the case where the λ_j are real and where there are no repeated or multiple roots. It is, of course, possible to have imaginary solutions to the characteristic equation. It turns out that the solution (Eq. 2.12) still holds in this case. The complex eigenvalues, if there are any, always occur in conjugate pairs. That is, if there is an eigenvalue $(a + bi)$ there will always be an $(a - bi)$ eigenvalue. This assures that the solution (Eq. 2.12) will always be real and that it will have some oscillatory component because exponentials with imaginary elements are connected by identities to trigonometric functions.

If there are repeated eigenvalues the general solution becomes more complicated, being:

$$\mathbf{x}(t) = \sum_{j=1}^{s} \sum_{k=0}^{r(\lambda_j)-1} e^{\lambda_j t}(\mathbf{A} - \lambda_j\mathbf{I})^k \frac{t^k}{k!} \mathbf{x}_j^0 \tag{2.14}$$

where s are the number of distinct eigenvalues, $r(\lambda_j)$ are the number of repeats of eigenvalue λ_j, and $\mathbf{x}_j^0$ are the eigenvectors $\mathbf{v}_j$ adjusted by multiplying by appropriate c_j's such as to satisfy the initial condition $\mathbf{x}(0)$. Notice that with no repeated roots Eq. 2.14 reduces to Eq. 2.12 because $(\mathbf{A} - \lambda_j\mathbf{I})^0 \, t^0/0! = \mathbf{I}$. For linear compartment models, like we are dealing with here, all eigenvalues will always have negative real parts, except, when the system is closed, there will be one or more zero eigenvalues, which determine the *steady state* conditions (the solution after a long period of time).

2.6.1   Solution to radionuclide model

Let us use the radiocesium flow model of Patten and Witkamp to illustrate the solution method using eigenanalysis. We will consider model (3) where radiocesium in the leaf litter flowed into the microflora and leachate. The flow matrix **A** for this model is:

$$\mathbf{A} = \begin{bmatrix} -0.036 & 0.0 & 0.0 \\ 0.017 & 0.0 & 0.0 \\ 0.019 & 0.0 & 0.0 \end{bmatrix}$$

Here compartment 1 is litter, compartment 3 is microflora, and compartment 5 is the leachate (see Fig. 2.1). The determinant of $\mathbf{A} - \lambda\mathbf{I}$ is:

$$\begin{vmatrix} -0.036 - \lambda & 0.0 & 0.0 \\ 0.017 & -\lambda & 0.0 \\ 0.019 & 0.0 & -\lambda \end{vmatrix}$$

When set equal to zero and expanded, this gives the polynomial characteristic equation $(-0.036 - \lambda)(-\lambda)(-\lambda) = 0$. The solutions are $\lambda_1 = -0.036$, $\lambda_2 = \lambda_3 = 0$. The repeated zero here creates a minor difficulty in setting the initial conditions because we have to use the general formula (Eq. 2.14). First we must solve for the eigenvectors, of which there are two. They are obtained by solving Eq. 2.13 using each eigenvalue, the same equation being used for repeated eigenvalues. For the -0.036 eigenvalue this vector equation is:

$$\begin{bmatrix} -0.036 & 0.0 & 0.0 \\ 0.017 & 0.0 & 0.0 \\ 0.019 & 0.0 & 0.0 \end{bmatrix} \begin{bmatrix} v_{11} \\ v_{12} \\ v_{13} \end{bmatrix} = -0.036 \begin{bmatrix} v_{11} \\ v_{12} \\ v_{13} \end{bmatrix}$$

Here the first subscript refers to the eigenvalue (e.g., the first eigenvalue) and the second refers to the variable—index 1 is for litter, 2 is for microflora, and 3 for leachate. The first equation is an identity and the second and third give a relationship between v_{11} and v_{12} or v_{13} (remember there are an infinite number of eigenvectors for each eigenvalue). We choose an arbitrary value for v_{11} of 1 and thus have $v_{12} = -17/36$ and $v_{13} = -19/36$.

The double zero eigenvalues result in a vector equation:

$$\begin{bmatrix} -0.036 & 0.0 & 0.0 \\ 0.017 & 0.0 & 0.0 \\ 0.019 & 0.0 & 0.0 \end{bmatrix} \begin{bmatrix} v_{21} \\ v_{22} \\ v_{23} \end{bmatrix} = 0.0 \begin{bmatrix} v_{21} \\ v_{22} \\ v_{23} \end{bmatrix}$$

This leads to all three equations implying that $v_{21} = 0$ because only 0 when multiplied by a nonzero constant gives 0. The other two elements of the eigenvector, v_{22} and v_{23}, are not defined by the above equation. This is because we have a double eigenvalue, which should have two independent eigenvectors (see the example in Appendix 2). We can combine the results of these two eigenvectors into a single eigenvector with $v_{22} = 1$ and v_{23} arbitrary.

Substituting the eigenvectors and eigenvalues into Eq. 2.14 (with $\mathbf{x}_j^0 = c_j\mathbf{v}_j$) gives as a general solution the following:

$$\mathbf{x}(t) = \left\{ \begin{bmatrix} 1 & 0 & 0 \\ 0 & 1 & 0 \\ 0 & 0 & 1 \end{bmatrix} \frac{t^0}{0!} + \begin{bmatrix} -0.036 & 0.0 & 0.0 \\ 0.017 & 0.0 & 0.0 \\ 0.019 & 0.0 & 0.0 \end{bmatrix} \frac{t^1}{1!} \right\} \begin{bmatrix} 0 \\ 1 \\ v_{23} \end{bmatrix} c_2 e^0$$

$$+ c_1 \begin{bmatrix} 1 \\ \dfrac{-17}{36} \\ \dfrac{-19}{36} \end{bmatrix} e^{-0.036t}$$

This solution would be complicated except that the second matrix inside the { } brackets when multiplied by the eigenvector gives a vector of zeros, which means that this term drops out of the solution. The coefficients c_1, c_2, and v_{23} can be solved for by setting this equation (simplified by omitting the second matrix inside the { } brackets) equal to the initial conditions (that is at time zero).

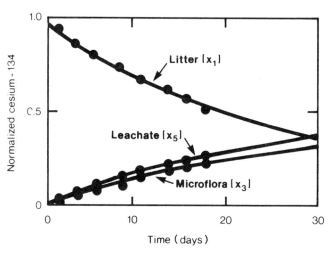

Fig. 2.8. Solution to the linear compartment model for cesium-134 flow in a laboratory microcosm containing litter, leachate, and microflora over a 30-day experiment. The closed circles are the experimental data. From "Systems Analysis of Radiocesium," by B. C. Patten and M. Witkamp, *Ecology*, 1966, *50*, 22–32. Copyright © 1966 by the Ecological Society of America. Reprinted by permission.

Because all the radionuclide was in the leaf litter at the start of this experiment our initial condition is (assuming a radionuclide dose scaled to 1):

$$\mathbf{x}(0) = \begin{bmatrix} 1 \\ 0 \\ 0 \end{bmatrix}$$

This gives us the equation:

$$\begin{bmatrix} 1 \\ 0 \\ 0 \end{bmatrix} = \begin{bmatrix} 1 \\ \dfrac{-17}{36} \\ \dfrac{-19}{36} \end{bmatrix} c_1 e^{-0.0360} + \begin{bmatrix} 0 \\ 1 \\ v_{23} \end{bmatrix} c_2 e^0$$

Solving these equations for c_1, c_2, and v_{23} at $t = 0$ ($e^0 = 1$) gives $c_1 = 1$, $c_2 = 17/36$, and $c_2 v_{23} = 19/36$. The final solution is:

$$\mathbf{x}(t) = \begin{bmatrix} 0 \\ \dfrac{17}{36} \\ \dfrac{19}{36} \end{bmatrix} + \begin{bmatrix} 1 \\ \dfrac{-17}{36} \\ \dfrac{-19}{36} \end{bmatrix} e^{-.036t}$$

This solution is shown graphically in Fig. 2.8, where it is compared with data from the analogous experiment. Notice that the eigenvector associated with the zero eigenvalue controls the steady state or equilibrium solution, whereas the eigenvector associated with the -0.036 eigenvalue is strongly influenced by the initial condition and controls the early behavior of the model.

Observation of Fig. 2.8 shows that this linear, donor controlled model provided an excellent fit to the data. You might be led to expect such fits for most ecological models. Do not be deceived. Three factors lessen the importance of the good fit obtained, although the fit is still remarkable and lends some credence to the donor controlled assumption in this case. First, this is a simple system. Second, and most important, this model was *calibrated* to the data to obtain the best fit. This process (calibration) of adjusting parameter values until the model adequately fits the data is commonly done in ecological modeling. Usually during calibration the parameters that are adjusted are constrained to be within certain limits suggested by the processes controlling the flows (e.g., maximum photosynthesis rate might be constrained to be within the range of values measured for it in controlled laboratory experiments). In the Patten and Witkamp case the parameter values were not so constrained. Finally, the parameters in this model have no biological meaning besides being flow rates. As such they are not amenable to independent measurement.

2.7 Difference equations: Leslie matrices

So far we have looked at differential equation models with time viewed as continuous. Alternatively, time can be looked at as discrete, with behavior of the system monitored only at integer time steps. Such a system is a *difference equation* model. The behavior of the system in the interval between these time steps is not considered in a difference equation model. A system of linear difference equations can be written:

$$\mathbf{x}(t + 1) = \mathbf{B}\mathbf{x}(t) \qquad (2.15)$$

where $\mathbf{x}(t)$ is the vector state of the system at time t and $\mathbf{B}$ is a matrix of the fractions of material that transfer in one time unit between the compartments.

Notice that the rate matrix in linear differential equations $\mathbf{A}$ has been replaced by a transfer fraction matrix $\mathbf{B}$ in difference equations. The Euler technique for approximating the solution to differential equations effectively replaces the differential equation by a linearized difference equation with a time step of Δ. If Δ is 1 in the Euler technique, then the Euler technique and the difference Eq. 2.15 are identical except that the matrices differ by $\mathbf{I}$ (i.e., the difference equation matrix $\mathbf{B}$ is $\mathbf{I}$ plus the differential equation rate matrix $\mathbf{A}$). This is apparent from Eq. 2.4 which, for $\Delta = 1$ and $\mathbf{B} = \mathbf{A} + \mathbf{I}$, is identical to Eq. 2.15.

It is important to distinguish between differential equations, where the rates of change of compartments are considered, and difference equations, where the fraction of material leaving a compartment is considered. If the time step is changed, the rates in a differential equation may be adjusted by simply multiplying by the ratio of the new to old time steps. This is not the case for difference equations where the fractional transfers used are time step dependent. As such, differential equations are more natural when results of experiments at one time step are to be extrapolated to another time step, whereas difference equations are more appropriate when little is known about what goes on during time intervals. For purposes of solution, however, the difference equation may be seen as equivalent to a differential equation solved by the Euler technique with a time step Δ equal to 1.

A special case of the difference equation is commonly used to represent the changes in animal population abundance and age distribution over time. This model was originally developed by Lewis (1942) and Leslie (1945) and has been graced with the name of *Leslie matrices*. The state variables in these models are the female population in each age class. The matrix of concern in this difference equation model is called the *population projection matrix* because it projects the population from one time step to the next. It consists of survival fractions on the lower subdiagonal (the elements just below the upper left to lower right diagonal of the matrix) and numbers of female offspring produced per female at each age in the top row of the matrix. It is assumed in this model that this projection matrix does not change over time.

The Leslie matrix difference equation is written as:

$$
\begin{bmatrix} n_0(t) \\ n_1(t) \\ \cdot \\ \cdot \\ \cdot \\ \cdot \\ n_k(t) \end{bmatrix} = \begin{bmatrix} f_0 & f_1 & f_2 & \cdot & \cdot & f_{k-1} & f_k \\ s_0 & 0 & 0 & \cdot & \cdot & 0 & 0 \\ 0 & s_1 & 0 & \cdot & \cdot & \cdot & \cdot \\ \cdot & 0 & \cdot & \cdot & \cdot & \cdot & \cdot \\ \cdot & \cdot & \cdot & \cdot & 0 & \cdot \\ 0 & 0 & 0 & \cdot & \cdot & s_{k-1} & 0 \end{bmatrix} \begin{bmatrix} n_0(t-1) \\ n_1(t-1) \\ \cdot \\ \cdot \\ \cdot \\ \cdot \\ n_k(t-1) \end{bmatrix}
$$

s_i denotes survival from age i to age $i + 1$, whereas f_i denotes fecundity (females per mature female) for age class i females.

Let $\mathbf{n}(t)$ denote the vector of age classes, $n_i(t)$. Then the above equation can be written in matrix notation as:

$$
\mathbf{n}(t) = \Lambda \mathbf{n}(t - 1) \tag{2.16}
$$

where Λ is the survival–fecundity matrix. The Leslie matrix can be used to project a population $\mathbf{n}(0)$ into the future any number of time steps by successive multiplications of the vectors $\mathbf{n}(t)$ by the population projection matrix Λ. Because the female population carries the reproductive potential for the entire population it alone is modeled and $\mathbf{n}(t)$ represents the age distribution of the female population. This can present a problem in cases where the entire population is of concern, such as when it is harvested, especially if the sex ratio changes with age (which it does naturally in some fish populations or as a result of sex selective harvest as with hunted deer populations). For most cases, however, a constant sex ratio can be assumed (or has been measured) and the female-only analysis can be easily extended to the entire population.

The Leslie matrix, as all other population models, approximates reality, where individual animals die when their time comes and breed either seasonally or at intervals throughout the year. If breeding is on a regular basis the appropriate Leslie model is to have the age class interval the same as the breeding interval. If breeding is irregular then the Leslie model approximates the true breeding pattern by a regular one—a source of inaccuracy. The Leslie model lumps all individuals in the population into age classes and treats the population in each age class as having average survival and fecundity.

Because the model looks at the system at discrete intervals of time some attention must be paid in using survival and fecundity estimates in the projection matrix. Usually the population is represented either just before or just after the breeding time because, for seasonally breeding populations this results in the minimum error in estimating survival and fecundity over discrete time steps. If the population is represented just before breeding then breeding takes place at the beginning of the time step as does recruitment to the next age class. Thus each age class is at the top of its age bracket at the time it is examined. Because of this the fecundity for an age i individual is actually fecundity as it enters the $i + 1$st year. Newborns are actually almost 1-year-old at the time the population is examined. Fecundities here should be adjusted by survival from age 0 to 1.

This involves replacing f_i by $f_i s_0$ and s_i by s_{i+1}. The adjustment in survival is necessary because age i individuals are actually almost age $i + 1$ when they are observed and a year later will be almost age $i + 2$. Thus s_{i+1} is their appropriate survivorship.

If the population is represented just after the breeding time, then breeding takes place at the end of the interval. Here fecundity must include the survival through the previous year. Thus f_i should be replaced by $s_i f_i$ in the first row of the survival–fecundity matrix. This is because not all the females at the start of the time interval breed but only those that survive to the breeding season, which is assumed to occur at the end of the time step. Because the females in age class i are almost age $i + 1$ when they breed, it may be necessary to use f_{i+1} instead of f_i for fecundity of age i females. Leslie assumed that all the individuals died by the end of the kth (oldest) age class. This is represented by having a 0 in the lower right hand corner of Λ. This assumption, however, is not necessary to using the Leslie matrix and an s_k could be placed in the lower right hand element of the diagonal of Λ. This is tantamount to lumping all age classes older than k into the kth age class.

2.7.1   Bird population dynamics

As an example of a Leslie matrix let us look at a particular breeding population of passerine (perching) birds. The maximum age of this population will be taken as 2 years old; they die at the end of their second year of life. The initial female population consists of 100 juveniles (1-year-old or less), 50 1-year-olds and 25 2-year-olds. The species begins breeding in its second year of life with younger adults (1-year-old) producing an average clutch of eight eggs. Two-year-olds produce an average clutch of 12 eggs. All eggs laid are assumed to hatch and the sex ratio at birth, and for all age classes, is 1:1. Survival is 50 percent for each of the age classes. For this analysis we will examine the population just before breeding.

The survival–fecundity matrix for this population (notice that f_i has been replaced by $f_i s_0$ and that the clutch sizes were reduced by 50 percent to account for only half the eggs being female) is:

$$\Lambda = \begin{bmatrix} 0.0 & 2.0 & 3.0 \\ 0.5 & 0.0 & 0.0 \\ 0.0 & 0.5 & 0.0 \end{bmatrix}$$

The initial condition is:

$$\mathbf{n}(0) = \begin{bmatrix} 100 \\ 50 \\ 25 \end{bmatrix}$$

By matrix multiplication we can obtain the solution at any time using Eq. 2.16 recursively starting with $\mathbf{n}(0)$. The population at the end of 1, 2, 3, and 5 years is:

$$\mathbf{n}(1) = \begin{bmatrix} 175 \\ 50 \\ 25 \end{bmatrix}, \ \mathbf{n}(2) = \begin{bmatrix} 175 \\ 87.5 \\ 25 \end{bmatrix}, \ \mathbf{n}(3) = \begin{bmatrix} 250 \\ 87.5 \\ 43.8 \end{bmatrix}, \ \mathbf{n}(5) = \begin{bmatrix} 381.25 \\ 140.63 \\ 62.50 \end{bmatrix}$$

Because the Leslie matrix is a difference equation the exact solution is obtained by matrix multiplication and no approximation technique like the Euler or Runge–Kutta techniques are necessary. Multiplication of the population vector by Λ is termed *population projection* because it projects the population into the future. The solution may also be obtained by eigenanalysis, although the solution is different than the differential equation solution Eq. 2.12 in that instead of having $e^{\lambda_j t}$ we have λ_j^t. We will not use eigenanalysis to solve the Leslie matrix example. We will instead develop the characteristic equation for matrix Λ and solve it for one particular eigenvalue—the largest one—and use this to give some information about the steady state solution to the example. The determinant of the matrix $(\Lambda - \lambda\mathbf{I})$ is first set equal to zero.

$$\begin{vmatrix} -\lambda & 2.0 & 3.0 \\ 0.5 & -\lambda & 0.0 \\ 0.0 & 0.5 & -\lambda \end{vmatrix} = 0$$

The characteristic equation is $-\lambda^3 + \lambda + 0.75 = 0$. There is only one real (noncomplex) positive solution to this equation, which is $\lambda = 1.263$. Leslie matrices have the property that there is only one positive eigenvalue. This eigenvalue, called the *dominant eigenvalue,* indicates what the steady state or intrinsic rate of growth of the population is. If it is larger than 1.0 the population will eventually grow without bound, whereas if it is less than 1.0 the population will eventually approach zero. A population with a dominant eigenvalue of 1.0 approaches a stable equilibrium. Despite the growth or diminution of the total population, the ratios of numbers of individuals in the age classes to each other becomes constant after a long time of population projection (steady state) and is called the *stable age distribution* of the population. This distribution is computed by solving for the dominant eigenvector associated with the dominant eigenvalue. This involves using Eq. 2.13 with the dominant eigenvalue and solving for the dominant eigenvector, which is the stable age distribution. Here the dominant eigenvalue is $\lambda = 1.263$. The equation is:

$$\begin{bmatrix} 0 & 2.0 & 3.0 \\ 0.5 & 0.0 & 0.0 \\ 0.0 & 0.5 & 0.0 \end{bmatrix} \begin{bmatrix} v_{11} \\ v_{12} \\ v_{13} \end{bmatrix} = 1.263 \begin{bmatrix} v_{11} \\ v_{12} \\ v_{13} \end{bmatrix}$$

This gives three equations in three unknowns, although one of them is redundant. The last two equations give $0.5v_{11} = 1.263v_{12}$ and $0.5v_{12} = 1.263v_{13}$. Arbitrarily setting $v_{13} = 1.0$ gives $v_{12} = 2.53$ and $v_{11} = 6.38$. This means that for every 2-year-old individual in the stable age population there will be 2.53 1-year olds and 6.38 juveniles. Furthermore, the model predicts that eventually the earth will be paved with these birds with the population each year being 1.263 times the population for the previous year. We might be interested in how fast a population approaches a stable age distribution in a Leslie matrix model. This depends in part on the matrix and in part on the initial population. In our example the age distribution at the end of 5 years is 6.15 : 2.25 : 1.00, which is close to the stable age distribution. Thus is because the population's initial age distribution was fairly close to the stable age distribution for this example.

2.8 Markov chains

Another type of linear difference equation model is the *Markov chain*. Markov chains are models that describe the changes in the state of a system when there are a discrete number of states and the changes occur at discrete time intervals. They are *stochastic* models in that they give the probability that the system is in each of the states. For example, a Markov chain can describe the probability that it rains on a given day (the states being rain and no rain) or the probability that the level of pollution is either low, moderate, or heavy. The state variables in a Markov chain model are a vector of probabilities $\mathbf{p}(t)$ of being in each of the states at time t, where time is treated as a discrete variable. The assumption is made that this probability vector depends only on the probability vector at the previous time, $t - 1$, $\mathbf{p}(t - 1)$, and the transition probabilities, which give the probabilities of transfers between states (and of remaining in the same state) during one time period conditioned on the starting state. This *probability transition matrix* $\mathbf{T}$ is assumed not to change over time in the Markov chain model. The property of the state depending only on the previous state and the transition matrix is called the *memoryless* property because the future can be projected in a probabilistic sense knowing the present state, without knowledge of the history of the system states.

An example will clarify the concept. Let $\mathbf{p}(t)$ be a vector of the probability of having a dry day, $p_d(t)$ and a wet day $p_w(t)$:

$$\mathbf{p}(t) = \begin{bmatrix} p_d(t) \\ p_w(t) \end{bmatrix}$$

where $p_d(t) + p_w(t) = 1$.

The transition matrix $\mathbf{T}$ then gives the probabilities of transition in one time step (1 day) between dry and wet days. For example, p_{dw} is the probability that tomorrow will be wet given that today is dry, whereas p_{dd} is the probability that a dry day will follow a dry day.

$$\mathbf{T} = \begin{bmatrix} p_{dd} & p_{wd} \\ p_{dw} & p_{ww} \end{bmatrix}$$

Columns of the probability transition matrix must sum to 1. For example, $p_{dd} + p_{dw} = 1.0$. Let us assume that the transition matrix $\mathbf{T}$ for Portland, Oregon (rain capital of the contiguous continental U.S.) in the winter is:

$$\mathbf{T} = \begin{bmatrix} 0.4 & 0.2 \\ 0.6 & 0.8 \end{bmatrix}$$

Then if it is raining today ($\mathbf{p}(0) = \begin{bmatrix} 0 \\ 1 \end{bmatrix}$), the state probability vector for tomorrow is:

$$\mathbf{p}(1) = \begin{bmatrix} 0.4 & 0.2 \\ 0.6 & 0.8 \end{bmatrix} \begin{bmatrix} 0 \\ 1 \end{bmatrix} = \begin{bmatrix} 0.2 \\ 0.8 \end{bmatrix}$$

There is an 80 percent chance that it will be raining tomorrow. If it is dry today then the probability vector for tomorrow is:

$$\mathbf{p}(1) = \begin{bmatrix} 0.4 & 0.2 \\ 0.6 & 0.8 \end{bmatrix} \begin{bmatrix} 1 \\ 0 \end{bmatrix} = \begin{bmatrix} 0.4 \\ 0.6 \end{bmatrix}$$

There is still a 60 percent chance that it will be raining tomorrow. If on Thursday night you are making plans for the weekend and the weather report says that there is a 50 percent chance of rain tomorrow you can figure out what the chances of rain over the weekend is. For Saturday you have:

$$\mathbf{p}(1) = \begin{bmatrix} 0.4 & 0.2 \\ 0.6 & 0.8 \end{bmatrix} \begin{bmatrix} 0.5 \\ 0.5 \end{bmatrix} = \begin{bmatrix} 0.3 \\ 0.7 \end{bmatrix}$$

and for Sunday:

$$\mathbf{p}(2) = \begin{bmatrix} 0.4 & 0.2 \\ 0.6 & 0.8 \end{bmatrix} \begin{bmatrix} 0.3 \\ 0.7 \end{bmatrix} = \begin{bmatrix} 0.26 \\ 0.74 \end{bmatrix}$$

Better plan for rain!

Knowing only $\mathbf{T}$ you can compute the *stationary* state probability vector $\mathbf{s}$. This is the vector of long-term or average probabilities for dry and wet days. It is calculated from the equation:

$$\mathbf{Ts} = \mathbf{s} \tag{2.17}$$

The reasoning behind this equation is that if a stationary vector $\mathbf{s}$ exists, then premultiplying it by $\mathbf{T}$ should result in that same state $\mathbf{s}$. This condition can be used to solve for $\mathbf{s}$ knowing only $\mathbf{T}$. In our rain example we have:

$$\begin{bmatrix} 0.4 & 0.2 \\ 0.6 & 0.8 \end{bmatrix} \begin{bmatrix} s_d \\ s_w \end{bmatrix} = \begin{bmatrix} s_d \\ s_w \end{bmatrix}$$

Although these two equations are redundant (they are not linearly independent) we also know that $s_d + s_w = 1.0$ (the probabilities for all states sum to 1.0). The equations represented in matrix form give $0.2s_s = 0.6s_d$ or $s_w = 3s_d$. Therefore $s_w = 0.75$ and $s_d = 0.25$.

2.9 Summary

We have introduced the major mathematical tools for solving linear donor controlled differential equation models. These include methods for solving these equations numerically (Euler and Runge–Kutta) and analytically (eigenanalysis). Besides giving examples of how to apply each of the techniques we also include the mathematical underpinnings for them. We do this because we believe that applying techniques without understanding their basis leaves us at the mercy of the technique rather than making us its master. Any modification or change would require us to throw up our hands unless we understand the basis for a method. Because modeling is a hands-on discipline there are often special cases in every system that require changes that are usually easily and properly made only if the basis for the method is understood.

Although the linear models in this chapter are not the major thrust of present day ecological modeling they do provide a forum for introducing many of the concepts and some of the methods that carry over to nonlinear and nondeterministic models. For example, the compartment diagram introduced here will be used again for nonlinear models—with added bells and whistles. So, too, for the Euler and Runge–Kutta techniques. Eigenanalysis will not be used again and is introduced mainly to show that linear models can be solved analytically and to give some indication about the error involved in the approximation technique. However, eigenanalysis is important for a wide range of applications including stability analysis of systems of nonlinear differential equations and for multivariate statistics. If you never use it again you will at least know how it can be used. If you use it again it will become an old friend in your bag of analysis tools.

Markov chains and Leslie matrices are two examples of the use of the matrix methods introduced initially for solving linear compartment models. They also show applications of difference equation models where no assumption is made about what happens between the time steps. Both these model types have a wide literature of their own. Leslie matrices are commonly used in managing fisheries and in setting quotas as well as in impact assessment, as examples in Chap. 9 will show. Markov chains are commonly used in weather generation for simulation models. We will be using them for this purpose in Chap. 7 to generate precipitation data.

In the next chapter we investigate a number of applications of donor controlled models by simulating them on the computer. We also use these examples to explore two important questions. The first is how good are linear models and the second is which solution method is preferable. We have already discussed some features of how good these models are in connection with the Patten and Witkamp model. Little more can be added to this discussion except by examining many of the applications of linear models in the literature. Often, the data are not available to evaluate the model's fit to data, as we could do with the Patten example. When such data do exist they are often much more variable than in the radiocesium experiment and thus the fit of the model is more equivocal. Most model applications assume nonlinearity and this in itself should settle questions about model realism. For initial investigation of a system, a linear model might provide some insights that could make the (limited) effort worthwhile.

We investigate the question of choice of a solution technique with some of the linear models in Chap. 3 and, again, with nonlinear models in Chap. 6. In general, the Runge–Kutta algorithm gives the most cost-effective solution, largely because round-off error in the computer becomes a factor in the accuracy of the Euler technique at small time steps.

2.10 Annotated Bibliography

Bailey, N. T. J.: *The Elements of Stochastic Processes with Applications to the Natural Sciences,* Wiley, New York, 1964.
Treats Markov chains as part of birth–death processes at a level of mathematical sophistication beyond the scope of this book.

Child, G. I., and H. H. Shugart, Jr.: Frequency response analysis of magnesium cycling in a tropical forest ecosystem, in *Systems Analysis and Simulation in Ecology, vol. 2,* ed. B. C. Patten, Academic Press, New York, 1972, pp. 103–135.
This book chapter reviews, in some detail, tools and techniques not discussed in this text for the analysis of linear compartment models—namely the use of frequency response analysis tools derived from electrical engineering.

Conte, S. D.: *Elementary Numerical Analysis, An Algorithmic Approach,* McGraw-Hill, New York, 1965.
Derives the Euler and second order Runge–Kutta techniques.

Leslie, P. H.: On the use of matrices in certain population mathematics. *Biometrika,* 33:183–212, 1945.
A pioneering paper for quantitative population dynamics. Introduces Leslie matrices.

Patten, B. C. ed.: *Systems Analysis and Simulation in Ecology, vol. 1,* Academic Press, New York, 1971.
The first chapter of this book, written by Patten, emphasizes tools for linear compartment model analysis. Some computer code is provided, although the emphasis is split between using digital and analog computers. Analog computers are no longer used for simulation because they are more difficult to learn and less flexible than digital computers. This, however, was not always the case and in 1971 the choice was still a toss up.

Patten, B. C., and M. Witkamp: Systems analysis of cesium 134 kinetics. *Ecology,* 48:813–824, 1967.
The simplicity of the cesium experiments and of the model and the closeness between the two make this an excellent introductory model for simulation study. We use it here.

Pielou, E. C.: *An Introduction to Mathematical Ecology,* Wiley, New York, 1969.
This is a classic in mathematical ecology. Well written and rigorous, Pielou covers Leslie matrix models more completely than we do here. She also deals with the analysis of spatial pattern in ecological systems.

Ralston, A: *A First Course in Numerical Analysis,* McGraw-Hill, New York, 1965.
Gives a complete derivation of the fourth order Runge–Kutta technique.

Schneider, H., and G. P. Barker: *Matrices and Linear Algebra,* Holt, Rinehart and Winston, New York, 1968.
Gives a more theoretical discussion of Markov chains, eigenanalysis, the theory of matrices, and linear algebra. Includes complex solutions to differential equations.

Searle, S. R.: *Matrix Algebra for the Biological Sciences,* Wiley, New York, 1966.
Develops matrix theory without linear algebra. Eigenanalysis (latent roots) is presented with applications to Leslie matrices.

Shields, P. C.: *Elementary Linear Algebra, 3rd ed.,* Worth Publishers, Inc., New York, 1980.

A good introduction to matrices and linear algebra. Applications include differential equations (eigenanalysis), Markov chains, and Leslie matrices.

Vandergraft, J. S.: *Introduction to Numerical Computations,* Academic Press, New York, 1978.
A simple derivation of the second order Runge–Kutta starting with a difference equation.

2.11 Exercises

Donor-controlled compartment models

2.1. A three-compartment system consists of dead grass, a herbivore that eats the dead grass, and a decomposer that decomposes both the dead grass and the dead herbivores. The system begins with 100 kg/ha of dead grass, and 5 kg/ha each of herbivore and decomposer biomass. The flow rates are as follows:

> from dead grass to herbivore: 0.01/day
> from dead grass to decomposer: 0.005/day
> from herbivore to decomposer: 0.001/day

Draw a compartment diagram depicting this system including flow rates as given. Write the differential equation model in matrix notation. Solve this model by eigenanalysis and compare the solution after 5 days with an Euler approximation with time step $\Delta = 1$.

2.2. The steady state solution to donor controlled differential equations may be sometimes calculated directly from an equation analogous to Eq. 2.17 for Markov chains. Such an equation simply assumes that at steady state the rate of change of state is zero. Namely:

$$\mathbf{A}\mathbf{x}_{eq} = 0$$

$$\mathbf{x}_{eq} = \text{equilibrium state vector at steady state}$$

a. Check that this equation is satisfied by the steady state solution for Patten's model used as an example in this chapter.

b. Try it as well for the three-compartment dead grass–herbivore–decomposer ecosystem in Ex. 2.1 but with the addition of a 0.002 flow rate from decomposers to dead grass. To find the steady state in this model you will need to solve the revised system by eigenanalysis.

c. Now try solving each of these equations directly for $\mathbf{x}_{eq}$ and show whether you get the same steady state solution. Remember that the Patten model is conservative of radiocesium and, as such, the sum of the radiocesium must equal the total amount added at the beginning of the experiment, and that the grassland model is conservative of area.

d. Under what conditions do you think you can use the above equation to solve for a steady state solution?

Recipient-Controlled Compartment Models

2.3. In this chapter we have focused on linear donor controlled differential equations. Many systems depart from the donor controlled assumption in that they have compartments that grow under their own steam. For a linear model this is accomplished by having a flow into those compartments that depends on the amount of material in them, rather than on that in the donor compartment. These are called *recipient controlled* compartments. As an example of such a system consider an extension to the system in Ex. 2.1 with a live grass compartment added. The live grass has flows to dead grass and herbivores and a flow from the decomposer compartment, presumably to represent the positive effect decomposition has on plant growth through nutrient recycling. Live grass growth can be represented as a flow from a source compartment (denoted by S).

   a. Draw a flow diagram of this system.

   b. Represent this system of differential equations in matrix form, with generic k_{ij} used to denote flow rates. Denote the flow from the source by k_{1s} (calling the live grass compartment x_1).

   c. How does this flow matrix differ from a donor controlled model flow matrix? Do the column sums still equal 0?

2.4. A tundra ecosystem has three major compartments, lichen, reindeer, and decomposers. The weekly flow rates between these compartments are as follows:

		from		
		lichen	reindeer	decomposers
	lichen			0.002
to	reindeer	0.02		
	decomposers	0.01	0.005	

These are all donor controlled flows. There is also a recipient controlled flow of 0.005 to the lichen from atmospheric carbon. Call the lichen, reindeer, and decomposer compartments x_1, x_2, and x_3, respectively, and treat atmospheric carbon as a source.

   a. Draw the flow diagram for this model

   b. Represent the differential equation in matrix form

   c. Ignore the source flow to lichen and solve for the steady state $\mathbf{x}_{eq}$ using $\mathbf{A}\mathbf{x}_{eq} = 0$.

   d. Omit the herbivores (reindeer), but include the source flow. Solve for the eigenvalues of this reduced model. Is there a steady state solution?

Leslie matrices

2.5. The dominant eigenvalue in a Leslie matrix model can be estimated using the method of *population projection,* and involves taking the ratio of the total population at time $t + 1$ to the population at time t. As t gets larger this gives a better and better approximation to the dominant eigenvalue. For the bird population dynamics problem in this chapter generate the population at time 10, $\mathbf{n}(10)$. Compute the total population $n_{tot}(10)$ and take its ratio to $n_{tot}(9)$. Compare this value with the dominant eigenvalue.

2.6. A population of *Microtus* (meadow voles) has the following survival–fecundity matrix:

$$\Lambda = \begin{bmatrix} 5p & 9 \\ p & 0 \end{bmatrix}$$

a. What value of p will lead to an equilibrium (nonchanging) population?

b. If $p = 0.05$ and $\mathbf{n}(0) = \begin{bmatrix} 10 \\ 5 \end{bmatrix}$ what is $\mathbf{n}(5)$?

Markov chains

2.7. The fire danger during the summer in Mount Baker National Forest, Washington is classified into one of three danger levels. These are $1 = $ low, $2 = $ moderate, and $3 = $ high (fire alert). The probability of daily transitions between these states is:

		to		
		1	2	3
	1	0.5	0.4	0.1
from	2	0.3	0.5	0.2
	3	0.0	0.5	0.5

a. Write the model in matrix form to project the fire danger probability from one day to the next.

b. Find the equilibrium probability of being in each state.

Answers to Selected Exercises

2.1. Euler's approximation gives:

$$\mathbf{x}(5) = \begin{bmatrix} 92.722 \\ 9.817 \\ 7.461 \end{bmatrix}$$

While the eigenanalysis solution gives:

$$\mathbf{x}(5) = \begin{bmatrix} 92.774 \\ 9.780 \\ 7.446 \end{bmatrix}$$

The eigenvalues are -0.015, -0.001 and 0.0.

2.2. The steady state solution for the grassland problem is:

$$\mathbf{x}_{eq} = \begin{bmatrix} 8.15 \\ 20.37 \\ 81.48 \end{bmatrix}$$

To obtain a solution there must be no absorbing state. All states must communicate with each other. This is not the case in the Patten model.

2.5. The dominant eigenvalue is 1.263 while $\dfrac{n_{tot}(10)}{n_{tot}(9)} = 1.266$.

2.6.

a. $p = \dfrac{1}{14}$.

b. $\mathbf{n}(5) = \begin{bmatrix} 14.90 \\ 0.69 \end{bmatrix}$

Chapter 3

Coding Linear Models

3.1   Introduction

In this chapter we discuss how to solve linear differential equations by simulation. The techniques are illustrated by examples of linear ecological compartment models. Although the programming information in this chapter may be familiar to some readers we encourage everyone to read it for the information specific to programming simulation models.

3.2   Computer languages

The choice of which computer language to present was difficult. At the time of this writing the most commonly used language for scientific computations is Fortran, however, it is being slowly replaced by more structured programming languages such as Pascal and C. Also the previously dominant centralized computer is now being replaced by personal computers and workstations for scientific programming. This leaves the preferred computer software anything but obvious. We have chosen to present our programs first in an easy to understand pseudocode, which is a mixture of formal programming language and informal English, and then in Fortran 77. Knowledge of some higher level programming language, such as BASIC, Fortran, Pascal, Algol, or C, will be assumed. This should be enough to understand our Fortran or at least the pseudocode.

Our programs adhere to the Fortran 77 standard so that all code should be portable to any machine with a Fortran 77 compiler. We emphasize structured programming, which contrary to popular belief, is not "goto-less" programming. Although Fortran 77 does not support all the program structures one would like (e.g., do while and do until), these structures can be translated from the pseudocode easily with logical if's and goto's. Some Fortran compilers may have these more structured features but we forego their use to maintain code portability.

There are computer languages written specifically for simulation modeling. Examples of these are DYNAMO, GASP, GPSS, Simula, and SLAM. These languages provide data structures that permit simple and efficient modeling. They also include many features for easy scheduling of events and simple input and output. All this makes for simple programming, with more time for model conceptualization. The principle disadvantage, however, is the lack of portability. Few of these languages are currently available for microcomputers. Also, debugging a model with these systems can be difficult. Their error messages are often cryptic and incomprehensible.

3.3   Program definition and design

Most of the time in any programming task should be spent defining and designing the program. There is no substitute for careful thought about what you want the computer to do before you go to the terminal to enter code. The aim here is to design clean, clear programs. Programs should be readable by other people as well as by the programmer 6 months after writing the program. A good test for readability is the telephone test: would someone understand the program if read to them over the phone? We do not try to be too clever in our examples or to optimize code because this usually leads to confusion for beginners. Efficiency is in the background here, we can come back later to improve model run time if necessary.

We suggest that programs be modularized. Each task should be assigned to a function or subroutine. The module should do one thing well and should have minimal interactions with others except through a well-defined connection, i.e., passing of variables or parameters. Of course a module can contain further submodules. This allows one to easily change a particular program and to build new ones out of these building blocks. It also makes debugging the program easier because each module can be run with test data. Then modules can be put together and tested in groups. Any errors arising at this point are then probably caused by problems in the communications between the modules. By restricting input and output to only a few of these modules, system specific features can be isolated, allowing portability of most of the code. This may appear to involve additional and unnecessary work for the example presented below, but it is a good habit to cultivate.

We write our programs using a *top-down programming* method. We start with a few skeleton modules that are described informally, often in pseudocode or even in English. Each of these modules is further subdivided into more basic modules until we get down to modules that consist of a few lines of code in a subroutine or function. We use pseudocode as an alternative to flowcharts to lay out the flow of the program.

The top or master module should be coded first and it should contain only the basic logic for calling the lower level modules. With the top module written one can easily test subroutines without having to write ''throw away'' code that is only used to test one subroutine.

3.3.1 Data types and structures

We assume that the reader is familiar with the difference between floating point and integer values and their arithmetic on computers [if not, a good reference is Kennedy and Gentle (1980)]. Variable types should be chosen with an eye to generality of the model. For example, although one may initially only want a time step of one time unit ($\Delta = 1$) for the Euler solution in a compartment model, someone using the program later may wish to use some other time step such as a half a time unit. Because the time step should be allowed to be changed from run to run and should permit decimal values (0.1, 0.5, and so on), declare the time step to be a floating point variable and not an integer. For variables, however, that can only be integer valued, such as the number of state variables in a model, use integers.

It is a good practice to declare all variable types at the start of each program and subroutine, even though this is not required if the Fortran convention that variables beginning with the letters from i to n are integer and all others are real values unless declared otherwise is used. This often allows more natural sounding mnemonic variable names such as *length* instead of *alngth*. Also, many Fortran compilers have a "declare variable option" that indicates at compilation those variables that are undeclared. This is an easy way to find variable name misspellings (variables accidentally spelled different ways in different parts of the program), which are a frequent programming error and often difficult to track down.

Arrays are an excellent way to group data together. They allow passing a single rate matrix to a subroutine as k, rather than as k11, k12, . . . , kpp. By storing similar variables in an array the same calculations can be performed on each variable by looping over the array.

3.3.2 Model input and output

Input and output are one of the most machine specific features of Fortran. To avoid problems we will keep our input and output simple; the standard input will be unit 5, standard output unit 6, and plot output will go to unit 7. For most Fortran compilers this will result in reading from the terminal, outputting to the terminal, and plot output going to a file called *fort.7* (or some similar filename). These units will be set equal to the integer variables IN, OU, and PL for input, text output, and plot output, respectively, and passed to all routines that need them via a *common* statement. This setup allows one to easily change the input and output units for all subroutines by changing the values of IN, OU, and PL only once.

We use free-format input wherever possible (denoted by an * rather than a format line number in a read statement). This prevents errors that invariably occur when counting columns for formatted read statements. Free-formatted reads also avoid having to match variable types with the proper format (i, f, or a). Unfortunately, to get nice looking output we have to use format statements. For real variables the g (general) or e (exponential) format is suggested to allow for a wide range of values without under or over flowing the format.

3.4   Linear compartment model code

We focus in this section on developing a computer simulation model to solve a system of linear differential equations, such as those described in Chap. 2, numerically using the Euler method. We describe a complete program that allows parameters and initial conditions to be changed from run to run and that has output directed both to a text output file and to a plotfile for later plotting.

The linear differential equation solver for compartment models must do the following tasks:

read in the simulation environment for the model (e.g., time step, number of compartments)
read in model parameters
read in model initial conditions
integrate the differential equation
produce model output

The following variables are needed (presented by type and code notation):

nstate	integer	number of compartments (state variables)
tzero	real	starting time
tend	real	ending time
dt	real	time step
tprint	real	print interval
tplot	real	plotting interval
time	real	time
k	real	parameter matrix

This is a list of the *global variables* that are used. They are used in the main program or top module and are also passed between modules as contrasted with *local variables* that are only used within a single module. Note that we use many mnemonic variable names. This makes reading the program easier than if we used a, b, c, d, and so on. The length of variable names is six characters or less for code portability.

From the outline above we can design a top module or main program that will call the following subroutines:

envin	read in the simulation environment
envout	print out the simulation environment
parin	read in the parameters
parout	print out the parameters
incond	read in the initial conditions

time loop
   euler   integration for one time step
   print   model output
   plot   model output for plotting

Notice that we have added some modules to the tasks outlined above, namely those for printing out (echoing) what is read in. It is good programming practice to check model input through printout because errors can result from reading in the wrong input or reading it in incorrectly. This output can also help identify different model runs. Although one can print the input directly in the input module we have chosen to include separate routines for input and output to allow for input–output flexibility.

Before writing the main program we define the interfaces between it and each subroutine, i.e., what global variables are passed between them. The *envin* and *envout* routines need:

   nstate   number of compartments
   tzero   time at start of simulation
   tend   time to end simulation
   dt   time step
   tprint   printing interval
   tplot   plot output interval

For the *parin* and *parout* routines we need to pass the number of compartments to know how many parameters to read in. We also pass the parameters:

   nstate   number of compartments
   k   parameter matrix (*nstate* by *nstate*)

The *incond* routine reads the initial conditions into the state variable x (vector) so it needs:

   nstate   number of compartments
   x   state array

The *euler* routine updates the state array x from *time* to *time* + *dt* and requires:

   nstate   number of compartments
   x   state array (*nstate* vector)
   k   parameter matrix (*nstate* by *nstate*)
   dt   time step

The *print* and *plot* routines write the time and the state at that time and use:

   nstate   number of compartments
   x   state array (*nstate* vector)
   k   parameter matrix (*nstate* by *nstate*)
   dt   time step

The time loop starts at time *tzero* and runs until time *tend* in increments of *dt* time units. At each increment the state variables are updated by the *euler* routine and the results are then sent to the output or plotfile if current time divided by *tprint* or *tplot* (respectively) results in an integer. In pseudocode:

> *initialize time to tzero*
> *while (time < tend)*
>   *increment time by dt*
>   *euler integration*
>   *if time to print*
>   *print*
>   *if time to plot*
>   *plot*
> *end while*

In our pseudocode notation, structured programming practice is followed. Activities occur in sequence and indentation is used to specify a group of activities that occur under some proviso, such as while or if a particular condition holds. This convention is apparent in the above example.

To determine if it is time to print or plot the modulus function (*amod*, for real variables in Fortran) and the absolute value function (*abs*, for real variables in Fortran) can be used. The modulus function computes the remainder from a division, e.g., modulus $(27,4) = 3$. If the modulus of *time* divided by *tprint* is zero, it is time to print. We never test floating point values just for equality, however, owing to machine specific aspects of how these numbers are stored and rounded off. Instead we test if modulus (*time,tprint*) is close to zero with the logical if:

$$if\ (abs(modulus(time,tprint)) < teps)$$

where *teps* is a machine-dependent epsilon small enough that printing will only occur at integer multiples of *tprint*.

To increment *time* we could use:

$$time\ =\ time\ +\ dt$$

However, there is always a problem in accumulating floating point values on a computer. Adding 0.0100 to 0.0000 one hundred times seldom results in 1.0000 and this error could throw off the above check for printing and plotting. A more numerically stable method is to use an integer increment that is initialized to zero outside the loop:

$$nstep\ =\ 0$$

and then inside the loop update *time* by:

$$nstep\ =\ nstep\ +\ 1$$
$$time\ =\ tzero\ +\ dt * nstep$$

We are now ready to write the main module in Fortran:

```
c-- main program for linear compartment model simulator
c
c Solves a system of linear, constant rate differential equations
c using the Euler method.
c
c Variables:
c    nstate -- integer -- number of compartments
c    maxst  -- integer -- maximum number of compartments allowed
c    nstep  -- integer -- time step accumulator
c    err    -- integer -- error code (0 => no error, 1 => error)
c    tzero  -- real    -- starting time
c    tend   -- real    -- ending time
c    dt     -- real    -- time step
c    tprint -- real    -- print interval
c    tplot  -- real    -- plotting interval
c    time   -- real    -- time
c    k      -- real    -- parameter matrix (nstate by nstate)
c    x      -- real    -- state variable array (nstate)
c
      real teps
      integer IN,OU,PL
      common /inout/ IN,OU,PL

      integer nstate,maxst,nstep,err
      real x(10),k(10,10),time,dt,tzero,tend,tprint,tplot

      data teps /1.0e-5/
      maxst = 10

      IN = 5
      OU = 6
      PL = 7

c read in and echo simulation environment checking for errors
      call envin(nstate,dt,tzero,tend,tprint,tplot,err,maxst)
      if (err .eq. 1) stop
      call envout(nstate,dt,tzero,tend,tprint,tplot)

c read in and echo out parameter matrix
      call parin(nstate,k)
      call parout(nstate,k)

c read in, print and plot initial conditions
      call incond(nstate,x)
      call print(nstate,tzero,x)
      call plot(nstate,tzero,x)
```

```
c   time loop
        nstep = 0
        time = tzero
   10 if (time .ge. tend) go to 20
            nstep = nstep + 1
            time = tzero + nstep * dt
            call euler(nstate,dt,k,x)
            if (abs(amod(time,tprint)) .lt. teps)
       +        call print(nstate,time,x)
            if (abs(amod(time,tplot)) .lt. teps)
       +        call plot(nstate,time,x)
            go to 10
   20 continue

        end
```

We have listed all variables at the top and have supplied comments but we do not comment the obvious. Loops, such as the time loop, are indented to allow for easier reading. Also, blank lines are used to separate appropriate blocks. The while loop from the pseudocode has been translated into Fortran with a logical if and goto's. The statement line numbers are in increasing order from top to bottom in increments of 10.

An error flag, *err*, has been added to the *envin* routine. The flag will be set to 1 (true) in *envin* if there is any error in the subroutine, otherwise it is set to 0 (false). On return from *envin* the error flag is checked and if true the program halts. Possible errors in *envin* are discussed below. One could put error flags in all the subroutines but experience shows that most errors occur in data input.

The variable *maxst* has been added and set equal to the declared dimension of the state array and parameter matrix (which is square). Because a common error in programming is exceeding array bounds, we can check for this problem by passing *maxst* to the *envin* routine as we do above.

The *envin* and *envout* routines are straightforward. Free-format input (denoted by an * in the read statement in *envin* below) is used wherever possible to avoid errors when counting columns when reading formatted input. The error code has been incorporated into the *envin* routine so that invalid input can be detected and appropriate action taken. Possible invalid input include having too many compartments (exceeding *maxst* as mentioned above), having *tzero* greater than *tend,* and having a negative *dt*. We handle these errors by printing an appropriate error message and then stopping the program upon return from the subroutine. The arrows in the subroutine header comments are used to indicate whether a variable is input ($\rightarrow$), output ($\leftarrow$), or both ($\leftrightarrow$) for that subroutine.

```
      subroutine envin(nstate,dt,tzero,tend,tprint,tplot,err,maxst)
c
c Read in the simulation environment from the input file skipping
c over the first line of each pair. Example:
c
c    nstate (this line skipped on input)
c    5
c    dt, tzero, tend (this line skipped on input)
c    0.10  0.0  30.0
c    tprint, tplot (this line skipped on input)
c    5.0 2.0
c
c    nstate <-- number of state variables
c    dt     <-- time step
c    tzero  <-- starting time
c    tend   <-- ending time
c    tprint <-- print interval
c    tplot  <-- plotting interval
c    err    <-- integer error switch, 0 => no error, 1 => error
c    maxst  --> maximum number of state variables
c

      integer IN,OU,PL
      common /inout/ IN,OU,PL

      integer maxst

      integer nstate
      real dt,tzero,tend,tprint,tplot

      err = 0

      read(IN,*)
      read(IN,*) nstate
      read(IN,*)
      read(IN,*) dt, tzero, tend
      read(IN,*)
      read(IN,*) tprint, tplot
```

```
c   check for errors
        if (nstate .gt. maxst) then
            write(OU,110) maxst
            err = 1
        end if
        if (dt .lt. 0.0) then
            write(OU,120)
            err = 1
        end if
        if (tzero .gt. tend) then
            write(OU,130)
            err = 1
        end if

        return

  110 format(1x,'error, too many state variables, maximum is ',i3)
  120 format(1x,'error, negative dt')
  130 format(1x,'error, tzero greater than tend')

        end




        subroutine envout(nstate,dt,tzero,tend,tprint,tplot)
c
c   Print out the simulation environment.
c
c       nstate --> number of state variables
c       dt      --> time step
c       tzero  --> starting time
c       tend   --> ending time
c       tprint --> printing interval
c       tplot  --> plotting interval
```

```
c

      integer IN,OU,PL
      common /inout/ IN,OU,PL

      integer nstate
      real dt,tzero,tend,tprint,tplot

      write(OU,110) nstate, dt, tzero, tend, tprint, tplot

      return

110   format(1x,'number of state variables: ',i3,/,
     +       1x,'time step: ',f9.5,/,
     +       1x,'time start: ',f9.3,3x,'end: ',f9.3,/,
     +       1x,'print interval: ',f9.3,3x,'plot interval: ',f9.3)

      end
```

The *parin* and *parout* routines are also simple. Again free-format input (*) is used.

```
      subroutine parin(nstate,k)
c
c  Read in the rate matrix from the input file in the following for
c
c     one line description (ignored on input)
c     first row of rate matrix
c     second row of rate matrix
c         .
c         .
c     last row of rate matrix
c
c     nstate --> number of state variables
c     k      <-- rate matrix, nstate by nstate, k(i,j) = flow
c     from j to i
```

```
c

      integer IN,OU,PL
      common /inout/ IN,OU,PL

      integer nstate,i,j
      real k(10,10)

      read(IN,*)
      do 10 i = 1,nstate
          read(IN,*) (k(i,j),j=1,nstate)
   10 continue

      return
      end


      subroutine parout(nstate,k)
c
c  Print out the rate matrix.
c
c    nstate --> number of state variables
c    k      --> rate matrix, k(i,j) = flow rate from j to i
c
      integer IN,OU,PL
      common /inout/ IN,OU,PL

      integer nstate,i,j
      real k(10,10)

      write(OU,110) (j,j=1,nstate)
      write(OU,120)

      do 10 i=1,nstate
          write(OU,130) i,(k(i,j),j=1,nstate)
   10 continue

      write(OU,140)

      return

  110 format(/,1x,'Rate Matrix:',/,15x,'from',
     +        /,3x,9(2x,i3,3x))
  120 format(1x,'to')
  130 format(1x,i2,1x,9(f7.3,1x))
  140 format(/)

      end
```

The routine to read in the initial conditions, *incond* is written in Fortran as follows:

```
      subroutine incond(nstate,x)
c
c Read in the initial conditions from the input file.
c Also write out a header for model output.
c Input is read in the following form:
c
c   one line description (ignored on input)
c   initial condition values (free format)
c
c   nstate --> number of state variables
c   x      <-- array of state variables at time tzero
c   of length nstate
c

      integer IN,OU,PL
      common /inout/ IN,OU,PL

      integer nstate,i
      real x(10)

      read(IN,*)
      read(IN,*) (x(i),i=1,nstate)

      write(OU,110) (i,i=1,nstate)
      write(OU,*)
  110 format(1x,'   Time      State Variables:',/
     +       8x,10(4x,i2,4x))

      return
      end
```

The *print* and *plot* routines are identical except that *plot* writes into PL unit instead of OU. Only the *print* routine is shown here.

```
      subroutine print(nstate,time,x)
c
c Print out the time and state variables.
c
c nstate --> number of compartments
c time   --> current time
c x      --> state variable array
```

```
c
        integer IN,OU,PL
        common /inout/ IN,OU,PL

        integer nstate, i
        real x(10)

        write(OU,110) time, (x(i), i=1,nstate)

        return

110 format (1x, f8.3,10(1x,g9.3))

        end
```

The *euler* routine is the most complicated of the modules. Recall from Chap. 2 that the Euler update formula for a constant rate model in matrix form is:

$$\mathbf{x}(t + \Delta) = \mathbf{x}(t) + \Delta\mathbf{A}\mathbf{x}(t)$$

where $\mathbf{A}$ is the constant rate matrix and Δ is the time step. Or for each x_i:

$$x_i(t + \Delta) = x_i(t) + \Delta \sum_{j=1}^{n} x_j f_{ij}$$

where n is the number of compartments. Converting this to pseudocode:

do i = 1, nstate
  zero sum
  do j = 1, nstate
    *sum = sum + k(i,j) * x(j)*
  end do
  *xnew(i) = x(i) + sum * dt*
end do

Of course we then must copy the array *xnew* to the array *x* because we return *x* as the updated state vector. From this the Fortran code is simple to write:

```
        subroutine euler (nstate,dt,k,x)
c
c Use the euler method to solve the system of linear
c differential equations.
c
c nstate --> number of state variables
c dt      --> time step
c k       --> rate matrix (nstate x nstate)
c x       <-> state variable array
```

```
c
      integer nstate,i,j
      real x(10),k(10,10),dt,sumkx,xnew(10)

      do 20 i = 1, nstate
         sumkx = 0.0
         do 10 j = 1, nstate
            sumkx = sumkx + x(j) * k(i,j)
10       continue
         xnew(i) = x(i) + sumkx * dt
20    continue

      do 30 i = 1, nstate
         x(i) = xnew(i)
30    continue

      return
      end
```

3.4.1 Radionuclide model example

We are now ready to use our program. Our first example is the Patten and Witkamp (1967) model of radiocesium flow through a microcosm system consisting of litter, soil, microflora, millipedes, and leachate. This model was discussed in Chap. 2 and a compartment diagram is presented in Fig. 2.1.

Although there are many combinations of compartments modeled in the original paper we consider only the most complicated system consisting of all five compartments:

$x(1)$ litter
$x(2)$ soil
$x(3)$ microflora
$x(4)$ millipedes
$x(5)$ leachate

This is a closed system with the leachate compartment as a sink.

The rate matrix for the models is:

$$\mathbf{A} = \begin{bmatrix} -0.1989 & 0 & 0.500 & 0.250 & 0 \\ 0.033 & 0 & 0 & 0.160 & 0 \\ 0.120 & 0 & -0.750 & 0.250 & 0 \\ 0.045 & 0 & 0.250 & -0.660 & 0 \\ 0.0009 & 0 & 0 & 0 & 0 \end{bmatrix}$$

The model will be run for 30 days with a time step of 0.01 days to simulate the experiment. The initial conditions are one unit of radiocesium in the litter and zero in the other compartments. The complete input file for our program applied to this example is shown in Fig. 3.1.

The output file is shown in Fig. 3.2. The model output plotted (using the file

```
                         nstate
                         5
                         dt, tzero, tend
                         .01 0.0 30.0
                         tprint, tplot
                         5.0 1.0
                         rate matrix:
                         -.1989 0.0 .50 .25 0.0
                         .033 0.0 0.0 .16 0.0
                         .12 0.0 -.75 .25 0.0
                         .045 0.0 .25 -.66 0.0
                         .0009 0.0 0.0 0.0 0.0
                         initial conditions
                         1 0 0 0 0
```

Fig. 3.1. Input file for running the ce-
sium-134 microcosm model with five
state variables. A line is provided before each group of variable for labels.

```
number of state variables:   5
time step:   0.01000
time start:    0.     end:     30.000
print interval:    5.000   plot interval:     1.000

Rate Matrix:
              from
       1      2      3      4      5
to
1  -0.199   0.    0.500   0.250   0.
2   0.033   0.    0.      0.160   0.
3   0.120   0.   -0.750   0.250   0.
4   0.045   0.    0.250  -0.660   0.
5   0.001   0.    0.      0.      0.


  Time    State Variables:
              1       2       3        4         5

  0.      1.00     0.      0.       0.        0.
  5.000 0.604    0.174   0.130    0.891e-01 0.334e-02
 10.000 0.477    0.330   0.109    0.784e-01 0.576e-02
 15.000 0.384    0.458   0.876e-01 0.635e-01 0.769e-02
 20.000 0.309    0.560   0.706e-01 0.511e-01 0.924e-02
 25.000 0.249    0.643   0.568e-01 0.412e-01 0.105e-01
 30.000 0.200    0.709   0.457e-01 0.332e-01 0.115e-01
```

Fig. 3.2. Output file generated by program *main* for the full cesium-134 microcosm
model using the input file in Fig. 3.1. Notice that the program reflects the input conditions
before giving values of state variables over time. The print interval for the model as well
as parameter values are set in the input file shown in Fig. 3.1.

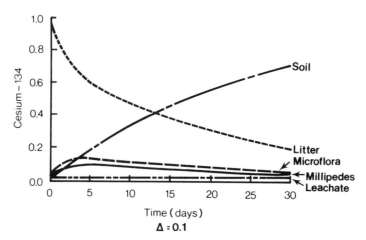

Fig. 3.3.   Graphical output from the full cesium-134 treated microcosm model for each of the five major state variables. This output was generated by the program given in this chapter using the input file in Fig. 3.1.

produced by subroutine *plot*) in Fig. 3.3 shows the litter gradually losing the cesium while the soil gradually picks it up. The microflora and millipedes initially gain some cesium and then gradually lose it. The leachate slowly picks up the cesium. Several simulation experiments for this model are suggested in the exercises at the end of this chapter.

3.4.2   Abandoned farmland succession example

For our second example we use the old field succession model by Bledsoe and VanDyne (1971). This model was discussed in section 2.4.2 and a compartment diagram is shown in Fig. 2.3. Recall that what is flowing in this system is the fraction of the total area covered by a particular successional stage after abandonment of farmland. The state variables are the five successional stages and are identified by the dominant or co-dominant grass species:

$x(1)$ *Helianthus annuus* (sunflower) and *Digitaria sanguinalis*
$x(2)$ *Aristida oligantha* (three-awn grass)
$x(3)$ *Aristida basiramea* (red three-awn grass)
$x(4)$ *Eragrostis secundiflora* (lovegrass)
$x(5)$ *Andropogon scoparius* (little bluestem)

The input file for the model that contains the flow rate matrix is shown in Fig. 3.4. The model is run for 10 years with a time step of 0.002 years. The initial conditions are 85 percent of the field in the first successional stage ($x(1) = 85$), 10 percent in the second stage, and 5 percent in the third stage. The model output is plotted in Fig. 3.5. The first successional stage has all but disappeared by 4 years. The middle three stages reach their maxima at about 1,

```
nstate
5
dt, tzero, tend
.002 0.0 10.0
tprint, tplot
1.0 .2
rate matrix:
-1.0 0.0 0.0 0.0 0.0
.8 -1.0 0.0 0.0 0.0
.2 1.0 -.7 0.0 0.0
0.0 0.0 .7 -.8 0.0
0.0 0.0 0.0 .8 0.0
initial conditions:
85 10 5 0 0
```

Fig. 3.4.   Input file for running the grazing land succession model for abandoned farmland in central Oklahoma.

2, and 3.5 years, respectively. Note that none of these intermediate stages ever dominate more than 35 percent of the total area. By year 8 after abandonment of the field it is almost completely in the climax little bluestem stage.

3.5   Comparison of Euler and Runge–Kutta accuracy

We have introduced the Euler and Runge–Kutta methods as alternative methods for numerically solving systems of linear differential equations. An obvious question is, given the increased complexity of the Runge–Kutta technique, why

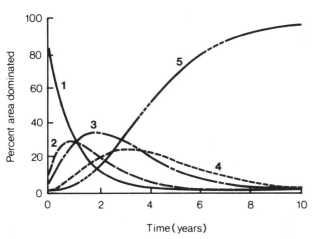

Fig. 3.5.   Graphical output for the grazing land succession model. Model output is generated using the program given in this chapter with the input file in Fig. 3.4. Stage 1 = *Helianthus anuus* and *Digitaria sanguinalis*, stage 2 = *Aristida oligantha*, stage 3 = *Aristida basiramea*, stage 4 = *Eragrostis secundiflora*, and stage 5 = *Andropogon scoparius* and *Bouteloua curtipendula*

bother using it at all if an Euler solution with a somewhat smaller time step Δ can give an equally accurate solution.

In running ecological simulations, especially nonlinear ones, we are concerned about model run time and, as such, a method that can give equal accuracy for a significantly larger time step is preferable. We can use linear models to give a preliminary estimate of the relative accuracy of the two methods and to give some indication about what Δ is appropriate. Obviously, we want as large a Δ as we can get away with and still have an accurate solution. Table 3.1 compares the Euler and Runge–Kutta methods with different time steps for the Oklahoma old field succession model. (The coding of the Runge–Kutta algorithm is left as an exercise). We are using three variables, x_1 (sunflower), x_3 (red three-awn), and x_5 (bluestem) at years 1, 3, 5, and 10 as measures of model performance.

The following conclusions are apparent from Table 3.1:

1. The Runge–Kutta technique is significantly more accurate than the Euler technique for a given Δ.
2. Simulating this model with an Euler technique and a step size greater than 0.1 is unacceptable.
3. The Runge–Kutta algorithm with $\Delta = 0.1$ is about as accurate as an Euler algorithm with $\Delta = 0.002$. In this case there are 50 iterations of the Euler algorithm for each iteration of the Runge–Kutta algorithm.

Table 3.1 Comparison of Euler and Runge–Kutta methods with different values of Δ using the grassland succession model of Bledsoe and VanDyne (1971).*

		Euler Δ				Runge–Kutta Δ			
		0.002	0.01	0.1	0.5	0.002	0.01	0.1	0.5
	time								
sunflower	1	36.8	36.6	34.9	25.0	36.8	36.8	36.8	37.5
	3	4.96	4.90	4.24	1.56	4.98	4.98	4.99	5.27
	5	0.670	0.657	0.515	0.098	0.674	0.674	0.676	0.742
	10	0.005	0.004	0.003	0.0	0.005	0.005	0.005	0.006
red three-awn	1	31.7	31.8	32.9	39.4	31.7	31.7	31.8	35.1
	3	32.7	32.7	33.1	34.5	32.6	32.6	32.6	33.0
	5	14.4	14.3	13.7	9.80	14.4	14.4	14.4	14.3
	10	0.737	0.725	0.596	0.174	0.740	0.740	0.741	0.745
bluestem	1	3.31	3.27	2.75	0.700	3.32	3.32	3.33	3.81
	3	37.4	37.4	37.3	35.9	37.4	37.4	37.4	37.5
	5	77.2	77.3	78.4	84.8	77.2	77.2	77.2	77.3
	10	112.0	112.0	113.0	114.0	112.0	112.0	112.0	112.0

*Numbers are acres dominated by the species.

4. Even though some of the rates in this model are large (e.g., 1.0), the Runge–Kutta algorithm with $\Delta = 0.5$ gives a surprisingly good approximation.

These results seem to suggest the Runge–Kutta algorithm to be preferable from a computational standpoint, and you can see how it came to be used. The proper step size to use depends on the flows in the matrix. A rule of thumb that we might derive from this example is that for the Runge–Kutta choosing Δ equal to half the reciprocal of the maximum flow rate would produce an accurate solution, whereas for the Euler method choosing Δ equal to 1/20 of the reciprocal of the maximum flow rate would produce less, though probably sufficiently, accurate result. These conclusions are also borne out by analogous simulations with the radiocesium flow model. It will be interesting to check this rule of thumb when we reexamine the accuracy of these methods with nonlinear models.

3.6 Summary

This chapter has brought our first exposure to the computer. We have busied ourselves with the logistics of computer code development such as to actualize the production of linear simulation models. Our language of choice, Fortran, reflects the emphasis in that language on computational ease. We have supplied the structure to make the code easy to understand, flexible to change and, as much as possible, general over computer systems.

We have used the examples introduced in Chap. 2 to demonstrate the functioning of our program, and also to provide a framework to evaluate the relative effectiveness of the Euler and Runge–Kutta methods for obtaining numerical solutions to the differential equations underlying the model.

As you will see in Chap. 6, much of the structure set up in this model will apply to nonlinear models as well. Some of the examples in the problem section explore extensions to the work in this chapter that will become part of the general package of tools used in simulating systems of nonlinear differential equations.

3.7 Annotated Bibliography

Anthony, T. F., and B. W. Taylor, III: Analyzing the predictive capabilities of Markovian analysis for air pollution level variations. *J. Environ. Manage.*, 5:139–149, 1977.
  This article is used as an example in the exercise section. The application, although peripheral to ecosystems themselves, gives some insight into how environmental conditions that effect ecosystems are themselves modeled.
Griffiths, P., and J. D. Hill: *Applied Statistics Algorithms*, Ellis Horwood Ltd., Chichester, 1985.
  A collection of program listings for statistical algorithms originally published in the *Journal of Applied Statistics*. Includes Fortran code for a plotting routine.

Hett, J. M.: Land-Use Changes in East Tennessee and a Simulation Model Which Describes These Changes for Three Counties, U.S. IBP Eastern Deciduous Forest Biome Report ORNL-IBP-71-8, Oak Ridge National Laboratory, Oak Ridge, Tenn., 1971, p. 56.

A linear donor controlled land-use-change model with different time periods having different rates reflecting particular temporal interventions in the history of the counties considered.

Hett, J. M., and R. V. O'Neill: Systems analysis of the Aleut ecosystem. *Arctic Anthrop.*, 11:31–40, 1974.

Used an example in the exercise section. This paper does a classical sensitivity analysis on the steady state solution, which is an approach not discussed here.

Kernighan, B. W., and P. J. Plauger: *The Elements of Programming Style, 2nd ed.*, McGraw-Hill, New York, 1978.

This little book teaches the style of computer programming. Rules are listed throughout the book to emphasize key style points. Although the examples are in Fortran and PL/1, the lessons are applicable to programming in any language.

Knuth, D. E.: *The Art of Computer Programming, vol. 2, 2nd ed.*, Addison-Wesley Publishing Co., Reading, Mass., 1981.

How to organize computer programs and to think them through

McKellar, H. N. Jr.: Metabolism and model of an estuarine bay ecosystem affected by a coastal power plant. *Ecol. Model.*, 3:85–118, 1977.

An application of linear models to studying the effect of increasing temperature on a coastal ecosystem in Florida.

O'Neill, R. V.: Examples of Ecological Transfer Matrices, U.S. IBP Eastern Deciduous Forest Biome Report ORNL-IBP-71-3, Oak Ridge National Laboratory, Oak Ridge, Tenn., 1971.

Includes a collection of transfer matrices that can be used as examples with the linear donor controlled simulator.

O'Neill, R. V., and O. W. Burke: A Simple Systems Model for DDT and DDE Movement in the Human Food-Chain, U.S. IBP Eastern Deciduous Forest Biome Report ORNL-IBP-71-9, Oak Ridge National Laboratory, Oak Ridge, Tenn., 1971, p. 18.

Yet another donor controlled model, this of pollutant flow through an ecosystem.

Parzen, E.: *Stochastic Processes*, Holden-Day, San Francisco, 1962.

A reference for Markov chains that has the derivation of the time to absorption problem discussed in the exercise section.

Patten, B. C.: A simulation of the shortgrass prairie ecosystem. *Simulation*, 19: 117–186, 1972.

This linear compartment model is noteworthy in that it is a direct simplification of a more complex nonlinear simulation model. To be able to mimic the behavior of the nonlinear model, Patten resorted to having time varying linear coefficients to reflect the effects of the environment on the model.

Patten, B. C., D. A. Egloff, and T. H. Richardson: Total ecosystem model for a cove in Lake Texoma, in *Systems Analysis and Simulation in Ecology, vol. 3,* ed. B. C. Patten, 1975, pp. 205–421.

Certainly the most complicated linear ecological simulation model ever presented. The work was so up to the minute that footnotes appeared correcting values in the tables.

Swartzman, G. L., and J. S. Singh: A dynamic programming approach to optimal grazing strategies using a succession model for a tropical grassland. *J. Appl. Ecol.*, 11:537–548, 1974.

A donor controlled succession model for a Benares, India grassland that includes the effect of grazing intensity on succession.

3.8 Exercises

Compartment models

3.1. Hett and O'Neill (1974) developed a model of an Aleutian Island eco-system. The model's objective was to explore the role the Aleut people play in carbon transfer throughout the ecosystem. It consists of a series of nine first order ordinary differential equations, and is a donor controlled model. The nine compartments are atmosphere, x_1, land plants, x_2, dead organic matter, x_3, man (the Aleuts), x_4, phytoplankton, x_5, zooplankton and other animals, x_6, dead aquatic organic matter, x_7, surface water, x_8, and deep sea, x_9. The transfer matrix for this model is:

$$
\begin{bmatrix}
-0.193 & 0.0223 & 0.0357 & 3.320 & 0 & 0 & 0 & 0.194 & 0 \\
0.050 & -0.0781 & 0 & 0 & 0 & 0 & 0 & 0 & 0 \\
0 & 0.0558 & -0.0357 & 0.458 & 0 & 0 & 0 & 0 & 0 \\
0 & 4.21e-8 & 0 & -3.778 & 0 & 5.08e-4 & 0 & 0 & 0 \\
0 & 0 & 0 & 0 & -8.0 & 0 & 0 & 0.080 & 0 \\
0 & 0 & 0 & 0 & 4.0 & -6.671 & 0 & 0 & 0 \\
0 & 0 & 0 & 0 & 4.0 & 6.67 & -0.01297 & 0 & 0 \\
0.143 & 0 & 0 & 0 & 0 & 0 & 0.01130 & -0.354 & 0.0013 \\
0 & 0 & 0 & 0 & 0 & 0 & 0.00167 & 0.080 & -0.0013
\end{bmatrix}
$$

All flow rates are in units of per year. Compartment units are in metric tons carbon.

Generate solutions for the Aleutian ecosystem over a 30-year period starting with the following sets of initial conditions and using the Euler approximation with a time step of 0.02 year:

$$
a)\ \mathbf{x} = \begin{bmatrix} 1.75e\ 8 \\ 1.12e\ 8 \\ 1.75e\ 8 \\ 1.02e\ 2 \\ 1.25e\ 6 \\ 7.5e\ 5 \\ 7.5e\ 8 \\ 1.25e\ 8 \\ 8.62e\ 9 \end{bmatrix}
\qquad
b)\ \mathbf{x} = \begin{bmatrix} 1.5e\ 8 \\ 5.0e\ 7 \\ 1.0e\ 8 \\ 1.0e\ 4 \\ 5.0e\ 5 \\ 3.0e\ 5 \\ 3.0e\ 8 \\ 1.0e\ 8 \\ 8.0e\ 9 \end{bmatrix}
$$

Do both these cases approach the same equilibrium? Which initial condition is closer to the equilibrium?

3.2. (continuation) Hett and O'Neill use an equilibrium analysis to show that land plants are more sensitive to changes in their growth rate (flow rate from the atmosphere to land plants) than zooplankton and animals, the major Aleut food sources in the sea, are to changes in phytoplankton uptake rates (the flow rate from surface water to phytoplankton).

a. Perform a sensitivity study to check this contention, which was used to show why the Aleuts are more dependent on marine than terrestrial food resources. The rationale here was that Aleuts would choose the more stable resource base. To do this analysis decrease each of the relevant flow rates by the same percentage. Several different percentages can be tried (e.g., 10, 50, and 80%). Remember that column sums must still equal 0.0. Use the first initial condition in the previous problem and run the simulation for 100 years.

b. Another claim made by the authors was that the Aleuts have a very minor effect on their food resources. Examine this statement by changing the flow rates into the Aleut compartment by 50 percent (both increase and decrease) and observe the effect on the Aleut's food compartments in a 30-year simulation beginning with the first initial condition.

3.3. John Teal (1962) did a study on energy flow in the salt marsh ecosystem of Georgia. Although Teal did not model his ecosystem, he did construct an annual energy flow diagram for the salt marsh. From this diagram, and the assumption that the equilibrium flow rates remain the same for nonequilibrium conditions and that the system is linear and donor controlled, a flow matrix can be constructed. The major compartments and flows are given in Fig. 3.6. Notice that the equilibrium amounts in each compartment are given in this figure. The flow rates can be calculated by dividing the flows out of each compartment by the amounts in the compartment. These flow rates are then assumed not to change over time.

a. Construct the flow matrix for this ecosystem. Use an atmosphere compartment from which all inputs come and to which all model outputs, except export, go. Assume that the atmospheric compartment initial conditions is 600,000 kcal/m^2.

b. Simulate the system using an Euler approximation to the differential equation having the flow matrix you have constructed with a time step of 0.1 years. Simulate over 100 years starting with (i) no biota, only atmospheric energy, (ii) twice the initial biota, and (iii) the initial biota but only half the initial atmospheric energy. What comments do you have about the realism of the model after these simulations? Why does this model not produce an equilibrium system? How might you modify the model to bring the system into equilibrium?

3.4. (continuation) Try the same simulation runs using a Runge–Kutta approximation with time steps of 0.25- and 1.0-year. This problem is a bit involved because you have not been given the code for the Runge–Kutta algorithm. Furthermore, you will notice when you look back at Chap. 2 that the fourth order Runge–Kutta algorithm has been given for a general and not necessarily linear, model. The general algorithm, to be introduced

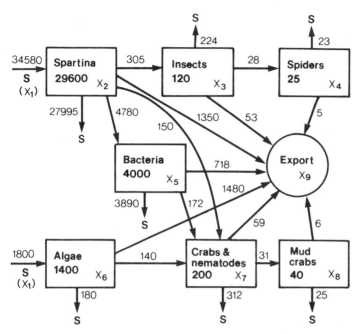

Fig. 3.6.   Flow diagram showing annual flows (kcal/m²/yr) between compartments in a salt marsh ecosystem. The amounts in the compartments are the annual averages (kcal/ m²) for that compartment. The source and sink compartments are the atmosphere, X_1.

in Chap. 6, is quite complex because it requires four calls to the model each time step. Significant time savings can be obtained for a linear model by substituting **Ax** for **f(x,t)** in Eq. 2.11. Do this first, then try coding the Runge–Kutta routine to replace the Euler routine. Which of the two approximation methods, Euler or Runge–Kutta, appear to be preferable for this system? Compare the number of computations and the solution accuracy between the various runs you made.

   Hint: You should obtain a polynomial in **A**, for **Φ**; which, because **A** does not change from one time step to the next, requires only that the higher powers of **A** be computed only once. You can do this in a subroutine *matmult,* which is called from the *main* program. The *euler* subroutine is then called within the time loop with **Φ** as an argument instead of **A**.

Simulating acute and chronic disturbances

3.5.   Material flow in an ecosystem often cannot be accounted for solely by the amount of material originally in the system and the flow rates. Usually some source of material adds to the total over time. Such is the case with the chronic addition of toxicants or nutrients to ecosystems. Let us investigate how such additions affect the dynamics of an ecosystem. We take as our example the experiment of Patten and Witkamp (1967). We use the full system with litter x_1, soil x_2, microflora x_3, millipedes x_4, and leachate x_5. The flow matrix for this experiment is:

$$A = \begin{bmatrix} -0.199 & 0 & 0.500 & 0.250 & 0 \\ 0.033 & 0 & 0 & 0.160 & 0 \\ 0.120 & 0 & -0.750 & 0.250 & 0 \\ 0.045 & 0 & 0 & -0.660 & 0 \\ 0.0009 & 0 & 0.250 & 0 & 0 \end{bmatrix}$$

a. Simulate this system assuming a spike of radiocesium of 1 unit at the start of the experiment to the litter using an Euler technique with a time step of 0.05 day. Do a 30-day simulation as in the original experiment.

b. Suppose that instead of the original spike that 1 unit were added over the entire 30-day period at a constant rate (1/30 per day). To simulate this with our donor controlled simulator we must add an additional subroutine—lets call it *inflow*—to allow for the addition of material to the model. This subroutine should be called in the main time loop just before subroutine *euler*. Write this subroutine and use it to simulate the above scenario. Compare the results with the original experiment. Are there any changes in where the radiocesium has gone? Try making the subroutine general in that any amount of material can be added at a constant rate. Note that this subroutine is not part of *euler* because the flow rates do not depend on the amount in the compartment. They are truly constant (not state dependent) flow rates.

c. How would you modify the code to include a constant inflow within subroutine *euler?*

3.6. Bledsoe and VanDyne's exploration of old field succession assumed that there are no disturbances to the field during the succession period. Often this is not the case. Simulate the following interventions over 10 years (including the first 5 undisturbed years) and plot each time trace as in Fig. 3.5.

a. After 5 years, 50 percent of the area is disturbed by a dirt bike rally and returns to the first succession stage. This area is assumed to come from each of the present succession stages in proportion to their area. Thus, half of each stage, except stage 1, is put into stage 1.

b. Chronic disturbance causes a 0.1 annual flow from each higher succession stage back to the first stage also beginning in year 5.

Markov chains

3.7. Air pollution has been a growing concern both in our cities, where we must breathe the air and suffer untoward consequences if the air is polluted, and also in ecosystems adjacent to urban areas, where plant growth and survival can be altered. We consider here a Markov chain model from Anthony and Taylor (1977) for forecasting the pollution condition in Atlanta, Georgia. The system state in this model is the probability that the air pollution index, a measure of particulate material in the atmosphere, is in a particular range. The six ranges are $x_1 = 0$–50 $\mu g/m^3$, $x_2 = 51$–75 $\mu g/m^3$, $x_3 = 76$–100 $\mu g/m^3$, $x_4 = 101$–125 $\mu g/m$, $x_5 = 125$–150 $\mu g/m^3$, $x_6 = $ greater than 151 $\mu g/m^3$. Our concern is with the probability that the system will be in state 6, an "alert" state, at some future time. A probability transition matrix, $\mathbf{T}$, has been estimated from data and is given by:

$$\mathbf{T} = \begin{bmatrix} 0.1852 & 0.1031 & 0.0471 & 0.0100 & 0 & 0 \\ 0.5926 & 0.3918 & 0.3647 & 0.2404 & 0 & 0 \\ 0.1852 & 0.3402 & 0.2471 & 0.3333 & 0.3333 & 0.2538 \\ 0.0370 & 0.1237 & 0.2000 & 0.2807 & 0.3888 & 0.3385 \\ 0 & 0.0103 & 0.0706 & 0.0806 & 0.2222 & 0.2308 \\ 0 & 0.0309 & 0.0705 & 0.0551 & 0.0555 & 0.1769 \end{bmatrix}$$

Assuming that the system begins in state 2 use the differential equation Euler routine with $\Delta = 1$ day to project the future of the population. Compute the state probability vector after 5, 10, and 20 days. How does the probability of an alert change over this time period? Note that when you use the Euler approximation with a 1-day time step to generate an *exact* solution to the Markov chain problem you must subtract 1.0 from each diagonal element of the $\mathbf{T}$ matrix because $\mathbf{T} = \mathbf{I} + \mathbf{A}$ where $\mathbf{A}$ is the equivalent flow rate matrix.

3.8. (continuation) We might be interested in the expected number of days to the next alert. This means how many days we can expect to have before we first arrive at an alert state given that we are now in some other state. To compute this from the above model you need to replace the $\mathbf{T}$ matrix by an identical matrix, except that the last column has all zeros except for the t_{66} entry, which equals unity. This is a matrix with state 6 being an *absorbing state.* By making this change we calculate the state probabilities given that once the alert state is entered the system remains in that state. We do this because we want to have the model enter the absorbing state only once in computing the average time to the first alert.

With this matrix the expected time to an alert state is the sum over all future time of the products of the time t times the difference in the probability of being in the alert state at time t minus the probability of being in the alert state at time $t - 1$. This latter term is the probability of arriving in the alert state for the first time at time t. In practice a good approximation can usually be had with a 200-day projection.

a. Start in states 1, 3, and 5 and compute the average time to absorption using the suggested method.

b. Now try the theoretical way to check your results. The expected time to absorption in an alert state starting from any state i is the ith row sum of the matrix $(\mathbf{I} - \mathbf{Q}^T)^{-1}$ (the sum of the elements in the ith row), where the $\mathbf{Q}$ matrix is the submatrix of $\mathbf{T}$ with the 6th row and column removed. $\mathbf{Q}^T$ is the transpose of the $\mathbf{Q}$ matrix; a matrix having the rows and columns reversed (transposed) from $\mathbf{Q}$. Technically speaking $\mathbf{Q}$ is the matrix of transition probabilities between nonabsorbing states.

Answers to Selected Exercises

3.1. Both cases do not approach the same equilibrium. The first initial condition was closer to its equilibrium. Land plants are more sensitive to changes in their growth rates than zooplankton and marine animals are to changes in phytoplankton growth rates. Changing the flow rate into the Aleut compartment has no significant effect on Aleut food resources but strongly affects the Aleuts.

3.4. The following table compares the solution after 100 years for each of the three initial conditions discussed, for the Euler with $\Delta = 0.1$ and the Runge–Kutta with $\Delta = 1.0$. The Runge–Kutta results in this case are identical with those for Δ of 0.25.

Teal Model after 100 Years

Variable	Case (i) Euler	Case (i) Runge–Kutta	Case (ii) Euler	Case (ii) Runge–Kutta	Case (iii) Euler	Case (iii) Runge–Kutta
x_1	318,550	318,600	352,370	352,430	176,190	176,220
x_2	15,793	15,795	17,469	17,472	8,375	8,376
x_3	64.15	64.16	70.96	70.97	35.48	35.48
x_4	13.43	13.44	14.86	14.86	7.43	7.43
x_5	2,145	2,145	2,372	2,373	1,186	1,186
x_6	746.7	746.8	825.9	826.1	413.0	413.0
x_7	107.2	107.2	118.6	118.6	59.28	59.29
x_8	21.60	21.60	23.89	23.89	11.95	11.95
x_9	262,560	262,500	297,520	297,460	148,760	148,730

In this case the Runge–Kutta with $\Delta = 1.0$ is preferable in that it gives a slightly more accurate solution than the Euler with $\Delta = 0.1$ with about 10 percent the run time.

The $\boldsymbol{\Phi}$ to be used in this case is given by:

$$\boldsymbol{\Phi} = \mathbf{Ax} + \frac{\Delta\mathbf{A}^2\mathbf{x}}{2} + \frac{\Delta^2\mathbf{A}^3\mathbf{x}}{6} + \frac{\Delta^3\mathbf{A}^4\mathbf{x}}{24}$$

3.5. The solution in the two cases after 30 days is:

Variable	Day 1 cesium spike	Constant cesium input rate
$x(1)$	0.2001	0.4256
$x(2)$	0.7096	0.4214
$x(3)$	0.0457	0.0860
$x(4)$	0.0331	0.0599
$x(5)$	0.0115	0.0071

3.7. The probability vector starting in state 2 after 5, 10, and 20 days is:

State	After 5 days	After 10 and 20 days
x_1	0.0611	0.0597
x_2	0.3199	0.3164
x_3	0.2964	0.2967
x_4	0.2006	0.2032
x_5	0.0671	0.0686
x_6	0.0550	0.0555

3.8. The expected times to an alert state calculated by the brute force method and by the theoretical method are:

Initial state	Simulation	Theoretical
x_1	22.45	22.57
x_3	20.54	20.65
x_5	20.41	20.57

Chapter 4

Model Processes

4.1 Introduction

This chapter is the meat and potatoes chapter of the book. It provides the substance of how to use data and hypotheses about how processes work to add to the tools for running simulation models developed in Chaps. 2 and 3. In this chapter the major challenge in simulation modeling is encountered—that of translating information about processes and supporting data into differential equations.

4.2 Variable definition

Most ecological simulation models may be expressed as a system of differential equations describing how the state variables change over time:

$$\dot{\mathbf{x}} = \mathbf{f}(\mathbf{x}, \mathbf{z}, t) \tag{4.1}$$

The $\mathbf{x}$'s are *state variables* and are of primary interest in the model. The $\mathbf{z}$'s denote *driving variables*. They are external to the system modeled, may change over time (denoted by t), and are not affected by the state variables.

Although Eq. 4.1 represents the general framework for most ecological simulation models, models are rarely written in that form. Ecologists commonly study processes that control the flow of material and nutrients in ecosystems, including such processes as photosynthesis, nutrient uptake, predation, translocation, and grazing. As such, the rates of flow ($\dot{\mathbf{x}}$) between compartments (state variables) are conventionally subdivided into component processes as are the rates of change ($\mathbf{f}$ in Eq. 4.1). For example, in a lake ecosystem model the rate of change of algal biomass (a state variable) depends on rates of photosynthesis, respiration, grazing, mortality, and sinking. Each of these rates in turn depends on other variables. For example, photosynthesis can be represented as depending on the temperature and solar radiation (driving variables), and on

68

algal biomass and internal nutrient concentration (state variables). Each of these dependencies is usually represented by a separate *subprocess equation*. By visualizing state equations, such as Eq. 4.1, as a hierarchical tree of process equations, which in turn include trees of subprocess equations, you can see how complicated Eq. 4.1 can get. Variables, such as process or subprocess rates, are identified in ecological modeling jargon as *intermediate variables*. These are usually denoted in this book by g. An example of an intermediate variable is the feeding rate of a predator on its prey given in Chap. 2, which depends on both the predator numbers x_j and the prey numbers x_i.

Intermediate variables represent *hypotheses* about how the process rates are affected at any time by the state of the system including the present (at time t) values of state, driving, and even other intermediate variables. The equations representing these processes are functional relationships that depend on the values of *parameters* or constants that do not change over time. Parameters in this book will be denoted by the letter k.

Sometimes functions of the state variables are desired as ecological indicators. Examples of these include the nitrogen to phosphorus (N/P) ratio used as a measure of limiting nutrient for plants, the fecundity (eggs per female) used as a measure of the health of zooplankton populations, and the diversity index used as a measure of the ecological health of ecosystems. In these cases separate *output variables,* which will be denoted by y, are computed in the model.

In summary, our simulation modeling environment consists of state, driving, and intermediate and output variables, which are *dynamic* in that they change over time and parameters, which are *static* in their temporal immutability. Parameters may change from one run of the model to another to represent changes considered in the shape of, but not the equation for, the process relationships. Changing parameters to improve model performance is a standard part of model development and is termed *calibration*. Many of the parameters are based on field or laboratory measurements of process rates that are not accurately measurable. Many uncontrollable factors may influence the measurement of a given process parameter from a laboratory experiment making the repeatability of the measured value for a parameter difficult. Often the parameter estimates are based on indirect evidence or on educated guesswork. Estimates from the literature on many parameters may not be available for the target biota and values from similar species are often used for initial parameter estimates. Because of these and other sources of parameter variability, such as the variance inherent in using the average rate for a population where the actual individual rates have some distribution of values, changing parameters between model runs must play an important role in model development and is usually justified.

4.3 Variable naming conventions

Because ecological models often include many variables, the naming of variables is an important consideration. The usual choice is between a *numbering system* with a key to identify the variables and a *mnemonic system*. Under the numbering system the only information given in the variable names are the type of variable; they are $-x$ for a state variable, z for a driving variable, and so forth. This is the convention used in Chaps. 2 and 3 for linear models. The mnemonic system presents a wide variety of choices. One alternative, which we prefer and use in this book, combines information about variable type with process or variable identifiers. In this notational scheme variable type is identified by the first character, being either z, g, x, y, or k for driving, intermediate, state and output variables, and parameters, respectively (sometimes t is used for temporary variables as a convenience notation and sometimes d is used for decision variables or management controls). The remainder of the variable name describes the variable in mnemonic terms. The mnemonic consists of up to three parts. The first part describes the main process or variable. For example *pht* is used for photosynthesis, *rsp* for respiration, *bio* for biomass, and *fd* for feeding. The second part, or suffix, modifies the variable or tells what material is involved. For example *wt* is used for weight, *n* for numbers, *tp* for temperature, and *p* for phytoplankton. The third part consists of subscripts that further specify a variable. For example *mx* is used for maximum, *tt* for total, and *rt* for ratio. The variable type and mnemonic main parts, suffixes, and subscripts are combined to make the full names of the variables. Using this scheme *gfd* is an intermediate variable representing the feeding rate of fish, kfd_{mx} is the maximum feeding rate parameter, a parameter in the *gfd* function, *xbiop* is phytoplankton biomass, and *ztp* is environmental temperature, a driving variable. Notice that not all variables require a main part, suffix, and subscript. Some have only a main part or a suffix. Some attempt was made to keep variable names relatively short, because long variable names make equations lengthy and cumbersome. This, however, was not always possible.

Often there are more than one compartment associated with a particular variable type. For example, there may be three groups of phytoplankton represented in a model, each having the same controlling processes, but with different parameters. It would be redundant to write over the same controlling equations for each species. Here numerical subscripts are combined with the mnemonic notation. For three zooplankton groups—say cladocerans, herbivorous copepods, and carnivorous copepods (labeled 1–3)—the weight per individual of each group would be xwt_1–xwt_3.

In this chapter subscripts on variables will be ignored when referring to different groups within a trophic level or species within a *functional group* (functionally similar species represented by the same processes and parameter values in the model). Only when we get to the models in Chap. 5 will we explicitly distinguish between multiple members of the same functional group by using subscripts.

Table 4.1 shows the parts of the variable names or *primitives* used in this book. It can serve as an aid in learning the notational scheme as well as a reference for "parsing" the name of a variable or parameter when it is encountered. We have denoted the parts of the name by roots (main parts), suffixes, and subscripts. The main parts are usually nouns that identify the variable. The suffixes are usually modifiers, although when a variable has no main part the suffix becomes a noun. The subscripts are always modifiers.

4.3.1   Notation justification

We have chosen the mnemonic notation for a variety of reasons. We wish to have a notation that is general over a wide range of ecosystems and that applies flexibly within a simulation framework. Trophic levels and species or functional groups are the structural currency of ecosystems, whereas processes are the functional currency. We prefer a notation that can identify the relevant process and characteristics for each variable. The first character of the variable name identifies the variable type and is needed to help the modeler to know what the variable is doing in the model—whether it is changing over time according to the differential equation, like a state variable, or not changing, like a parameter, or not affected by model behavior but affecting it, like a driving variable. Our mnemonic scheme provides both the type of variable and information about the variable that places it in its proper ecological context. The other, numbering notation is succinct and is compatible with arrays often used in Fortran but allows no way to identify what ecosystem or biota we are dealing with. All models look the same to it. Because it is the preferred notation for mathematics we will use it in appropriate places, to give a shortened equation form so that the functional curve form can be easily seen, or when we derive an equation to make substitutions so that equation manipulations can be easily followed.

One important feature of our notation is that it is general to a wide variety of ecosystems and biota. This makes it preferable as a *general ecological simulation model notation* to alternative notational schema that, although more standard for any particular process or type of biota, do not generalize over types of ecosystems or biota. Thus, whereas I_0 is commonly used for incident solar radiation in models and measurements of the effect of light on phytoplankton growth, it is not commonly used in forest canopy models or experiments. As another example z is used as depth in aquatic models, whereas h is used to denote depth in forest canopy models. If we were just considering aquatic ecosystem models we could assemble a more standard notation than ours, but it would not apply to forest canopy models, where a different notation would be used for the same variable. We found this unacceptable, and for most types of processes and ecosystems there is a lack of any standard notation—something that is sorely needed.

With our notation we always can look to the first character to identify the type of variable, the next characters to identify the process, the suffix for modifiers or indicators of the material involved, and the subscript for additional description. We have opted for a mnemonic notation that is unique. *lt* is always light, *fd* is always feeding, and *n* is always numbers no matter what biota are involved. Those who have used this scheme have found it possible to read the equations with an eye to both the variables and the ecological process information.

Table 4.1   Notation primitives classified according to location in the variable name.

Notation Roots (nouns)	Meaning
area	area
assim	assimilation rate
bio	biomass
birth	birth rate
C	carbon
cap	prey capture
Chl	chlorophyll *a*
chl	challenge (lion)
dbh	diameter at breast height
deg	°C
dig	digestion rate
dis	discharge rate
eat	eating (consumption) rate
emig	emigration rate
enctr	encounter rate (of prey)
energ	energy content
evap	evaporation
exc	excretion
ext	extinction (light)
fd	feeding rate
flow	flow rate
grow	growth rate
gut	gut contents
han	handling time (of prey)
hatch	hatching (egg)
immig	immigration rate
insol	incident solar radiation
intak	energy intake rate
lay	laying rate (egg)
len	body length
lodger	lodger lions
mix	water mixing rate
mois	soil moisture
mrt	mortality rate
pht	photosynthesis rate
pred	predation
pref	feeding preference (for prey)
prey	prey density
pride	pride lions
pup	seal pups
q	Q_{10}
recrt	recruitment (to a fishery)
rsp	respiration rate
sex	sex (ratio)
space	space (effect)
srch	search rate (for prey)

Table 4.1   *(cont.)*

Notation Roots (nouns)	Meaning
stock	adult fish stock
stren	strength
suc	success
surv	survival
swim	swim speed
umb	umbrella (medusa)
upt	uptake rate
vol	volume

Suffixes (adjectives or nouns if no adjectives are used)

age	age
bs	basal (area)
dy	day
eff	efficiency
egg	egg
ent	entangled
ht	height
lf	leaf
lg	large
lt	light
n	numbers
nit	nitrogen
p	phytoplankton
prp	proportion
py	prey
rad	radius
sda	specific dynamic action
tim	time
tp	temperature
wt	weight
z	zooplankton

Subscripts (secondary modifiers)

av	average
eq	equilibrium
hs	half saturation
mn	minimum
mx	maximum
op	optimum
pr	probability
rt	ratio
th	threshold
tt	total
vr	variance
10	in Q_{10}

From Starfield, A. M., et al., A model of lion population . . .
In *Dynamics of Large Mammal Populations*, Fowler and Smith,
eds., Wiley, 1981. Reprinted by permission.

Despite all these benefits of our notational scheme, we recognize that it is new to most readers of this book. As such, we italicize all model variables so that the notation can stand out. The glossary in the back of the book can be used to check the notation for any equation. Each equation will have a list of the new variables introduced in that equation just below the equation. Despite all these aids to using the notation we expect that you may have to read some equations more than once before the notation sinks in.

4.4 Steps in model formulation

Ecological simulation model building, like any building, is best conducted in a methodical fashion. The plans must be drawn before a materials list is made. The materials must be purchased before building starts. The foundation must be laid before the building is framed. The first step in building models—conceptualization or planning—begins with setting model objectives. Simulation models are developed to meet specific objectives. These can be placed into the following classes:

1. To replicate ecosystem behavior under ordinary conditions and predict behavior in the future.
2. To further understanding of ecosystem behavior and to determine the degree to which our present understanding of the system is adequate to replicate ecosystem behavior.
3. To organize and use, in a whole ecosystem framework, data and information about an ecosystem and its component processes.
4. To identify areas for future research.
5. To generalize research on an ecosystem beyond a single site.
6. To investigate the effect of changes, either accidental, planned, or resulting incidentally from our activities on ecosystem behavior.

These objectives are general, and can be made more specific for any particular model. They are included to give an idea of the breadth of model objectives. The use of models for management purposes can target several of the above objectives. For example, although the managers would like a model to predict the effect of changes on the ecosystem, they are now also recognizing some of the other benefits of modeling, such as providing research direction and organizing data and information into a whole ecosystem framework as valid objectives in their own right.

Model objectives can play as important a role as the ecosystem itself in determining the model structure. For example, several ecological models have been made for the Great Lakes having objectives ranging from understanding the effect of power plant thermal loading, to studying lake eutrophication, to examining the effects of fish species introductions on native species, to serving as an aid to better understand the ecosystem. The state variables were different among these models. The fish introduction models centered on fish species and considered plankton in the most cursory fashion if at all. The eutrophication models ranged in complexity from those including only plankton and nutrients to including all trophic levels. A thermal loading model confined the ecosystem boundaries to the area around the power plant, whereas the eutrophication models ranged in spatial boundaries from lake segments (e.g., Saginaw Bay) to the entire lake, depending on the degree of detail required about point source release of phosphates. We cannot generalize about how the objectives of the model and the system modeled interact to produce a particular model structure. We will show instead, through examples scattered throughout this book, how this occurred in various case studies.

The second conceptual step in model development is defining the variables of interest, the boundaries of the ecosystem, and the major ways that the variables interact. This step is aided by the development of a flow diagram showing the major variables, the quantities flowing, and the major connections between the variables (the flows) and, sometimes, the factors that control these flows. Several examples of flow diagrams for ecological models will be given in the next chapter. We have already seen simple examples of such diagrams in the linear model discussion of Chap. 2.

The third conceptual step is to specify the functional mechanisms—the factors that control the rates of flow between variables or control the events that cause changes in the variables. This step is the heart of model development. At this point, data and information are reviewed and decisions are made about how much detail to include in representing each process. A balancing act is required between using the best information available and having limited data to work with, between not including a process because of insufficient data for the study species and using data from another species, and between having a simple model and also having realistic components.

Models often compromise by using scanty data about one process to allow another process, expected to be very important, to be represented in as much detail as possible. For example, decomposition is often represented in a simple fashion owing to scanty data, yet without it an important source of recycled nutrients, essential to finance plant growth, would be omitted from the model. Sometimes processes are represented in an extremely simplified fashion (e.g., linear) for lack of any substantive data or information about the process. A simpler alternative may be preferred to a complicated yet realistic process representation because no data are available for several of the parameters in the complicated model or because the detailed output from this process is incompatible with the level of detail in companion processes. For example, a great deal is known about nutrient cycling within plants, but the value of including much of this detail is lost because of imprecision in the decomposition process. Another example is with the photosynthesis process in models of long-term changes in forest stand species composition (*forest succession models*). Although the photosynthesis process has been studied in cuvette experiments (field enclosures where the environment can be controlled) and its immediate and marked response to changes in solar radiation and temperature has been established, many of these effects average out over the growing season and it is possible to establish a good empirical relationship between photosynthesis and annual indexes of temperature and solar radiation that does not require the detail explicit in the cuvette experiments.

In this way the weighing and measuring continues, allowing each process mechanism to find its natural level. Because the factors determining these decisions are often not all available at the start of model building, you can expect to change detail in some processes as further information is gained about the relative importance of different processes and about the availability of data.

After specifying the model boundaries, variables, processes, and equations, model development moves from the conceptualization to the *implementation* phases. Implementation steps include estimating parameter values from data, coding, and debugging the model and calibrating the model. This last step is the process of fine tuning parameters to make model output better fit our known understanding of ecosystem behavior based on data time traces from samples of ecosystem biota.

An essential, and often overlooked, part of model development is model documentation—the recording of the model structure, assumptions, parameter values, and resulting behavior to make them clear to a person not involved in model development. Keeping an updated record of model structure and behavior makes it possible to have an accurate portrayal of model development as well as of the model itself. Because models are frequently altered during development, keeping accurate records of process equation and parameter changes makes it easier to benefit from lessons learned in previous changes. We cannot teach this fact; it must be learned by experience. In Chap. 5 we describe some documentation techniques used to aid model development as well as in preparing a complete model document.

The next stage of model development is model *evaluation*. Several techniques and measures are used, the most common of which, termed model *corroboration* (validation), compares dynamic model behavior with data time traces of ecosystem state variables, preferably using data independent from those used to calibrate the model. These techniques are treated in Chap. 8.

Simulation experiments can be made with a model. Here the model is used in scenario mode to do experiments that would be difficult or too expensive to do on the real ecosystem. Examples include stocking a lake with fish at several levels of intensity, removing a fish species, increasing the rate of input of nutrients to a lake, or raising the temperature 5°C to simulate the greenhouse effect on a forest succession model. The model can also be used as a gedanken (thought) experiment to investigate the relative importance of various processes to model behavior. Examples of such experimentation for management purposes are given in Chap. 9.

4.5 Process representations—three examples

To give some flavor of the process representations that form the heart of simulation models let us examine three different kinds of processes and some of the subprocesses that comprise them. The first is the fish feeding rate represented in units of grams consumed per gram of fish biomass per day (or / day). The second is the photosynthesis of terrestrial plants and aquatic algae in units of grams per square meter per day ($g/m^2/day$). The third represents the challenges between territorial pride lions and challenger male lions in the Kruger National Park in South Africa. Each of these processes can be represented in Eq. 4.1 as a function of environmental and state variables that change over time. The first two are rate processes, whereas the third is a discrete time process occurring at regular intervals.

4.5.1   Fish feeding rate

Fish feeding rate or daily ration has most commonly been represented as the product of a weight-related factor, a temperature-related factor, and a prey density-related factor, each of which are subprocesses or intermediate variables. We denote these by *gfdwt, gfdtp,* and *gfdpy,* respectively. The effect of each of these factors on fish feeding rate has been well studied in the laboratory and, therefore, much information is available about them. Separating them implies that their effects on feeding rate are independent of each other. This means that, for example, if the prey density is reduced to a level such that the daily feeding rate is halved then this same percentage reduction (50%) will occur at this prey density at 20°C as well as at 10°C, although the actual feeding rate will be different owing to the different effect of temperature in the two cases.

The weight change or growth rate of individual fish is represented by the difference between food assimilated and respiration:

$$xwt = kassim \times gfdwt \times gfdtp \times gfdpy \times xwt \qquad (4.2)$$
$$- grspwt \times grsptp \times xwt$$

xwt = fish weight (g)
$gfdwt$ = effect of weight on feeding rate (/day)
$gfdtp$ = effect of temperature on feeding rate
$gfdpy$ = effect of prey density on feeding rate
$grspwt$ = effect of weight on per-unit-weight respiration (/day)
$grsptp$ = effect of temperature on respiration rate
$kassim$ = assimilation efficiency

The daily growth rate of individual fish, *xwt,* is in units of g/day. The feeding rate part includes a weight effect in units of /day, and temperature and prey density effects, both of which are unitless. The weight effect units are grams consumed per gram of body weight per day. This is multiplied by body weight in Eq. 4.2 to give a rate in g/day. Separating the process into a per-unit-time rate times the compartment biomass (here fish weight) is the same as was done in the donor controlled models, where the rate k_{ij} was multiplied by the donor compartment x_j. It has the benefit of having the process rates in units of /*time.* This makes the rates free of the units of what is flowing (biomass or nutrients or whatever). It also can create some difficulties, as we will see when we look at *gfdwt* in detail in the next section by resulting in parameters that are in different units and, therefore, have different values from those commonly presented in the literature. Once these difficulties are recognized and taken into consideration, having rates in /*time* units is a convenience, both for coding the model and for having dimensional consistency between flow rates. The units for the respiration part of the equation are analogous to those used for the feeding rate and are not discussed further.

Weight effect on feeding

The weight effect *gfdwt* has been represented by Winberg (1956), Ursin (1967), and Kitchell et al. (1977) by the *allometric consumption function:*

$$gfdwt = kfdwt_1 \times xwt^{kfdwt_2} \qquad (4.3)$$

$kfdwt_1$ = daily feeding rate by a 1-g fish $(g^{-kfdwt_2}/\text{day})$
$kfdwt_2$ = exponent in effect of fish weight on feeding

This equation is a power function of the form:

$$g(x) = k_1 \times x^{k_2}$$

$kfdwt_1$ has units of g^{-kfdwt_2}/day. These unconventional units result from the need for dimensional balance. Because *gfdwt* is in units of grams per gram per day (/*day*) and *xwt* is in grams we have from Eq. 4.3:

$$1/day = g^{kfdwt_2}g^{-kfdwt_2}/day$$

This equation dimensionally balances. If possible, units such as those of $kfdwt_1$ should be avoided owing to the difficulty of interpreting a parameter having such units. Here the historical development of models about the process has carried with it an equation with awkward units for one parameter.

Most fish species have a value of $kfdwt_2$ between 0.0 and -1.0. This results in large fish eating a proportionally smaller fraction of their body weight daily than smaller fish. Let us illustrate this with an example. Suppose, in the equation for *gfdwt*, $kfdwt_1$ has a value of 0.25 and $kfdwt_2$ has a value of -0.33. Here $gfdwt = 0.25 \times xwt^{-0.33}$. A 1-g fish would feed at a daily rate equal to 25% of its body weight. A 27-g fish, on the other hand would feed at a rate equal to $0.25 \times 27^{-0.33}$ or 0.25/3 g/g body weight/day. The 1-g fish feeds at three times the rate per gram of body weight as the 27-g fish.

Notice that for a 1-g fish the per unit weight feeding rate is the same as the feeding rate of the fish. To obtain the feeding rate per fish we must multiply *gfdwt* by *xwt*. In our example we would have $gfdwt \times xwt = 0.25 \times xwt^{0.67}$. The exponent is 1.0 greater than the exponent in *gfdwt*. We now have the effect of weight on the feeding rate of the fish (rather than per gram of fish weight). A 27-g fish would have this rate equal to $0.25 \times 27^{0.67}$ or 2.25 *g/day*. From this we can see that although the per unit gram feeding rate goes down, the feeding rate per fish goes up with increasing fish weight. In the laboratory it is the feeding rate per fish (not per weight) that is measured. The weight effect parameter presented in the literature based on these experiments is, therefore, $1.0 + kfdwt_2$. When we present a value of -0.33 the equivalent literature value is 0.67. This is the difficulty alluded to before. Once you understand the difference you have mastered a major stumbling block to the compatibility of models—the difference in units chosen. The take home message is to always check the units in a model for dimensional consistency (a process known as *dimensional analysis*) and beware of differences in units between parameter values presented in the literature.

The equation for *gfdwt* is supported by a host of empirical evidence, which will be presented in a later section of this chapter. Several physical interpretations have been given to suggest why the relationship between weight and feeding rate should have weight to some power with a value usually close to -0.33 (0.67 for per fish feeding rate). One rationale is that feeding rate is controlled (assuming food is not limiting) by the rate that food can pass through the gut. This may be proportional to the gut surface area, because food absorption into the bloodstream is through the gut walls. Under this assumption feeding rate would be controlled by the gut surface area, which would be related to fish volume (which is directly proportional to weight) by a power function with an exponent about 0.67. For a sphere the volume is proportional to r^3 and the surface area to r^2, where r is the radius. Their ratio would be of the order of $r^{0.67}$. Although neither the fish gut nor the fish is spherical, the power analogy extends to less regular surfaces.

Temperature effect on feeding

The temperature and prey density effects are modeled as dimensionless, multiplicative factors. Temperature influences fish largely through their metabolism. Because fish are poikilotherms (their temperature changes with changes in environmental temperature), their metabolic processes are highly temperature dependent. As such, their feeding rate must also be temperature driven to reflect their temperature driven metabolism and activity level. It is generally accepted that fish adapt their metabolism to their environmental temperature and that at some temperature maximum feeding rate will be achieved, if other factors such as prey density are equal. A gamma density function is most often used to describe the rise in consumption rate with increasing temperature at lower temperatures and its drop above the optimal temperature $kfdtp_{op}$. The general form of the gamma density function is:

$$g(x) = k_1 \times x^{k_2-1} \times e^{-k_3 x} \qquad (4.4)$$

A modified version of this equation, commonly used in the literature to represent the temperature effect on feeding rate (O'Neill et al., 1972; Kitchell et al., 1977), is:

$$gfdtp = t_1^{t_2} \times e^{t_2(1 - t_1)} \tag{4.5}$$

where
$$t_1 = \frac{kfdtp_{mx} - ztp}{kfdtp_{mx} - kfdtp_{op}}$$

and
$$t_2 = (kfdtp_{mx} - kfdtp_{op}) \times \left(\frac{\log(kfdq_{10})}{10}\right)$$

$kfdtp_{mx}$ = temperature above which feeding ceases (°C)
$kfdq_{10}$ = the change in feeding rate with 10°C change in temperature
$kfdtp_{op}$ = temperature at which feeding rate is maximum (°C)
t_1, t_2 = temporary variables

The temperature effect curve is shown in Fig. 4.1 for different temperature optima between 15 and 27°C with $kfdq_{10} = 2.0$ and $kfdtp_{mx} = 30°C$. You can see the general form of the gamma density function as a unimodal (one peaked) curve. Although the equation for *gfdtp* may not seem to be similar in form to the gamma equation, closer inspection will show that they are similar. t_1 depends only on the temperature *ztp* and parameters, and is closely related to x in Eq. 4.4. t_2 depends only on parameters and is, therefore, a parameter itself, much like k_2 in the gamma equation. The $e^{-t_2 t_1}$ term is like $e^{-k_3 x}$, and e^{t_2} is a constant that could be brought out in front of the equation like k_1.

$kfdq_{10}$ is also called the Q_{10} for feeding rate. The Q_{10} is a concept that arose from experiments measuring process rates in the laboratory. It is defined as the ratio of the process rate over a 10°C increase in temperature. A Q_{10} of 2 implies a rate doubling over 10°C. Temperature effect *gfdtp* increases with increasing temperature at a rate close to Q_{10} for low temperatures but as the temperature approaches $kfdtp_{op}$, the rate of increase diminishes until *gfdtp* reaches a maximum at $kfdtp_{op}$. Above this temperature, consumption drops until $kfdtp_{mx}$, at

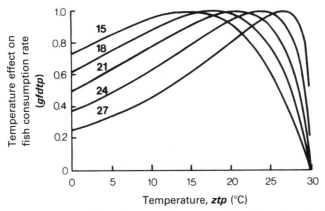

Fig. 4.1. The effect of temperature *ztp* on daily fish ration (feeding rate) *gfdtp* based on the gamma-like equation. The family of curves have different values of optimal temperature for feeding $kfdtp_{op}$ ranging from 15 to 27°C.

which temperature consumption rate is zero. Notice that $kfdtp_{op}$ has a significantly greater effect on *gfdtp* near the optimal temperature than it does in the lower temperature range where $kfdq_{10}$ has the major influence.

We see in Fig. 4.1 how changing parameter values can change the relative shape of the process rate function. It is useful to be able to identify over what range of controlling variable (in this case temperature) each parameter exercises the major control, because this will help in understanding how a parameter change affects model behavior. Therefore, good process representations have their curve shape controlled by parameter values in a clearly defined manner. Later in this chapter we will suggest other measures for evaluation of process equations.

Prey density effect on feeding

The prey density effect is the least understood of the processes controlling feeding rate. Many factors can play an important role here, such as the spatial distribution of the prey (random, regular, or patchy), the feeding pattern of the predator (schooling, solitary searching for prey, or solitary stationary), the selectivity of the predator for different prey, the need of the predator to avoid predation itself, the existence of prey refugia (areas where the prey are protected from predation), and predator switching between prey as relative prey abundance changes. To consider all these factors would require a model beyond the present level of understanding of predator–prey interactions. The function we present here is a simplification of the Holling disk equation (Holling, 1965), originally developed to represent predation by praying mantis and now widely applied to other species:

$$gfdpy = \frac{gprey}{gprey + kfdpy_{hs}} \qquad (4.6)$$

gprey = total prey density weighted by relative selectivity (g/m^3)
$kfdpy_{hs}$ = prey density at which feeding rate is half the maximum (g/m^3)

You may recognize this equation, despite the fancy notation, as identical to the hyperbolic feeding function given as an example in Chap. 2. The general functional form is:

$$g(x) = \frac{x}{x + k}$$

$kfdpy_{hs}$ is sometimes called the *half saturation* parameter. This equation form is identical to the Michaelis–Menton function, commonly used both in chemical kinetics (the rates of chemical processes) and to represent nutrient uptake (absorption rate of nutrients into algal cells) rates in phytoplankton. Fig. 4.2 shows a family of curves of *gfdpy* graphed against prey density *gprey*, with different values of $kfdpy_{hs}$. The lower the half saturation value the higher the feeding rate at a given prey density.

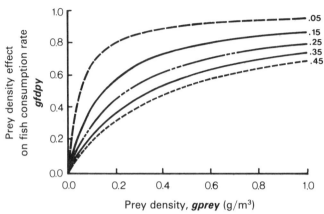

Fig. 4.2. The effect of prey density *gprey* on daily fish feeding rate *gfdpy* based on the Holling equation. The family of curves have different values of prey density half saturation parameter *kfdpy*$_{hs}$ ranging between 0.05 and 0.45.

Because fish feeding is controlled by fish weight, temperature, and prey density, the combined effect is multidimensional. The composite equation for fish feeding rate has many parameters and many curve forms can be obtained. Each of the effects on fish feeding rate assumes that the other two effects are independent of them. There is no interaction in their combined effect on fish consumption rate and the overall effect is the product of the individual effects.

4.5.2 Light effect on photosynthesis

The effect of light on photosynthesis is represented in a similar fashion for both trees and phytoplankton. This may at first appear unusual because of their size difference and because one is on land and the other in water. A gamma-like function was proposed by Steele (1962) to represent photosynthesis by phytoplankton. In this formulation the instantaneous photosynthesis rate *gpht* in units of grams per gram per day (/day) responds to incident light *zlt* by increasing until an optimal light intensity *kphtlt*$_{op}$, beyond which further increase in light will reduce photosynthesis. Reduction of photosynthesis at high light intensities is termed *photoinhibition* and is found in most terrestrial as well as aquatic plants.

This equation has also been used to represent photosynthesis of canopy trees. A major common factor between trees and phytoplankton is that both live in an environment that is often light limited. For both systems light is attenuated by the biota such that light intensity decreases with depth. Depth in the canopy is the distance below the top of the forest foliage, whereas for plankton it is the distance below the water surface. Light attenuation (reduction in light levels with increasing depth below the surface) is often represented through Beer's law, which states that light decreases exponentially through the water column:

$$zlt = zinsol \times e^{-kextlt \times z} \tag{4.7}$$

zinsol = solar radiation incident to the surface (ly/day)
z = depth (m)
kextlt = extinction coefficient (m)
zlt = light at depth *z* (ly/day)

Often the amount of foliage changes through the year, which requires making the extinction coefficient depend on the foliar (or phytoplankton) biomass. In this case the units of the extinction coefficient would be m^2/g. Equation 4.7 would be modified by replacing *kextlt* by *gextlt* = *kextlt* × *xbiop*, *xbiop* being the plant biomass in g/m^3. The product is still in units of /m, which assures the dimensional correctness of Eq. 4.7 (i.e., *gextlt* × *z* is unitless). The phenomenon of light attenuation due to plant biomass in a canopy or water column is termed *self-shading*.

Steele's equation for the effect of light on photosynthesis is given by:

$$gphtlt = kpht_{mx} \frac{zlt}{kphtlt_{op}} e^{1 - \frac{zlt}{kphtlt_{op}}} \tag{4.8}$$

gphtlt = effect of light on photosynthesis
$kphtlt_{op}$ = light intensity at which photosynthesis is a maximum (ly/day)
$kpht_{mx}$ = maximum photosynthesis rate (/day)

This equation is graphed in Fig. 4.3A as a function of light intensity *zlt* (langleys per day) with different values of optimal light intensity $kphtlt_{op}$ and with $kpht_{mx} = 1.0$. The equation is an exponential function of the form:

$$g(z) = k \times \left(\frac{z}{z_0}\right) e^{1 - \frac{z}{z_0}}$$

The parameters in Fig. 4.3A have been chosen to have biological meaning. $kpht_{mx}$ is the maximum photosynthesis rate, which occurs when light *zlt* equals the optimal light intensity $kphtlt_{op}$. Because light changes with depth and photosynthesis with light we can combine Eqs. 4.7 and 4.8 to obtain the total depth-averaged effect of light on photosynthesis rate $gphtlt_{tt}$. To do this we must integrate Eq. 4.8 over depth when Eq. 4.7 has been substituted for *zlt*. Thus:

$$gphtlt_{tt} = \frac{kpht_{mx}}{kd} \int_0^{kd} \frac{zinsol \times e^{-kextlt \times z}}{kphtlt_{op}} e^{1 - zinsol \frac{e^{-keztlt \times z}}{kphtlt_{op}}} dz \tag{4.9}$$

In this equation *zinsol* and $kphtlt_{op}$ are both in units of langleys per day. Depth *kd* is the depth of the *euphotic zone;* the depth at which net photosynthesis is zero. Integration of Eq. 4.9 is greatly simplified by making the variable substitution $s = e^{-kextlt \times z}$. If we also substitute $L = \dfrac{zinsol}{kphtlt_{op}}$, $g = gphtlt_{tt}$,

Model Processes

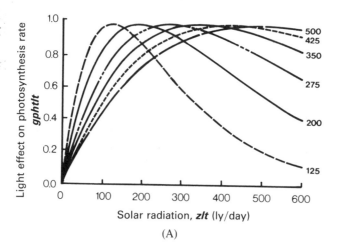

(A)

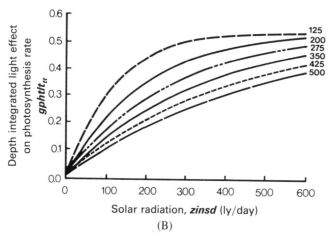

(B)

Fig. 4.3. Family of curves for the effect of solar radiation on phytoplankton photosynthesis rate based on Steele's equation. Curves are for values of optimal light intensity for photosynthesis $kphtlt_{op}$ ranging between 125 and 500 ly/day. Curve A. gives photosynthesis rate $gphtlt$ as a function of light intensity zlt incident to the phytoplankton. Curve B. gives photosynthesis rate $gphtlt_{tt}$ over the entire water column as a function of solar radiation incident to the surface ($zinsol$).

$k_1 = kpht_{mx}$, and $k_2 = kextlt$ for notational simplicity we have:

$$g = \frac{k_1 \times L}{kd(-k_2)} \int_1^{e^{-k_2 \times kd}} e^{1-L \times s}\, ds$$

Completing the integration gives:

$$g = \frac{k_1 \times e}{k_2 \times kd}[e^{-L \times e^{-k_2 \times kd}} - e^{-L}]$$

Substituting $\dfrac{zinsol}{kphtlt_{op}}$ for L gives, as well as reversing the other substitutions gives:

$$gphtlt_{tt} = \frac{kpht_{mx} \times e}{kextlt \times kd}(e^{-\frac{zinsol \times e^{-kl \times tlt \times kd}}{kphtlt_{op}}} - e^{\frac{-zinsol}{kphtlt_{op}}}) \qquad (4.10)$$

This equation is graphed as a function of *zinsol* in Fig. 4.3B using different values of optimal light intensity $kphtlt_{op}$ with $kd = 1.2$ and $kextlt = 40/m$. Notice that the depth-integrated light effect on photosynthesis increases even when surface light intensity *zinsol* is greater than the optimal light intensity $kphtlt_{op}$. This is because light attenuation reduces the light over most of the water column to below the optimal light intensity, even though the light intensity near the surface is above it. A drop in photosynthesis is possible with increased surface light intensity above the optimum light intensity only if light remains above the optimal intensity over most of the water column, a case that rarely occurs except in shallow Arctic lakes and thermal pools.

Another relationship between light and photosynthesis used for both trees and phytoplankton is the Michaelis relationship:

$$grphtlt = \frac{kpht_{mx} \times zlt}{zlt + kphtlt_{hs}} \qquad (4.11)$$

grphtlt = effect of light on photosynthesis
$kphtlt_{hs}$ = the light at which photosynthesis is half the maximum rate (*ly day*)

This relationship does not consider photoinhibition, but instead assumes photosynthesis increases with increasing light intensity. It has been used in models of tree and shrub photosynthesis (Holm and Kellomaki, 1984) and of phytoplankton photosynthesis (Larsen et al., 1974). As with the previous relationship the Michaelis or hyperbolic relationship can be combined with Beer's law and integrated to give a total depth-averaged photosynthesis rate:

$$gphtlt_{tt} = \frac{kpht_{mx}}{kextlt \times kd} \log\left[\frac{zinsol + kphtlt_{hs}}{kphtlt_{hs} + zinsol \times e^{-kextlt \times kd}}\right] \qquad (4.12)$$

4.5.3   Lion behavior and population dynamics

Let us turn to a different type of model process. We will consider the encounters in Kruger National Park in South Africa, between lion (*Panthera leo*) prides (those in breeding family groups) and those in "lodger" groups. Lodger groups are groups of male and female lions that are not mating and whose males are looking to join with a group of mature females. This must be done by lodger males challenging the dominant males in a pride.

The basic model (Starfield et al., 1981; Starfield and Bleloch, 1986) is a discrete time, lion population dynamics model with behavioral rules that govern the migration between territories and the interaction between lion prides and lodgers. Lodger groups are formed by expulsion from existing prides when they become subadults (age 2 to 3). These groups exist within the home territories of prides and migrate between territories based on food availability—migration being certain if food availability is poor.

The particular part of the model we wish to emphasize is what happens after migration when a lodger group encounters the territorial pride in that new territory. Females in the lodger group will attempt to join the pride, which they may do on a first come, first served basis if the number of pride females is below the "female capacity" for that territory. Others remain with the lodger group. Lodger males may challenge existing pride males. Success of the challenge depends on the mean age of both the challenger and the pride males and on the numbers of males in each group.

This relationship is quantified in terms of a challenge ratio $gchl_{rt}$, which determines the probability of success in a challenge. The challenge ratio is defined as the ratio of the threshold strength of challenge $gchl_{th}$, which in turn depends on the average ages of both pride and lodger males (Table 4.2), and the strength of challenge *gstren*, which is the number of challengers (*xlodger*, standing for lion lodgers) divided by the number of males in the pride (*xpride*). Thus:

$$gstren = \frac{xlodger}{xpride}$$

gstren = lodger challenge strength in a specific challenge
xlodger = the number of lodger males in a specific challenge
xpride = the number of pride lion males in a specific challenge

$$gchl_{rt} = \frac{gchl_{th}}{gstren}$$

$gchl_{rt}$ = challenge ratio, a measure of the challenge success probability
$gchl_{th}$ = challenge threshold, an age-dependent chance of challenge success

Table 4.2   Threshold strength for challenges in the lion model.

| | | Age of challenger males (yr) | | | |
		5	6	7	8
	5	1.50	1.00	0.90	1.00
	6	2.00	1.40	1.50	1.80
Age of pride males	7	2.00	1.40	1.40	1.80
(yr)	8	1.30	1.20	1.25	1.40
	9	1.25	1.00	1.05	1.35
	10	0.90	0.30	0.30	1.00

The higher the challenge ratio the smaller the chance of a successful challenge. For challenge ratios below 0.8 the challenge is sure to succeed and above 1.2 the challenge is hopeless and is not attempted. The probability of challenge success $gsuc_{pr}$ is a linearly decreasing function of the challenge ratio:

$$gsuc_{pr} = \begin{cases} 1.0 & \text{for } gchl_{rt} \leq 0.8 \\ 1.0 - \dfrac{gchl_{rt} - 0.8}{0.4} & \text{for } 0.8 < gchl_{rt} \leq 1.2 \\ 0.0 & \text{for } gchl_{rt} > 1.2 \end{cases} \tag{4.13}$$

$gsuc_{pr}$ = probability of challenge success

This relationship is depicted in Fig. 4.4. It is a linear relationship with a couple of thresholds. The workings of this challenge module can best be illustrated with an example. If six lodgers of average age 6 challenge five pride males of age 5, $gchl_{rt}$ is 0.83 (challenge threshold = 1.0 and challenge strength = 6/5 = 1.20) and the challenge is almost certain to succeed, whereas if the lodgers are of average age 5, $gchl_{rt}$ is 1.25 and there is no chance that the challenge will succeed. This shows how sensitive lion challenge dynamics are to the age of the lodgers.

If the challenge succeeds the old pride males are killed, reflecting the observation that ousted pride males die within a short time. The challengers then kill a proportion of the cubs, this proportion depending on the number of challengers. Challenges that fail result in mortality to both the lodger and pride lions, the mortality depending on the challenge ratio. The challenge function is a stochastic function in that challenge success is represented as a probability. In implementing such a model, a *Monte Carlo* method is used in which the value of a

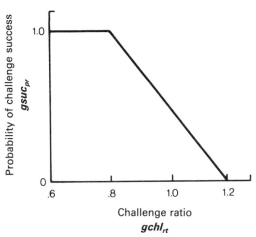

Fig. 4.4.   Probability of a successful challenge by lodger males on territory holding pride male lions, $gsuc_{pr}$, as a function of the challenge ratio, $gchl_{rt}$, a measure of the relative numbers and ages of lodger and pride lion males involved in the challenge.

generated random number is compared with the success probability $gsuc_{pr}$. Success occurs if the random number is smaller than the probability of challenge success $gsuc_{pr}$. Details of Monte Carlo methods in ecological simulation are discussed in Chap. 7.

4.6   Rationale and estimation for parameters

We now turn to the question of parameter estimation for ecological processes. This task represents a special set of problems for model building because the data base used is often incomplete and always has more detail in some processes than in others. Sometimes parameter estimation can be done using statistical techniques such as least squares regression. More often the data available are not conducive to this approach, either because there are too many parameters changing at the same time in the data to separately estimate the parameters or because insufficient data are available to obtain a useful estimate. Some parameter values are not estimable directly from data and must be estimated from data on another species or from a value taken from a similar model. Often data from several experiments conducted by several researchers are available and these are used to suggest a range of estimates for a parameter that are later refined through the calibration process. Because the models are *mechanistic* (have process equations based on some physical or biological rationale) we often have some idea what the parameter values ought to be. An example is with the effect of weight on fish respiration, where hypothesizing that the respiration rate is proportional to fish surface area results in an exponent for the weight effect of around $\frac{2}{3}$ (the surface area to volume ratio produces a $\frac{2}{3}$ in the exponent).

Ecological process parameter estimation is a melange of methods ranging from classical regression to extrapolation from other ecosystems or models to plain educated guesswork. It is best presented through examples. Let us return to discuss parameter estimation for each of the examples of the previous section, with an eye to the biological supporting rationale and equations for them.

4.6.1   Weight effect on fish feeding

Many process equations evolve over a time period far beyond the time span of a single modeling project. A good example is the effect of weight on fish consumption rate. Figure 4.5 shows several models that used Eq. 4.3, the allometric consumption function, and who they cited. As can be seen, the initial suggestion of this function dates back to 1920, long before it was used in any simulation model. The evidence for use of this function is largely empirical, with regression analyses by Winberg (1956) and Ursin (1967) collating data from many laboratory studies and weight-at-age data on a wide variety of fish species. These data analyses support the hypothesis that fish growth is a power function of weight and that the exponent $kfdwt_2$ in the consumption expression is remarkably similar from one species to the next, being between -0.33 and -0.44. Remember that consumption rate $gfdwt$ is in units of /day or gram

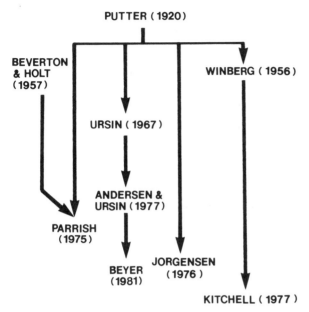

Fig. 4.5.   Hierarchical tree showing researchers doing work leading to the allometric consumption function for fish ration as a function of body weight given in Eq. 4.3.

consumption per gram body weight per day. To convert this into a consumption rate per individual *gfdwt* must be multipled by body weight *xwt*. Table 4.3 summarizes the results of a review by Ursin (1967) of regressions using length at age data for 81 species. The relatively narrow range of parameter values in this table is strong evidence for the applicability of Eq. 4.3 over a wide variety of fish species.

The estimates for $kfdwt_2$ in Table 4.3 were obtained by Ursin in a complicated way from length at age data. A few process studies on fish consumption and respiration were available to him. One estimate for $kfdwt_2$ of -0.31 was obtained from data of O_2 consumption by fed and fasting cod (*Gadus morhua*). These values were not used with Ursin's length at age data, however. Instead, the assumptions were made that:

$$x\dot{w}t = [kfdwt_1 \times xwt^{kfdwt_2} - krspwt_1 \times xwt^{krspwt_2}]xwt \qquad (4.14)$$

and

$$xwt = klenwt \times xlen^3 \qquad (4.15)$$

$krspwt_1$ = respiration rate of a 1-g fish ($g^{-krspwt_2}$)
$krspwt_2$ = exponent in the effect of fish weight on respiration rate
$klenwt$ = parameter in the fish length-weight allometric relationship (g/mm^3)

The weight differential equation does not consider temperature or food density effects. Ursin corrected the data for temperature differences where possible and used experiments where food was not limiting.

Table 4.3   Summary of the effect of fish weight (*xwt*) on fish consumption rate.

Species group	No. species	Range	Mean	Standard deviation
Clupeidae (herring, sardine)	10	$(-0.58, -0.24)$	-0.40	0.08
Gadoidae (cod, pollock, hake)	10	$(-0.64, -0.28)$	-0.43	0.10
Pleuronectoidae (sole, halibut)	15	$(-0.48, -0.26)$	-0.37	0.08
Scombridae (tuna, mackerel)	8	$(-0.52, -0.28)$	-0.40	0.08
Perciformes (perch, bass, mullet)	24	$(-0.67, -0.26)$	-0.44	0.12
Scorpinoformes (rockfish, sablefish)	6	$(-0.70, -0.31)$	-0.45	0.14
Other fish	8	$(-0.57, -0.38)$	-0.39	0.13

Respiration exponent $krspwt_2 = 0.832$. Fish are organized into groups with example species listed in parentheses. (Adapted from Ursin, 1967.)

First *klenwt* was estimated for each species from length versus weight data. Sometimes, where weight data were not available, a *klenwt* value for a similarly shaped fish was used. The allometric relationship, Eq. 4.15 was substituted into the weight change differential equation, Eq. 4.14 using a change of variables to convert the weight differential equation into a length differential equation. This gave the differential equation:

$$\dot{xlen} = \frac{kfdwt_1}{3} \times klenwt^{kfdwt_2} \times xlen^{3kfdwt_2+1}$$
$$- \frac{krspwt_1}{3} \times klenwt^{krspwt_2} \times xlen^{3krspwt_2+1} \tag{4.16}$$

A researcher at present would probably approach this estimation problem using nonlinear regression. In the late 1960s, however, when this work was done, linear regression was the state of the art and this highly nonlinear equation is not amenable to linear regression. Ursin approached the problem by assuming that $krspwt_2 = 0.0$ and $kfdwt_2 = -\frac{1}{3}$, giving a linear regression of $\dot{xlen}$ versus $xlen$ from which initial estimates of $kfdwt_1$ and $krspwt_1$ could be obtained (the differences in length between adjacent age classes was used for values of $\dot{xlen}$). The rationale for using these values were that they were Putter's original values (Fig. 4.5).

Data on O_2 consumption from several fasting fish experiments led to an estimate for $krspwt_2$ of -0.17. This suggested that the estimate of 0.0 for $krspwt_2$ was not acceptable. The value of $krspwt_1$ from the above linear regression of *xlėn* versus *xlen* was retained in further estimates of other parameters. Ursin was still left with three parameters to estimate in Eq. 4.16. Another simplifying assumption, that $krspwt_2 = -0.33$ was made. This made Eq. 4.16 amenable to linear regression analysis by dropping *xlen* entirely from the respiration term in the equation (second term on the right), because the exponent on *xlen*, $3krspwt_2 + 1$, is now zero. This leaves a constant for the second term on the right and makes the equation depend only on $kfdwt_1$, *klenwt*, and $kfdwt_2$. The constant was added to estimates of *xlėn* and logarithms of both sides taken to give a linear equation for *xlėn* (plus a constant) in terms of *xlen*. $kfdwt_1$ and $kfdwt_2$ were estimated by regression on this equation.

The 81 species regression gave a $kfdwt_2$ of -0.44. Estimating a rate by an annual difference in lengths, as was done for *xlėn* can introduce some bias into the regression estimates, especially when the initial length is used for *xlen*. When $kfdwt_2$ was adjusted for this factor by using average length rather than the initial lengths a $kfdwt_2$ of -0.41 was obtained.

Table 4.3 shows that there is no significant difference between taxonomic fish groups in this parameter, or for that matter, between fish having different habitats, feeding styles (e.g., active or passive, schooling or solitary). Other estimates from controlled feeding experiments (Stewart et al., 1983), where feeding rates were measured directly by keeping track of the food removed, gave values for $kfdwt_2$ between -0.24 and -0.35. The estimates of $kfdwt_2$ from Table 4.3, which were about the best available in 1967, are significantly lower (more negative) than the values estimated from direct experiments, which may suggest a general bias in the assumptions and method adopted in using length at age data for estimating $kfdwt_2$. This example poignantly illustrates the large number of assumptions needed to estimate parameters when direct experimental evidence is unavailable. Ursin had to make strong statistical assumptions to force the data into a mold for allowing parameter estimation. The bad news here is that such machinations are frequently necessary in ecological modeling. Many parameter values used in the literature are based on considerably less evidence than $kfdwt_2$.

In addition to the empirical evidence there are two hypotheses that provide a rationale for the weight effect of feeding function. The first, from Andersen and Ursin (1977), alluded to earlier, is that food absorbing surface area is proportional to xwt^{kfdwt_2} and that consumption is proportional to this. Surface area to volume ratio depends somewhat on the shape of the fish. Body weight can be assumed to be directly proportional to body volume. Thus, the surface area to volume ratio can be equivalenced to the ratio of surface area to body weight. For a sphere or a cylinder surface area would be proportional to $xwt^{0.67}$ and the per unit weight (volume) ratio would be proportional to $xwt^{-0.33}$. This, incidentally, is often used to justify a respiration exponent of -0.33, because individual fish respiration rate can be seen to depend on the surface area and the per gram fish body weight respiration rate to depend on the surface area to volume ratio.

The other rationale, reported in Kitchell et al. (1974), is based on changes in stomach volume and digestion rates in relation to change in body weight. Ex-

perimental evidence that maximum food intake is linked to body size via stomach
volume and digestion rates comes from laboratory experiments by Brett (1971)
on sockeye salmon (*Oncorhynchus nerka*) and Elliott (1976) on brown trout
(*Salmo trutta*). In these experiments stomach volume is a function of body
weight xwt^{kfdwt_2}.

4.6.2 Photosynthesis

Two light–photosynthesis relationships were presented. The first was a gamma-
like function from Steele (1962), whereas the other was a hyperbolic function
used by Larsen et al. (1974) and Caperon (1967) for phytoplankton and by Holm
and Kellomaki (1984) for tree and shrub photosynthesis.

The main difference between the two is that Steele's equation postulates the
existence of photoinhibition, or the reduction of photosynthesis at high light
intensities. This phenomenon has been commonly observed for phytoplankton
(Ryther, 1956; Ryther and Menzel, 1959; Lassiter, 1975). For trees, photoinhibition has not been commonly observed. This is in part because macrophytes
(larger plants) can orient the stomates inside their leaves to reduce exposure to
high light intensities. Also, many tree leaves have wax layers that are thought
to filter ultraviolet wavelengths, the major frequencies that can damage the
stomates. Adaptation to cope with high light intensities was a requisite part of
the evolution of terrestrial plants.

Table 4.4 gives some values of optimal light intensity for photosynthesis
$kphtlt_{op}$ for different groups of phytoplankton, and the data sources for estimating
them. These are based on laboratory experiments under controlled light conditions. Table 4.5 gives the half saturation parameter $kphtlt_{hs}$ for the effect of light
on photosynthesis for both terrestrial plants and phytoplankton. The terrestrial
plant values are based on cuvette experiments where leaves were kept in controlled temperature, CO_2 and O_2 environments and photosynthesis rate (as CO_2
uptake rate) was measured along with light intensity.

Light extinction coefficients through the canopy or water depend on the density of biomass. In water, they also depend on such factors as turbidity. Technically, light attenuation should be computed separately for the photosyntheti-

Table 4.4   Optimal light intensity, $kphtlt_{op}$ (ly/min) for different
groups of phytoplankton.

Group	Experimental range	Typical value	Species (genus)
Diatoms	(0.130, 0.200)	0.160	*Navicula*
Blue-greens	(0.067, 0.183)	0.106	*Anabaena*
Small greens	(0.050, 0.264)	0.066	*Chlorella vulgaris*
Colonial greens	(0.055, 0.099)	0.055	*Scenedesmus*
Filamentous greens	(0.019, 0.079)	0.019	*Ulothrix*

(After Rose, 1985.)

Table 4.5   Solar radiation level at which photosynthesis is half maximum $kphtlt_{hs}$ (ly/day) reported for both terrestrial and aquatic plants. Photosynthesis is assumed to follow Eq. 4.11.

Producer type	$kphtlt_{hs}$	Reference
Oak saplings	200*	Holm and Kellomaki (1984)
Warm season grasses	255	Sauer (1978)
Cool season grasses	200	Sauer (1978)
Forbs	190	Sauer (1978)
Shrubs	200	Sauer (1978)
Cactus	225	Sauer (1978)
Aquatic macrophytes	50–259*	Cosby et al. (1984)
Phytoplankton	163	Larsen et al. (1974)

*Conversion factors assume 1 *lux* = 5.6×10^{-5} ly/min and a 12-hour day.

cally active wavelengths because plants do not use all light wavelengths in photosynthesis, and because attentuation is wavelength specific. Hutchinson (1957) found that below the first meter, use of an average light extinction (attenuation) coefficient over all light wave lengths was a good approximation. Self-shading, or light attenuation due to biota is important both in the sea and forest. Models for the effect of self-shading on the light extinction coefficient *gextlt* have been developed by Riley (1956) and Chen (1970) for phytoplankton and by Horn (1971), Botkin et al. (1972), and Reed et al., (1976) for trees.

4.6.3   Lion challenge

The lion model challenge module was based entirely on observation of lion challenges in the field. For example, the challenge thresholds in Table 4.2 were obtained by a series of statements on the evenness of match-ups obtained from park rangers and others doing lion behavior studies. Thus the statement that two 6-year-old pride males and three 7-year-old lodger males are evenly matched would result in a challenge threshold strength for challenges between 7-year-old lodgers and 6-year-old pride males of 1.5 (1.5 lodgers are needed for each pride male to have a challenge success probability of 0.5).

Other aspects of the challenge, such as the increase in mortality, after a successful challenge, in proportion to the number of challenging males, were also based on observation and hypothesis. Here it was hypothesized that with more males in the new group, increased competition for females results in more adverse adult male contact with cubs and subsequently greater cub mortality.

4.7   Discussion

The following discussion of parameter estimation and rationale for various processes gives a picture of the present state of the art. Although many processes are based on controlled laboratory or field experiments, others are based on general observation or anecdotal information. The examples we chose are representative of the range of ecological simulation model process formulations.

The following generalizations can be made regarding parameter estimation and equation rationale for simulation models.

1. The higher the trophic level, the less accurate are the process representations and the parameter estimates. Exceptions include studies on domestic herbivores, for which a plethora of data are available.
2. Abiotic and producer processes are most amenable to study under controlled conditions or are based on physical principles.
3. Processes relating to animal behavior are in general based on observation and theory and not on experiments. Parameters for these processes are often not estimable directly from data.
4. Parameters for higher trophic level organisms are often obtained as empirical estimates from data on a wide variety of species and often are not available for the species of interest in the model.
5. One important objective of ecological simulation is to find model equations that apply to a wide variety of species with only the parameter values in the equations changing from species to species.
6. Just as ecologists characterize species by their life history or their taxonomic relationship to others, they may also characterize them by the parameter values they have for processes common to a group of species. For example, phytoplankton may be grouped according to their optimal light intensities for photosynthesis $kphtlt_{op}$. These parameters can be used to define in a more quantitative way qualitative concepts that exist in ecology. Shade tolerant trees, in this sense, would be trees having lower light optima so that they reach their maximum growth potential at low light intensities, whereas shade intolerant trees have high light optima and show little or no growth under low light conditions.
7. Biota are often classified into functional groups of species with similar parameters for the processes. Biota in the same functional group would be looked at as identical in the model.
8. There are no general standards for parameter estimation. Methods tend to be ad hoc and differ from model to model depending on the system being modeled and the biota of concern.

4.8 Summary

We have given a feeling for the model formulation process through examples taken from three very different kinds of model processes. We have carried these examples through the equation stage, giving their rationale and also giving alternative formulations where these exist. We have also traced the data estimation procedures for selected parameters. To communicate clearly model equations we use a mnemonic notational scheme that is introduced here and is used throughout the book. As we will show later this scheme conflicts somewhat with the structure of Fortran, which is array oriented, and becomes inefficient when a repetitive computation procedure must be written separately for each case. We have found that initially the mnemonic scheme appears clumsy when compared with an array notation. When several models are developed, however, it is wise to have notation that is specific to the models or confusion between variables in different models arises. Our scheme is a compromise between comprehension and succinctness.

Chapter 4 is the breakwater chapter in this book. It opens the door to the burgeoning field of ecological simulation modeling and provides the keys to the specification of process functions and justification for the choices made. We also consider the use of data to estimate model parameters. Most simulation models stand or fall on how well this work is accomplished. Using available evidence and data in constructing these models is an activity which, although common to all types of simulation, has a unique flavor in ecological simulation. This is because of the complexity of the systems, the relative paucity of data about the processes, and the lack of a unified theory for ecosystems. Because of this, process formulation has more of the flavor of an art than of a scientific endeavor. Like the artist we are frequently mixing data from many sources. These data are filtered through experiments often conducted on different biota and at different locations than those modeled. It is not surprising if the results look somewhat "Impressionistic."

4.9   Annotated Bibliography

Bowie, G. L., W. B. Mills, D. B. Porcella, C. L. Campbell, J. R. Pagenkopf, G. L. Rupp, K. M. Johnson, P. W. H. Chan, S. A. Gherini, and C. E. Chamberlin: *Rates, Constants, and Kinetics Formulations in Surface Water Quality Modeling, 2nd ed.,* EPA Environmental Research Laboratory Report EPA/600/3-85/040, p. 455, Athens, Georgia, 1985.
   A comprehensive review of process equations and parameter values for all aspects of water quality modeling ranging from detailed uptake and growth processes to general hydrological flow processes.
Eppley, R. W.: Temperature and phytoplankton growth in the sea. *U.S. Nat. Mar. Fish Serv. Fish. Bull.,* 70:1063–1085, 1972.
   The first convincing argument for the use of functional groups in simulation models based on the way the growth rate of a multispecies plankton population responds to temperature.
Holling, C. S.: The functional response of predators to prey density and its role in mimicry and population regulation. *Can. Entomol.,* 41:385–398, 1965.
   First treatment of and rationale for the use of a Michaelis-type function for the effect of prey density on feeding rates of predators.
Hutchinson, G. E.: *A Treatise on Limnology, I. Geography, Physics and Chemistry,* Wiley, New York, 1957.
   All books by Hutchinson are classics as information sources for limnology.
Jorgensen, S. E.: *Handbook of Environmental Data and Ecological Parameters,* International Society of Ecological Modelling, Copenhagen, Denmark, 1979.

Chock full of parameter values from laboratory and field studies and models. This can be very valuable for indicating the range of values reported in the literature for a particular parameter. We have found it difficult to locate a particular parameter.

Jorgensen, S. E. ed.: Modeling primary productivity. *Ecol. Model.,* 23:1–184, 1984.
Nine articles on modeling primary production in all kinds of ecosystems in this special volume of Ecological Modelling. These models provide further reading on the photosynthesis review in this chapter.

Leidy, G. R., and R. M. Jenkins: *The Development of Fishery Compartments and Population Rate Coefficients for Use in Reservoir Ecosystem Models,* Final Contract Report Y-77-1, Environmental Effects Laboratory, U.S. Army Engineer Waterways Experiment Station, Vicksburg, Miss., 1977, p. 134.
A review of fish energetics processes and parameters with an emphasis on fish species in southern U.S. reservoir ecosystems.

Leidy, G. R., and G. R. Ploskey: *Simulation Modeling of Zooplankton and Benthos in Reservoirs: Documentation and Development of Model Constructs,* Technical Report E-80-4, Environmental Effects Laboratory, U.S. Army Engineer Waterways Experiment Station, Vicksburg, Miss., 1980, p. 221.
A review of parameters and data sources for zooplankton and benthos processes. The review is by process and gives a systematic treatment of the various relevant parameters using a standard notation and units.

Scavia, D., and A. Robertson, eds.: *Perspectives in Lake Ecosystem Modeling,* Ann Arbor Science, Ann Arbor, Mich., 1979.
A modeler's guide to lake ecosystem modeling. The book treats improvements in process representations from those in existing models. It also discusses the role of models in setting research priorities and serving as a management tool.

Sokal, R. R., and F. J. Rohlf: *Biometry, 2nd ed.,* W. H. Freeman and Co., San Francisco, 1981.
Gives a good, ecologically oriented review of basic statistics including such relevant topics as ANOVA, correlation, and simple linear regression.

Steele, J. H.: Environmental control of photosynthesis in the sea. *Limnol. and Oceanogr.,* 7:137–150, 1962.
This is the paper from which Steele's equation, the most commonly used for the effect of light on temperature, came.

Sullivan, P. G., Swartzman, and A. Bindman: *Process Notebook for Aquatic Ecosystem Simulations, 2nd ed.,* U.S. Nuclear Regulatory Commission Report NUREG/CR-3392, 1983, p. 182.
A manual for fish simulation modeling. It contains process by process comparisons of different hypotheses in the literature, reviews parameter values, and gives data sources. All equations are given in standard notation and units are standardized between the different resource models reviewed.

Swartzman, G. L., and R. Bentley: A review and comparison of plankton simulation models. *ISEM Journal,* 1:30–78, 1979.
An extensive review of the processes affecting plankton growth, feeding, and population dynamics that translates other models into a standard notation.

Ursin, E.: A mathematical model of some aspects of fish growth. *J. Fish. Res. Board Can.,* 24:2355–2453, 1967.
An impressive review of the data evidence supporting fish energetic process parameter estimates.

4.10 Exercises

Derivations

4.1. Derive Eq. 4.12 using Eq. 4.11, Beer's law and integration of the resultant equation over depth. This should be done in a manner analogous to the derivation of Eq. 4.10 from Eq. 4.8 through Eq. 4.9 and direct integration.

Parameter estimation from data

4.2. Beer's law assumes light attenuates with depth in the water column. Phytoplankton in the water column act as self-shaders, with the light attenuation coefficient depending on the phytoplankton density. Water also attenuates light. Generally light attenuation because of water is distinguished from that caused by phytoplankton. The following data give the percent penetration of light at different depths (m) in a lake during the winter when minimal biota are present in the water column.

depth	% light
0.0	100.0
0.5	44.0
1.0	30.0
2.0	23.0
3.0	18.0
4.0	12.0
5.0	9.5
6.0	8.0
7.0	5.0
8.0	4.0
10.0	2.5
12.0	1.4

Use these data to estimate the clear water light extinction coefficient *kextlt*. These data were collected under cloudy ice, which accounts for the relatively small light penetration through the first 0.5 m. To fit Beer's law you should start with the light at 0.5 m as if it were the incident light intensity to the water column.

4.3. The following average length and weight at age were computed from catch data for an Atlantic Ocean rockfish *Sebastes mentella:*

age (yr)	length (cm)	weight (g)
1	5.2	2
2	8.5	8
3	11.5	21
4	14.3	38
5	16.8	69
6	19.2	117
7	21.3	148
8	23.3	190
9	25.0	264
10	26.7	293
11	28.2	318
12	29.6	371
13	30.8	455
14	32.0	504
15	33.0	518
16	34.0	537
17	34.9	651
18	36.4	719
19	37.1	726
20	37.7	810

These data can be used to support an empirical relationship between body length and weight, which has been found to be accurate for most fish species growing under normal (not starvation) feeding conditions. This so-called *allometric length–weight relationship* has the form in Eq. 4.15. Use these data to obtain an estimate for *klenwt* assuming first that the exponent is $+3$ and then estimate both *klenwt* and the exponent. How far off is $+3$ as an exponent for these data? How much change is there in *klenwt*?

4.4. Stewart et al. (1983) considered the following model for lake trout consumption:

$$gfd = kfdwt_1 \times xfw^{kfdwt_2} \times e^{kfdtp \times ztp}$$

where, xfw = fish weight (g)
ztp = temperature (°C)
$kfdwt_1$, $kfdwt_2$, and $kfdtp$ are constants

a. Use the following fish feeding experiment data (simulated) to estimate $kfdwt_1$, $kfdwt_2$, and $kfdtp$ using multiple regression. (*Hint:* a log transformation would help here).

xwt	ztp	gfd	xwt	ztp	gfd
40	3	0.0408	600	12	0.0113
100	3	0.0207	900	12	0.0294
250	3	0.0236	1200	12	0.0478
600	3	0.0095	40	15	0.1816
900	3	0.0125	100	15	0.0943
1200	3	0.0076	250	15	0.0584
40	6	0.0366	600	15	0.0640
100	6	0.0411	900	15	0.0444
250	6	0.0279	1200	15	0.0414
600	6	0.0167	40	18	0.0677
900	6	0.0154	100	18	0.0550
1200	6	0.0196	250	18	0.1016
40	9	0.0657	600	18	0.0483
100	9	0.0525	900	18	0.0336
250	9	0.0323	1200	18	0.0324
600	9	0.0323	40	21	0.1282
900	9	0.0197	100	21	0.3156
1200	9	0.0305	250	21	0.2012
40	12	0.0552	600	21	0.1163
100	12	0.0566	900	21	0.1024
250	12	0.0742	1200	21	0.1219

b. Is this a good model for these data?

c. Which factor, weight or temperature is more important for modeling consumption?

Comparing curve types

4.5.

a. Graphically compare the gamma (Eq. 4.5) with the Q_{10} formulations for the effect of temperature on fish feeding. The Q_{10} formulation is given by:

$$gfdq_{10} = kfdtp \times kfdq_{10}^{\frac{ztp - 10}{10}}$$

$kfdtp$ = temperature effect at 10°C
$kfdq_{10}$ = Q_{10} for feeding

Make the graphs as a function of temperature over the 0 to 30°C range. Set $kfdq_{10}$ = 2.0 for each curve. Make the optimal temperature for feeding in the gamma function $kfdtp_{op}$ 25°C and the upper feeding temperature $kfdtp_{mx}$ 30°C. Adjust the $kfdtp$ parameter so that both curves are equal at the optimal temperature for growth in the gamma function. How do they compare at low temperatures (0–10°C)?

b. If instead we make the two agree at low temperatures how far off are they near the optimal temperature?

c. If you were to model a lake that had average temperatures between 0 and 22°C, what Q_{10} curve would you choose that would give the best approximation to the gamma curve with a $kfdq_{10} = 2.0$ over this temperature range? Assume that you can change both parameters in the Q_{10} curve for this problem.

4.6. The Holling equation (Eq. 4.6) is the most commonly used equation for the effect of prey density on feeding rate. An alternative was developed by Ivlev and is given by:

$$gfdpy = 1.0 - e^{-(kfdpy \times gprey)}$$

Graph both curves as a function of prey density choosing parameters such that both curves have half the maximum feeding rate at a total prey density *gprey* of 0.3 g/m³. Which curve is more sensitive to a 50 percent increase or decrease in its parameter?

Answers to Selected Exercises

4.2. $kextlt = 0.2890/\text{m}$.

4.3. Regressing weight on length³ through the origin gives $klenwt = 0.0148$ g/mm³. To estimate both the exponent and *klenwt* the logarithm of weight is regressed on the logarithm of length to get $klenwt = e^{-4.37} = 0.0126$ and the exponent estimate is 3.05.

4.4. Multiple regression using the logarithms of *gfd* and *xwt* along with *ztp* gives $kfdwt_2 = -2.874$, $kfdtp = 0.1057$ and $kfdwt_1 = e^{-2.765} = 0.063$. The coefficient of determination for the logarithm model is $R^2 = 0.775$. The model fits the data well. Temperature appears to be more important than weight as the correlation between the logarithm of consumption and temperature is higher than its respective correlation with the logarithm of weight.

4.6. The Ivlev curve is more sensitive to parameter change.

Chapter 5

Nonlinear Models

5.1 Introduction

In this chapter we look at simple, yet full blown, ecological simulation models. To avoid getting your heart rates up into the danger zone, we begin with a simple case study of passerine (perching bird) population dynamics and energetics (Swartzman, 1969). We then move up in scale to describe a model of plankton in a river delta (Di Toro et al., 1971) and finally a forest succession model (Botkin et al, 1972). Our descriptions are mainly of the model equations and their rationale. We show how the choice of state variables and processes depend on the ecosystem, the model objectives, and the underlying assumptions about the system. Coding and analyzing model output are deferred to Chap. 6. Before beginning our discussion of these models we describe some protocols for model documentation. We illustrate these with the description of each of the three models exemplified in this chapter.

5.2 Model documentation

As mentioned earlier, model documentation needs to portray the model objectives, the conceptual basis for the model, how it was implemented and how it is performed, to persons not connected with the modeling project. This is no mean task. We suggest several approaches that we have found helpful, both in documenting models and in reviewing models from the literature.

The document for the conceptual model must describe the following:

1. Underlying model assumptions (e.g., spatial boundaries, time scale)
2. Major state variables
3. Driving variables
4. Flows between state variables (compartments)
5. information about and rationale for equations controlling flows
6. Model parameters, units, and values in the model
7. Data sources and estimation methods for model parameters

The implementation document must contain (or provide on request):

1. Model code with adequate code documentation (comments) to clarify the code
2. Initial conditions for state variables
3. Initial and calibrated parameter estimates
4. Graphs of or reference to driving variable data

Tools for model evaluation are discussed in Chap. 8, along with methods for documenting model evaluation.

5.2.1 Model diagrams

We have found flow diagrams helpful in communicating the major state variables and flows in the model. Several approaches are commonly used. The simplest of these is the compartment diagram, introduced in Chap. 2. Although this diagram gives the major state variables and flows, it gives no information about what controls these flows. When several types of material flow are represented in the model, separate diagrams are required as is seen in Fig. 5.1 for a coniferous forest stand growth model CONIFER (Swartzman, 1979a).

Two alternatives to the compartment diagram are H. T. Odum's (1983) energy circuit language diagram and Forrester's (1961) flow control diagram. We in-

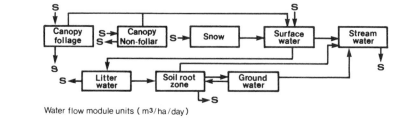

Water flow module units (m3/ha/day)

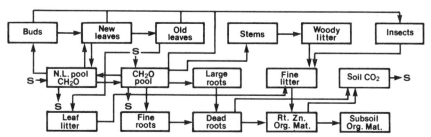

Carbon flow module (tons C/ha/wk, 1 ton = 1000 kg)

Fig. 5.1. Compartment diagram for the CONIFER model, showing the flow of carbon and water in a coniferous forest stand in the northwestern United States dominated by douglas fir, *Pseudotsuga menziesii.* Water flow is in units of m³/ha/day and carbon flow is in units of tons C/ha/wk. Sources and sinks of water and carbon are labeled with directed arrows and the symbol S.

clude these approaches here because they, or minor variants, are commonly used in the ecological literature and because both require some introduction, i.e., they are not self-evident like the compartment diagram is. We will first illustrate these diagrams as they represent a simple model of diurnal patterns in blue-green algal microcosms (Odum, 1983). These microcosms have a mat of blue-green algae growing underwater on a sand substrate.

Odum's approach is based on the theory that energy is the currency of eco-systems and that all ecosystem processes are energy transactions. He builds compartments out of energy transaction elements, energy storage elements, energy combining elements, and energy processing elements. Figure 5.2A gives an example of an Odum energy flow diagram that indicates how ecosystem models are built from energy processing modules. The diagram symbols come from electric circuit diagrams and analog computers. Respiration losses are represented by the electrical ground symbol ($\doteq$). Triangles show energy combiners ($\rightarrow\!\!\!\triangleright$) and arrow-like hexagons are energy processers or *work gates* ($\boxtimes$). Passive energy storers are in compartments that look like water tanks ($\diamondsuit$), whereas driving variables are represented by circles. Odum represents producers by cutoff ovals ($\supset$) and consumers by regular hexagons ($\bigcirc$). It is difficult to separate different materials flowing in these diagrams because all flows are perceived as energy transactions.

Notice how every transaction results in an energy loss to respiration, which, according to thermodynamic principals, is the case. Also, notice how major compartments in the simple flow diagram include several energy processing or storing subunits. For example, producers consist of a storage element, two energy combiners, and two energy processers. The energy processers are the photosynthesis and respiration processes. The first energy combiner adds pho-tosynthesis products to oxygen in the water to produce dissolved oxygen. The second combiner takes photosynthate and oxygen and produces cell material. Although the energy flow nature of this diagram is unambiguous, there is am-biguity in combining the oxygen in the water molecules and the oxygen dissolved in the water into the same compartment. Also, the detail considered in the energy processing within the producers is greater than that needed to model the oxygen and carbon flow in this simple system. Such detail, despite its revelation-like quality, can be a hindrance in modeling the system. We are reminded of Rilke's *Notes from Malte Laurids Brigge* where the protagonist is so sensitive that he becomes aware of the motion of the earth. His sensitivity causes him to be unable to get out of bed because of dizziness. He envies the poor fools who lack his sensitivity, for they are able to function while he is not.

Forrester's approach (Fig. 5.2B) embeds a flow diagram in an information or control network. Flows between compartments or state variables are controlled by valves, represented by two triangles point to point symmetric around the flow line that they control ($\bowtie$). This valve is connected via dashed lines to those factors controlling the flow. The dashed lines can be seen as shutting or opening the valves that control the flows. Controlling factors include state vari-ables (inside rectangles), driving variables (inside either small ovals with lines

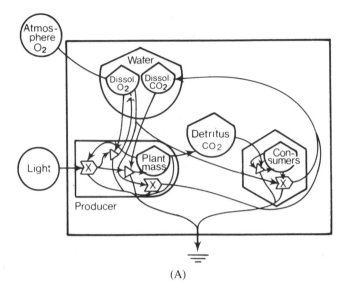

(A)

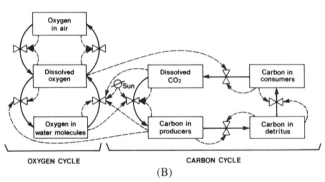

(B)

Fig. 5.2.   Representation of a model simulating the diurnal pattern of blue-green algal mats using A. Odum's energy diagram, and B. Forrester flow control diagrams. Both oxygen and carbon are flowing in this model. From Odum, H. T., *Systems Ecology,* Wiley, 1983. Reprinted by permission.

through them or hexagons), and parameters (inside circles). Sources and sinks of energy or material are represented in these diagrams by amorphous masses. You can see from Fig. 5.2B that donor controls are denoted by a dashed arrow originating from the donor compartment and going to the valve controlling the flow from that compartment. This implies that the flow rate out of the donor compartment is controlled by the amount in that compartment. Also notice the separation of flows in the Forrester diagram into oxygen and carbon flows. The Forrester diagrams combine the work gates and energy combiners in Odum's diagrams into valves with controls. They emphasize a different aspect of the processes than the Odum diagrams.

For further comparison of the three diagram alternatives let us take a single model: the eutrophication model by Di Toro et al. (1971), to be presented later in this chapter. The diagrams are given in Fig. 5.3A, B, and C. An advantage of the Odum diagram is that it distinguishes the major building blocks in the model from each other, for example the autotrophs (primary producers) from the heterotrophs (consumers), through different diagrammatic symbols. The energy focus of these diagrams is emphasized by observing that all interactions between compartments require work gates, which lead to respiration as an energy loss. The Odum diagram, however, mixes the nutrient and biomass flows together as energy equivalents, which misses the role that nutrients play in photosynthesis as a flow moderator. Also the Odum diagram emphasizes the flows instead of the equations. Notice that the Odum diagram in Fig. 5.3B does not have the internal detail on the producer and consumer contained in the Odum diagram in Fig. 5.2A. The focus of the model makes this detail unnecessary and, as you can see, it is easy to collapse or expand the amount of detail in the Odum diagram. Finally, because of its energy focus, the Odum diagram does not have the ability to uniquely represent some processes such as nutrient control or behavior. This limits the perspective we can have on such processes.

The Forrester diagram adds details and realism by separating the flows from the controls. Although nutrient and biomass flows are not distinguished, the

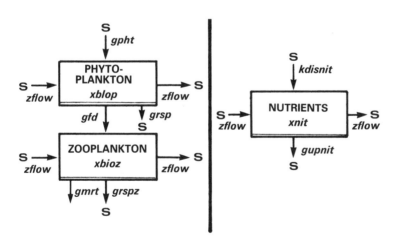

Process List

gpht—photosynthesis	*gmrt*—mortality
grspz—respiration	*gupnit*—nutrient uptake
gfd—grazing	*kdisnit*—nutrient discharge
zflow—inflow/outflow	

(A)

Fig. 5.3. Flow diagrams showing a model of the San Joaquin estuary plankton and their response to nutrient enrichment. A. Compartment diagram showing both biomass and nutrient flows,

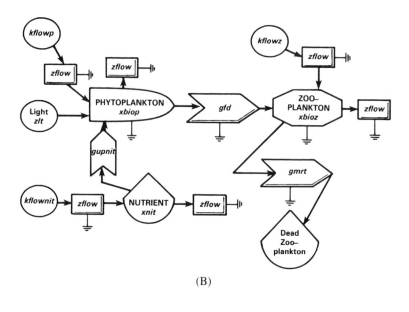

(B)

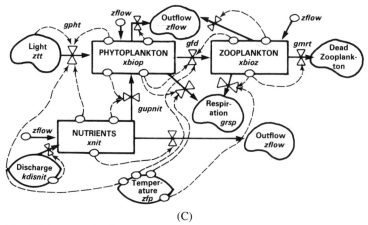

(C)

Fig. 5.3 *(cont.).*   B. Odum's energy diagram, and C. Forrester flow control diagram.

control of nutrient level over photosynthesis is apparent. The inclusion of controls is the real strength of the Forrester diagrams. It leads, however, to a large number of connections that are difficult to separate and to clearly diagram. The figures look like spaghetti, even for a model as simple as the San Joaquin model. Also, this diagram seems to spawn sources and sinks. Sources and sinks are more simply handled in the Odum and compartment diagrams.

The compartment diagrams are the most straightforward. They have processes labeled by notation that is compatible with the code. They distinguish nutrient from biomass flow by having separate diagrams. They give no control information and all compartments look the same in them, unlike the Odum diagrams.

We think the compartment diagrams are most useful when combined with a complete documentation of the flow equations and controls on them. For simple models we prefer the Odum and Forrester diagrams because they give some information about factors controlling the flows between compartments.

The Odum diagrams are especially valuable in representing the ecosystem as a series of energy transactions, which they are, even if we cannot always study them at this level because of a lack of data or an inability to measure the necessary conversion factors. The Forrester diagrams emphasize control and, for simple models, can include a complete documentation to the model. This is not true for a model of the complexity of the Di Toro model. The Odum diagrams are difficult to learn, but provide a consistent picture of the energy basis underlying flows. The Forrester diagrams can often appear cluttered because they create a compartment for each intermediate variable and parameter. For large ecosystem models, using Forrester's approach requires subdividing the model into modules that can be individually examined. It has the advantage of introducing all the model notation directly in the diagram. Whichever approach is used, additional documentation is required to give model equations and their rationale. We suggest that you return to these three diagrams again after you have read the complete San Joaquin model document to see which one you prefer.

5.2.2   Model document organization

For documenting the model equations we use a ''top down'' approach. Here differential equations for the rates of change of each of the state variables are presented first. The processes controlling these rates of change are given as intermediate variables on the right hand side of these equations. These controlling processes are then described, each process description identifying subprocesses—other intermediate variables—that in turn are described completely before the next process is considered. We usually start with the producer differential equations, then primary consumers, secondary consumers, and finally detritivores. The position of the abiotic processes is variable. Some models have abiotic processes such as temperature modeled explicitly. For example, heat balance equations can be used to model temperature changes in the soil over time. In such cases, when the abiotic components are separate from the biotic components, they are usually described first. In other cases, such as nutrient cycling in an ecosystem, where the abiotic components depend explicitly on biotic state variables, the abiotic components are described last.

The top-down approach can lead to some redundancy and some misplacing of processes, although it is usually the most ecologically natural order. For example, the consumption process for primary consumers will appear first as the producer process of *predation mortality* (herbivory) even though it is primarily driven by consumers. The process must then be described again as the herbivore's consumption process. Although the two processes are not identical, they have enough overlap that reference to one can be made as an aid in describing the other. Because the consumption process is associated primarily with the consumer, but appears first as a producer process, it is useful to refer forward in the documentation to the consumer section when first documenting the process. The process description should also give the rationale for each of the processes, with references given to supporting evidence in the literature.

Documentation of model parameters should give the notation, units, values, and how the values were estimated, with references to supporting data. Examples of this type of documentation were given for several processes in Chap. 4. In the rest of this chapter we give "concept" documents for each of the three models presented in this chapter following the precepts outlined above. We also give examples of model output from each of the models. The *implementation document* for the eutrophication and forest succession models are used as examples in Chap. 6 for how ecological model code is written.

5.3   Bird energetics model

Our first model considers the energetics and population dynamics of passerine birds in short grass prairie environments. These birds spend only part of the year in these breeding sites. We wanted to develop a model that reflected the weight and population change of these birds on the breeding site in terms of a few simple process rates, including consumption, respiration, assimilation, birth, mortality, and migration. This model could then provide insight into the relative importance of these processes in controlling weight and population changes in the grassland. This work was part of the International Biological Program (IBP) grassland biome, a project concerned with understanding the structure and function of grassland communities. Models were used there to promote interaction between the modeler and the field biologist and specifically to identify areas where future research was needed.

The bird community was chosen for an initial model effort and the most common species, the lark bunting, was used as an example. These birds are summer residents on the Pawnee National Grassland in northeastern Colorado, where the field work supporting this model was done. They arrive beginning on May 1 and start leaving August 15. Data on weight change over the season and on population totals throughout the season were available, as were data on the rate and timing of egg laying and hatching and the survival of eggs. Because there were no data on the age of the birds, it was decided to represent the birds as a single nonage-structured population represented by an average weight xwt (g), population density xn (no./100 ha), and biomass $xbio$ (g/100 ha). The model time step is daily, although the time unit is a week.

5.3.1   Population dynamics and energetics

The rate of change of the bird population xn depends on immigration, emigration, natality (birth), and mortality:

$$x\dot{n} = gimmig - gemig + gbirth - gmrt$$

$gimmig$ = immigration rate in the spring (no/wk)
$gemig$ = emigration rate in the fall (no/wk)
$gbirth$ = birth rate (natality) (no/wk)
$gmrt$ = mortality rate (no/wk)

Immigration and emigration are time varying rates reflecting the average observed lark bunting migration pattern.

The bird's weight is controlled by the following differential equation:

$$x\dot{w}t = kassim \times kfd - grsp \qquad (5.1)$$

kfd = feeding rate (g/wk)
$kassim$ = assimilation efficiency
$grsp$ = respiration rate (g/wk)

Respiration rate changes over the season as the birds arrive, mate, breed, and feed their young. The biomass $xbio$ is represented as the product of weight times population density.

In this simple formulation all the intermediate variables (except those connected with birth) are represented either as time varying rate coefficients or as constants. Birth rate is represented as the product of the egg laying rate lagged in time by the hatching period times the fraction of eggs that survive the hatching period times the female population at the egg laying time (the product of the total population times the sex ratio).

$$gbirth(t) = glayegg(t - khatch) \times ksurvegg(khatch) \times$$
$$ksex_{rt}(t - khatch) \times xn(t - khatch)$$

$glayegg(t)$ = egg laying rate at time t (no/wk)
$khatch$ = time period for eggs to hatch (days)
$ksurvegg\ (khatch)$ = fraction of eggs laid that survive to hatch
$ksex_{rt}(t)$ = sex ratio (female fraction) of bird population at time t

The model is simple in its mechanisms. All the rates are either constants or are donor-controlled coefficients multiplied by state variables (as with natality and mortality).

An interesting feature of this model is that it ties the parameters to observed patterns over time of bird migration and nesting and feeding behavior. These parameter changes implicitly include hypotheses about how strongly the various behavioral modes of the lark bunting (immigrating, emigrating, breeding, or brooding) affect their feeding and mortality rates. For example, an increased respiration rate from June 2 to July 10 reflects increased activity required by the adults to feed nestlings. Parameter values and their data sources are given in Table 5.1.

Table 5.1   Parameter values for bird population dynamics and energetics model applied to lark bunting population on the Pawnee National grassland in Colorado.

Notation	Description	Units	Value		Data source
gimmig	Immigration rate	no./100 ha/wk	60 0	May 1–15 Else	Observation
gemig	Emigration rate	no./100 ha/wk	50 0	Past Aug 15 Else	Observation
gmrt	Mortality rate	no./100 ha/wk	0.015 0.025	May 1–June 10 Past June 10	Dead bird count
kassim	Assimilation efficiency	—	0.7		Literature
kfd	Feeding rate	g/bird/wk	41		Stomach content analysis
grsp	Respiration rate	g/bird/wk	28.0 29.5 29.0 26.0	May 1–June 1 June 2–July 10 July 1–30 Past July 30	Literature
khatch	Time to egg hatch	Days	9		Observation
glayegg	Egg laying rate	egg/female/wk	1.2 0.4 0.0	June 1–21 July 10–21 Else	Nest survey
ksurvegg	Egg survival fraction	—	0.4		Nest survey
$ksex_{rt}$	Sex ratio	—	0.45		Sample of bird kills

5.3.2   Model output

Model output for population density and average weights are shown in Fig. 5.4A and B, along with data values for these variables. This model replicated the pattern of weight and population change in the Pawnee site lark bunting population. The main parameters requiring more accurate estimates are the respiration rates for the lark bunting and how they change with different activities through the season. This would require working with captive birds and perhaps even a breeding program for birds in captivity. This conclusion derives from the scanty data available about respiration and from the fact that the weight change is as sensitive to respiration as it is to consumption and assimilation processes (see Eq. 5.1); processes for which more data are available.

There are some serious problems with this model, yet there are some strengths as well. The major unrealistic assumption is that the lark buntings are one nonage-specific pool. This has the ludicrous result of making newly hatched

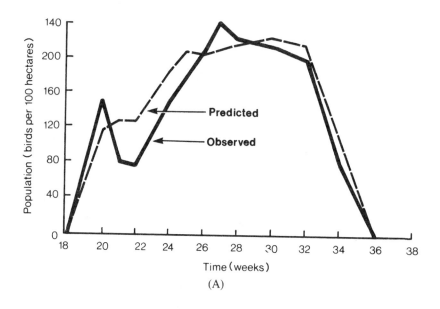

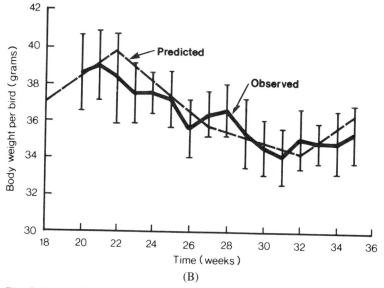

Fig. 5.4.   Graphical model output for A. lark bunting population and B. lark bunting weights from a model of bird population dynamics and energetics compared with data collected on the Pawnee National Grassland in northeastern Colorado. Population estimates are based on direct counts and have no variance estimate, whereas weight estimates are from direct samples and are graphed with two standard error of the mean bounds.

111

birds weigh the same as adults. Actually, the hatchlings grow within 2 weeks to weigh about as much as the adults, making the common weight assumption not too bad. More serious is the artifact that newborn chicks can participate in the second part of the breeding season as adults. We do not know the full implications of this factor, although it is certainly considerable.

Another weak assumption is that energetics changes over the season affect respiration but not consumption rates. Because adults must feed the young there must be a considerable increase in consumption during the fledgling period. However, because most of this food goes to the young and it is the average consumption of both young and adults that we model, the constant consumption assumption may not be too bad. One important weakness of this model is that increasing (or reducing) both respiration and consumption can produce identical weight changes. The two processes cannot be separated.

The choice of variables and processes in this model reflect the data and the objectives, which were to identify areas for future research. The high mortality of eggs suggests that the breeding process is pivotal to lark bunting population dynamics. Because much data were available about the breeding process, its role in population dynamics is well known and it was not identified as an important area for future focus. Yet the source of egg mortality and its variation from year to year can be important to population dynamics even though this particular model would not show that. The strengths of this model are its focus on a single species and its relatively complete data base about that species. A more complicated and realistic model would not have been warranted by the data.

5.4   Eutrophication model

Eutrophication is the phenomenon observed in bodies of water that receive large influxes of nutrients due to agricultural runoff or urban waste disposal. It is characterized by blooms of either green or blue-green algae (often noxious smelling) and by a drastic reduction in dissolved oxygen that often makes it impossible for many species of fish and zooplankton to live in the water.

Simulation models have been a primary tool in the study of eutrophication in lakes (see Chap. 9). Many eutrophication models have been developed both to predict the effect of nutrient additions on lake biota and water quality and to examine how effective various nutrient diversion alternatives might be in improving water quality. For this chapter we based our model on a eutrophication model by Di Toro et al. (1971); the first model of its type developed and a prototype for freshwater plankton simulation models to this day. The model was applied to the Sacramento–San Joaquin Delta in California and will, henceforth, be referred to as the San Joaquin model. The primary objective was to predict the effect of direct discharge of nutrients into the San Joaquin Delta, from sewage and agricultural runoff, on plankton dynamics and the possible effect of reducing this nitrogen input. Because the model addressed changes in plankton density after nutrient addition and reduction, its state variables include the biomass density (mgC/L) of phytoplankton, *xbiop*, lumped into a single taxonomic group, the biomass density of zooplankton *xbioz* (mgC/L), and the concentration of inorganic nitrogen *xnit* (mgN/L) in the water. The latter was included because nitrogen was judged the primary limiting nutrient in the San Joaquin ecosystem and release of nitrogen into the water of all the nutrients would result in the most noticeable change in plankton dynamics.

Because of the large size of the San Joaquin river system the model was subdivided into segments. We deal in this discussion with the segment around Mossdale. Flow of nutrients and biota into this segment from upriver was assumed to be controlled by the river flow rate, with densities upriver assumed to be constant. This assumption was made instead of modeling the dynamics of the spatial segments separately (including their interactions) and represents a simplification of the spatial dynamics of the system. The time scale (run time) of the model of 1 to 2 years suggested a time unit of 1 day, resulting in rates given on a daily basis.

Carbon (mgC) was used as a currency in this model as opposed to total dry weight. This is often done in models that also consider nutrient concentrations because the ratio of carbon to nutrients is commonly and easily measured. When considering multiple elements flowing in parallel pathways, attention must be paid to the *stoichiometric ratio* of these elements (the average ratio in a particular biota of the amount of one element to another). This may change from one trophic level (or even species) to another. Because dry weight is not a particular element it does not fit neatly into the stoichiometric classification. Dry weight units can be calibrated, however, to carbon units (i.e., carbon to dry weight ratio) and the two can be used interchangeably. Driving variables in this model are ambient water temperature *ztp* (°C), flow rate into and out of the water segment under consideration, *zflow* (L/day), and the average daily solar radiation at the water surface *zinsol* (ly/day).

The model is expressed as a series of three differential equations for the rates of change of phytoplankton, zooplankton and nitrogen.

$$\dot{xbiop} = (gpht - grsp) \times xbiop - gfd \times xbioz$$

$$+ \frac{zflow}{kvol}(kflowp - xbiop) \tag{5.2}$$

$$\dot{xbioz} = (gfd \times kassim - grspz - gmrt) \times xbioz$$

$$+ \frac{zflow}{kvol}(kflowz - xbioz) \tag{5.3}$$

$$\dot{xnit} = -knitC_{rt} \times gpht \times xbiop + \frac{kdisnit}{kvol}$$

$$+ \frac{zflow}{kvol}(kflownit - xnit) \tag{5.4}$$

gpht = daily photosynthesis rate (/day)
grsp = daily phytoplankton respiration rate (/day)
gfd = daily grazing or feeding rate (/day)
zflow = flow rate into and out of the water segment modeled (/day)
kvol = volume of water segment modeled (L)
kflowp = density of phytoplankton in the upriver segment (mgC/L)
grspz = daily zooplankton respiration rate (/day)
gmrt = daily zooplankton mortality rate (/day)
kflowz = zooplankton density in the upriver segment (mgC/L)

$knitC_{rt}$ = phytoplankton $\dfrac{nitrogen}{carbon}$ ratio (mgN/mgC)

kdisnit = discharge rate of nitrogen (mgN/day)
kflownit = nitrogen density in the upriver segment (mgN/L)
kassim = zooplankton assimilation efficiency

$\dfrac{zflow}{kvol}$ = per unit volume flow rate (/day)

The flow of nutrients and plankton into the Mossdale segment is assumed to be proportional to flow rate through the segment, with a constant amount per unit flow rate entering per unit time. Flow out of the compartment is proportional to the amount in the compartment and the flow rate. A tacit assumption in this model is that all phytoplankton species and all zooplankton groups (dominated in this system by rotifers) have constant weights that do not change over time.

Nitrogen is reduced through phytoplankton uptake, which is assumed to be driven by photosynthesis. The (assumed) constant phytoplankton nitrogen to carbon stoichiometric ratio $knitC_{rt}$ is used to convert photosynthesis in carbon units to uptake in nitrogen units. Nitrogen loading caused by direct discharge into the river is assumed constant (*kdisnit*). Notice that although the grazing rate on phytoplankton by zooplankton equals *gfd* × *xbioz*, the rate of assimilation of phytoplankton into zooplankton biomass equals *gfd* × *xbioz* × *kassim*, the assimilation efficiency determining the fraction of food eaten actually assimilated. All state variables and differential equations are expressed on a per unit volume basis.

Let us now examine the intermediate variables of Eqs. 5.2 to 5.4 categorized by trophic level.

5.4.1 Phytoplankton equations

The photosynthesis rate *gpht* was calculated as the product of a time and depth integrated total light effect, $gphtlt_{tt}$, a temperature effect, *gphttp*, and a nutrient (nitrogen) effect *gphtnit*. Because nitrogen is considered to be the limiting nutrient in the San Joaquin river system, it was the only nutrient considered in this model. Having photosynthesis as the product of three factors is tantamount to assuming these factors to act on photosynthesis rate independently. For example, if the light effect is half its maximum and the nutrient effect is half its maximum the combined effect will be a quarter. Some scientists think that this is too severe a limitation on photosynthesis. An alternative is *Liebig's law of the minimum,* which asserts that the combined effect of, let us say, light and nutrients is the minimum of the effect of light and of nutrients. This assumption results in less severe restriction on photosynthesis. Other possible hypotheses are that the overall effect is some other combination of the individual effects or that the effects do not act independently but rather that there is interaction between the light and nutrient effects.

The light effect, $gphtlt_{tt}$, was represented by Steele's equation, given in Chap. 4, and integrated over depth and time, assuming that light is a square wave with constant light intensity during the day and no light at night. This assumption leads to some bias in computing photosynthesis because light is not a square wave in natural ecosystems. The instantaneous effect of light on photosynthesis cannot be integrated, however, in closed form to give a daily total photosynthesis unless the light function over time has a simple form such as a square wave. Kremer and Nixon (1978) obtained an estimate of the bias by integration over small time segments of the day using a step function approximation to the light curve. They found that the square wave approximation overestimated photosynthesis by about 15 percent. The result is further complicated by light changes caused by clouds and light reflection. We hold with the simple approximation here. The equation for the depth-integrated $gphtlt_{tt}$ is given in Eq. 4.10. The light extinction coefficient, which was a constant (*kextlt*) in Eq. 4.10, was replaced in this model by an intermediate variable *gextlt*, which depends on the density of phytoplankton chlorophyll *a* in the water column. Riley (1956) developed the following expression:

$$gextlt = 0.0088 \times \frac{xbiop}{kCChl_{rt}} + 0.053 \times \left[\frac{xbiop}{kCChl_{rt}} \right]^{2/3} + kextlt$$

$kCChl_{rt}$ = phytoplankton carbon to chlorophyll *a* ratio (mgC/mgChl)

A simple linear function of temperature was used for the effect of temperature on photosynthesis.

$$gphttp = kphttp \times ztp$$

$kphttp$ = photosynthesis temperature effect parameter (°C)

This equation is based on a linear fit to data by several investigators and is graphed in Fig. 5.5. These data are a composite of many phytoplankton species. The term *saturated growth rate* in Fig. 5.5 means the photosynthesis rate when both light and nutrients are at their optimal (nonlimiting) levels. The middle line in this figure is the least squares regression line through the data. The upper and lower lines are plus and minus one standard deviation around the mean slope and were treated by Di Toro et al. (1971) as measures of the expected parameter range.

Other models (e.g., Lehman et al., 1975) have used the fact that for most phytoplankton species, or even functional groups, photosynthesis rates drop at higher temperatures. This effect, although marked for most species, only occurs at temperatures substantially higher than those encountered in nonthermally

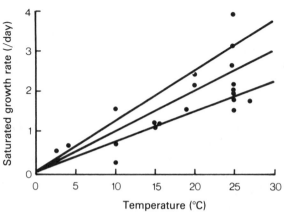

Fig. 5.5.   Temperature effect on phytoplankton photosynthesis rate *gphttp* in the San Joaquin model showing data supporting the linear relationship and parameter values used in the model. The three lines have the average slope, 0.1, and plus and minus 0.025 for this; slopes that can be seen as a possible range for the average slope. Data were collected from laboratory experiments on a variety of species. From Di Toro et al. (1971). Reprinted with permission from *Advances in Chemistry Series*. Copyright 1971 American Chemical Society.

loaded natural environments (where temperatures are not artificially elevated by an external, and usually man-derived, source). Because this model is not examining the effect of elevated temperatures on phytoplankton, and was used in a region of moderate temperature range, it is probably adequate to use the original linear relationship of Di Toro et al. (1971). Gamma-like or exponential (Q_{10}) relationships are used more commonly than the linear temperature effect assumed here (Swartzman and Bentley, 1979). Eppley (1972) has argued that when phytoplankton are represented by a single assemblage, the drop in photosynthesis at higher temperature that occurs for each species is masked by a shift in species composition with increasing temperatures over the growing season toward species that are better adapted to higher temperatures. This produces an overall increase in photosynthesis rate with increasing temperature; more like a Q_{10} than a gamma relationship.

Nutrients were assumed to affect photosynthesis as a Michaelis function of external or ambient nutrient level given by:

$$gphtnit = \frac{xnit}{xnit + kphtnit_{hs}}$$

gphtnit = nitrogen effect on photosynthesis
kphtnit$_{hs}$ = half saturation photosynthesis parameter for nitrogen (mgN/L)

The half saturation parameter is the nitrogen level at which photosynthesis is reduced to half the maximum rate (given that light and temperature are nonlimiting). Empirical evidence suggests that nutrient uptake rate is driven by external nutrient concentrations (external to the cells). One example is the data of Ketchum (1939) shown in Fig. 5.6. The fit to the Michaelis model is good.

If the assumption is made that photosynthesis in a nutrient-limited environ-

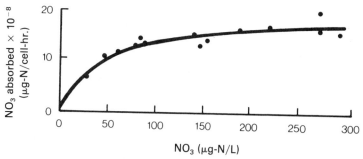

Fig. 5.6.   Ambient nitrate concentration effect on phytoplankton uptake of nitrate show-ing data due to Ketchum (1939) supporting a Michaelis–Menten function with a maximum uptake rate of 20.0×10^{-8} and a nitrate half saturation of 45.1. Fits of the Michaelis function to data like these were used in the San Joaquin model to provide a half saturation parameter estimate and confidence interval. Reprinted with permission from *Advances in Chemistry Series*. Copyright 1971 American Chemical Society.

ment will be controlled by the nutrient uptake rate, then photosynthesis rate can be linked directly to the external nutrient concentration. This was the assumption made in the San Joaquin model. More recently, models have recognized that, depending on the algal species and environmental nutrient status, uptake and photosynthesis rates can be more or less tightly coupled. Most models represent this coupling through explicit modeling of phytoplankton internal nutrients. Up-take is seen as the source of the nutrients, whereas the internal nutrient levels themselves are assumed to control photosynthesis.

The respiration rate of phytoplankton is the rate at which the phytoplankton oxidize their organic carbon to carbon dioxide. Based on data for several species presented by Riley (1949), and shown in Fig. 5.7, this model uses a linear

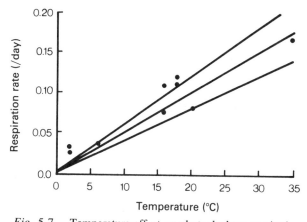

Fig. 5.7.   Temperature effect on phytoplankton respiration rate *grsptp* used in the San Joaquin model showing data supporting the linear relationship and parameter values used in the model. Only part of the data used to conduct the regression for estimating the slope of the line are shown in this figure. From Di Toro et al. (1971). Reprinted with permission from *Advances in Chemistry Series*. Copyright 1971 American Chemical Society.

relationship between temperature and respiration. The equation is:

$$grsp = krsptp \times ztp$$

$grsp$ = phytoplankton respiration rate (/day)
$krsptp$ = temperature parameter for phytoplankton respiration (/°C/day)

5.4.2 Zooplankton equations

Zooplankton grazing on phytoplankton, *gfd*, was affected in this model by both temperature and prey density. The grazing rate is equal to the maximum grazing rate kfd_{mx} times the prey density effect *gfdpy* times a temperature effect *gfdtp*. Data were presented (Fig. 5.8) for several zooplankton species indicating that grazing rate is affected by temperature. This model used a linear function of temperature:

$$gfdtp = kfdtp_1 + kfdtp_2 \times ztp$$

$kfdtp_1$ = zooplankton grazing rate temperature coefficient
$kfdtp_2$ = zooplankton grazing rate temperature coefficient (/°C)

Notice that the units for grazing rate in Fig. 5.8, (L/mg-dry wt/day) are different from the units of *gfdtp*, which is unitless. A direct conversion can be obtained by multiplying by prey density *xbiop*. Because prey density might have changed from one experiment to another, the slopes of the curves in Fig. 5.8 would

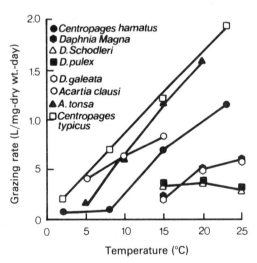

Fig. 5.8. Temperature effect on zooplankton grazing rate *gfdtp* for various studies reported in the literature. The lack of a consistent pattern in these curves was used to support a linear hypothesis for the relationship between temperature and grazing rate in the San Joaquin model, but with the relationship not going through the origin. From Di Toro et al. (1971). Reprinted with permission from *Advances in Chemistry Series*. Copyright 1971 American Chemical Society.

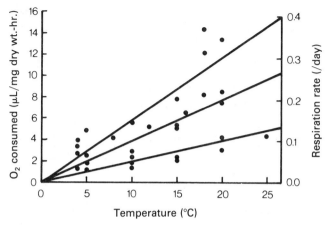

Fig. 5.9. Zooplankton respiration rate, *grspz*, as a function of temperature. Data for a number of cladoceran and copepod species were used to justify a linear relationship in the San Joaquin model. Respiration rates in L O_2/mg dry wt./hr were converted to mg dry wt./mg dry wt./day by assuming a constant dry weight to oxygen ratio for respiration. From Di Toro et al. (1971). Reprinted with permission from *Advances in Chemistry Series*. Copyright 1971 American Chemical Society.

change when units are converted, but the relative shapes of the curves would not.

The prey density effect was based on the Holling equation, presented in Chap. 4 in Eq. 4.6. Because the notation is a trifle different we will give the equation again:

$$gfdpy = \frac{xbiop}{xbiop + kfdpy_{hs}}$$

$kfdpy_{hs}$ = prey half saturation for zooplankton grazing (mgC/L)

Loss of zooplankton biomass results from respiration and mortality. Collation of a large amount of data by Di Toro et al. (1971) on several species suggested that endogenous respiration rate (rate of respiration produced internally by the biota) is temperature dependent (Fig. 5.9). A linear function of temperature was used because the data showed no clear nonlinearity:

$$grspz = krsptpz \times ztp$$

$grspz$ = zooplankton respiration rate (/day)
$krsptpz$ = zooplankton respiration rate temperature coefficient (/°C/day)

Mortality due to natural causes and predation from carnivorous zooplankton and fish are assumed to be a constant. Thus, $gmrt = kmrt$.

5.4.3 Nitrogen dynamics

Nitrogen is taken up by phytoplankton by an active process that apparently is linked to photosynthesis rate. In this model this linkage is carried a step farther to assume that photosynthesis *drives* nutrient uptake. Thus, uptake rate *gupnit* is represented as the photosynthesis rate times the phytoplankton biomass multiplied by the phytoplankton nitrogen to carbon ratio *knitC_{rt}*. The primary nitrogen source in this model is from external additions or discharge. Other sources of nitrogen, which can be important in oligotrophic (low nutrient) or mesotrophic (average nutrient) systems, are nitrogen addition to the water because of zooplankton excretion, and endogenous respiration by both zooplankton and phytoplankton; nitrogen fixation by blue-green algae can be important in nutrient rich bodies of water. In the eutrophic San Joaquin system these inputs are presumed to be negligible in comparison with the primary external inputs and are neglected.

5.4.4   San Joaquin delta parameters

Parameter values used in simulating the San Joaquin River ecosystem at Mossdale bridge, California, are given in Table 5.2. Table 5.2 also gives the type of data source for the parameter estimates. Several of the parameters were estimated from data obtained from the literature. These have been presented in this document by figures that summarize much of the data. References to the original data are in Di Toro et al. (1971). For several other parameters *calibration* was used as a data source. Parameters were estimated during simulation by adjustments made in their values between simulation runs to improve the fit of model output to field data. To some extent this lessens the validity of the model, because the same data cited as a measure of success for the model are also used for parameter estimation. The use of calibration has been justified by the authors of this model by pointing out that many of the parameter estimates are unavailable, that comparison of the final parameter values were made with ranges of parameter values taken from the literature on other related species or on the same species in other systems and that the calibrated values are reasonable by this measure of comparison. We discuss this topic in more detail in Chap. 8.

5.4.5   Model output

Driving variables for temperature, flow rate, and solar radiation from 1966 and 1967 were used to run the model and are given in Fig. 5.10. Notice that the two years had different river flow patterns. This had a significant effect on model output as can be seen from Fig. 5.11, which compares phytoplankton, zooplankton, and nitrogen dynamics from the model with data from the Mossdale site for the 2-year period. The system appears to have been dominated by the flow in the second year and by the predator–prey dynamics between the zooplankton and the phytoplankton in the first year. The high flow in the second year kept the nitrogen level lower in the spring and kept flushing out the plankton, which prevented the spring phytoplankton bloom observed in the first year. The lack of a spring bloom provided insufficient food for zooplankton growth (which is, in part, driven by prey density).

Table 5.2   Parameter values for a model of the San Joaquin estuary by Di Toro et al. (1971).

Parameter	Description	Units	Value	Data source
$kphttp$	Photosynthesis temperature effect	/day/°C	0.1	Literature
$kphtlt_{op}$	Optimal light for photosynthesis	ly/day	300	Literature
$kextlt$	Light extinction coefficient	/m	4.0	Experiment
kd	Depth of euphotic zone	m	1.2	Measurement
$kphtnit_{hs}$	Half saturation nitrogen effect on photosynthesis	mgN/L	0.025	Literature
$krsptp$	Phytoplankton respiration rate	/day/°C	0.005	Literature
$kflowp$	Influent phytoplankton concentration	mgC/L	0.25	Measurement
kfd_{mx}	Maximum zooplankton feeding rate	/day	0.39	Calibration
$kassim$	Zooplankton assimilation efficiency	—	0.6	Literature
$kfdpy_{hs}$	Half saturation prey feeding effect	mgC/L	3.0	Calibration
$kfdtp_1$	Temperature effect on zooplankton feeding	—	0.0	Measurement
$kfdtp_2$	Temperature effect on zooplankton feeding	/°C	0.2	Measurement
$krsptpz$	Zooplankton respiration rate	/day/°C	0.01	Literature
$kmrt$	Zooplankton mortality rate	/day	0.075	Calibration
$kflowz$	Influent zooplankton concentration	mgC/L	0.05	Measurement
$knitC_{rt}$	Phytoplankton nitrogen to carbon ratio	mgN/mgC	0.17	Measurement
$kCChl_{rt}$	Phytoplankton carbon to chlorophyll ratio	mgC/mgChl	50	Measurement
$kflownit$	Influent nitrogen concentration	mgN/L	0.1	Measurement
$kdisnit$	Nitrogen direct discharge rate	mgN/day	5.67×10^9	Measurement
$kvol$	Segment volume	L	2.75×10^{10}	Measurement

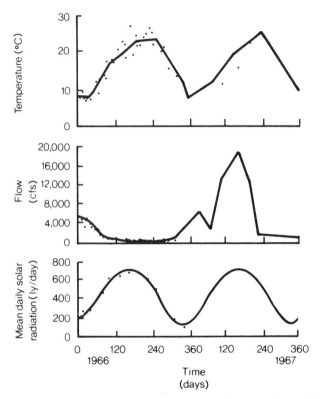

Fig. 5.10.   Model driving variables for the San Joaquin model. These include temperature, flow rate, and mean daily solar radiation incident to the water surface. The solid lines used for driving variables in the model were derived from data points collected at Mossdale Bridge on the San Joaquin estuary for 1966 and 1967. From O'Connor, D. J., D. M. DiToro, and R. V. Thomann, "Phytoplankton Models and Eutrophication Problems." *In* Russell, C. S., ed., *Ecological Modeling in a Resource Management Framework,* RFF Working Paper QE-1 (Washington, D.C., Resources for the Future, Inc., 1975) pp. 174, 176, 179. Reprinted by permission of Resources for the Future, Inc., copyright holder.

In the first year the phytoplankton biomass buildup during the spring (when temperature was rising, light was increasing, and nutrients were high) was followed by a phytoplankton drop and an accompanying zooplankton rise caused by rapid zooplankton growth and high grazing rates. This is the classical picture of predator–prey dynamics characterized in the *Lotka–Volterra* equations, to which this model is similar (if you ignore light, temperature, and the Michaelis prey density effect on grazing). The autumn phytoplankton bloom observed in the second year is smaller than the spring bloom of the first year and apparently is not grazed back by the zooplankton as the first year bloom was, but instead drops owing to reduced temperatures and solar radiation. Results of this model show that nutrient loading is more of a problem in the low water flow years than in the high years. The contribution of eutrophication models to water quality management is discussed in Chap. 9.

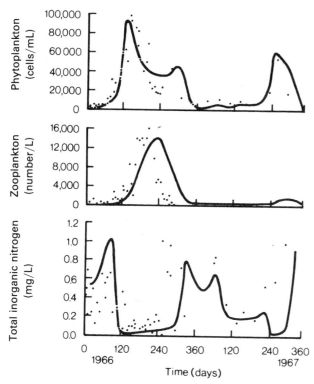

Fig. 5.11.   Model output for the San Joaquin model for phytoplankton, zooplankton, and total inorganic nitrogen compared with field data values collected in 1966 and 1967 at Mossdale Bridge, California. Model output has been converted to the same units as the data. From O'Connor et al., 1975. Reprinted by permission of Resources for the Future, Inc., copyright holder.

5.5   Forest succession model

A modeling approach to forest succession was originally developed for a north-eastern United States deciduous forest at Hubbard Brook, New Hampshire, by Botkin, Janak, and Wallis (1972), and has been expanded and applied abundantly to areas as different as a subtropical rain forest (Shugart et al., 1980), a coniferous forest ecosystem (Reed and Clark, 1979), and an Appalachian mountain forest (Shugart and West, 1977). The ecosystem considered here is a 10 × 10 m forest plot within the Hubbard Brook experimental forest in New Hampshire.

The objective of this model is to represent the change in species composition and biomass in a forest over time. The approach is to represent how tree growth both responds to and changes the external environment over time. A good example to illustrate this concept is sunlight. The forest canopy is composed of many trees, often of several species that have their foliage at different levels of the canopy depending on the time they sprouted, the environment they sprouted in, the growth potential of each tree, and the environment that the trees experienced over time (both the external environment, e.g., temperature, and the *intrinsic* or internal environment that has been created by the trees themselves). As the trees grow, their height in the canopy changes and they contribute additional shading both to themselves and to other trees. In this way they can be seen as altering their own environment.

The approach in this model is somewhat different from the San Joaquin model. Long-term succession is assumed to take place in an environment in which the contribution of each tree to both crowding and shading other trees is seen as important, but where the exact locations of the trees themselves is not explicitly considered. Thus, only the height and diameter of the trees affect the succession sequence on the modeled plot. Differences in location of individual trees is seen as a source of growth variability, which is represented as a random component of the model (not included in the version implemented in Chap. 6).

The model flow diagram is shown in Fig. 5.12. It consists of four modules: a growth module, a birth or initialization (establishment) module, a mortality module, and a site specification module. The model time step is a year. Each tree in the growth module is characterized according to its growth potential, which depends on the size (diameter) of the tree, its leaf area, age, light conditions, temperature, and crowding. Competition for nutrients and soil moisture, which can also play a major role in forest succession, were not explicitly included in this version of the model but were instead represented by the crowding function.

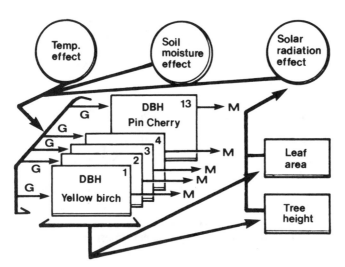

Fig. 5.12.   Flow diagram for the forest stand succession model of Botkin, Janak, and Wallis (1972). The flow structure is simple, with diameter at breast height (DBH) being the only major state variable. Tree height and leaf area are additional state variables. Driving variables are shown inside circles. The full model deals with 13 species of trees and considers the growth (G) and mortality (M) of individual trees of each species.

5.5.1   Tree growth

The model represents growth on an annual basis for each tree. Light, temperature, and space competition are all factors that reduce growth in a multiplicative fashion below over that attained by a tree growing under optimum conditions—the largest tree of its kind. Growth is couched in terms of tree volume $xdbh^2 \times xht$ (tree diameter at breast height $xdbh$ squared times tree height xht). The annual growth increment in volume is computed as a differential equation:

$$x\dot{v}ol = kgrow_{mx} \times xarealf \times \left[1.0 - \left(\frac{xdbh \times xht}{kdbh_{mx} \times kht_{mx}} \right) \right] \quad (5.5)$$

$x\dot{v}ol$ = annual volume growth rate (cm³/yr)
$kgrow_{mx}$ = maximum (species specific) tree growth rate (cm/yr)
$xarealf$ = tree leaf area (cm²)
$xdbh$ = tree diameter (cm)
$kdbh_{mx}$ = maximum diameter observed for a forest grown tree of this species (cm)
kht_{mx} = maximum height observed for a forest grown tree of this species (cm)

Several assumptions are made to reduce Eq. 5.5 to a differential equation in $xdbh$, the annual change in the diameter alone. This is desirable because several model variables, such as leaf area, are directly linked by forest mensurationists (forest statisticians who relate the dimensions of trees to each other to estimate timber production) to the diameter at breast height (DBH). These assumptions involve expressing tree height as a function of diameter:

$$xht = kdbhht_1 + kdbhht_2 \times xdbh - kdbhht_3 \times xdbh^2 \quad (5.6)$$

$kdbhht_i$ = parameters in the relationship between tree height and diameter
xht = tree height (cm)

Also leaf area is assumed proportional to leaf weight, which is in turn proportional to tree diameter squared:

$$xarealf = karealf \times xdbh^2 \quad (5.7)$$

$karealf$ = parameter in the relationship between tree leaf area and diameter

Equations 5.6 and 5.7 assume that the form of the relationship between leaf area, height, and diameter does not change with the size of the tree. The leaf weight (area) to diameter ratio actually varies over the life of a tree with the exponent (2) in Eq. 5.7 actually varying between 1.5 and 3. Botkin et al. (1972), however, state that the overall succession pattern is not sensitive to the value of this exponent over the range of 1 to 3 and that the lack of accuracy of Eq. 5.7 can, therefore, be ignored.

Because tree volume $xvol$ is given by $xdbh^2 \times xht$ and both leaf area $xarealf$ and tree height xht are connected to DBH $xdbh$ by Eqs. 5.6 and 5.7 we can express Eq. 5.5 entirely in terms of *xdbh*. The only tricky part is expressing the rate of change of volume *xvol* in terms of *xdbh*. This can be done using the chain rule for differentiation and is left as an exercise. The results after substitution and differentiation are as follows:

$$\dot{xdbh} = \frac{kgrow_{mx} \times karealf \times xdbh \times \left[1.0 - \dfrac{xdbh \times xht}{kdbh_{mx} \times kht_{mx}}\right]}{[2 \times kdbhht_1 + 3 \times kdbhht_2 \times xdbh - 4 \times kdbhht_3 \times xdbh^2]}$$

(5.8)

Factors owing to solar radiation, temperature, and soil moisture serve to reduce the growth below its maximum rate. The solar radiation factor is given by:

$$ggrowlt = kgrowlt_1 \times \left[1.0 - e^{-kgrowlt_2 \times (glt - kgrowlt_3)}\right]$$ (5.9)

ggrowlt = the effect of solar radiation on annual tree diameter growth
glt = available solar radiation (normalized)
kgrowlt_i = parameters controlling effect of light on tree growth

The available solar radiation depends both on the incident solar radiation and the leaf foliage above the tree in question. As mentioned earlier, it is an example of how the environment of the forest is changed over time by tree growth. Available solar radiation is computed using an application of Beer's law given in Eq. 4.7 and is given by:

$$glt = zinsol \times e^{-kextlt \times garealf}$$

garealf = leaf area sum for all trees higher than the present tree (cm^2)
zinsol = total annual insolation (normalized)

The function for the effect of temperature on growth assumes that each species grows in response to an annual accrued number of degree days above some threshold temperature. Temperature *ztp* is expressed as annual degree days over 40°F (4.4°C) (an arbitrary baseline temperature for tree growth, having some relevancy to observed tree growth at the Hubbard Brook site). The temperature effect equation is:

$$ggrowtp = \frac{4 \times (ztp - kdegdy_{mn}) \times (kdegdy_{mx} - ztp)}{(kdegdy_{mx} - kdegdy_{mn})^2}$$

ggrowtp = effect of temperature on tree growth
ztp = annual temperature index (degree days above 4.4°C)
kdegdy_{mx} = maximum annual temperature producing growth (degree days)
kdegdy_{mn} = minimum annual temperature producing growth (degree days)

This function is symmetric around $\dfrac{kdegdy_{mx} + kdegdy_{mn}}{2}$ (the optimal annual number of degree days for growth). The parameter values are obtained by examining the temperature extremes over the range of each species. Competition for moisture and nutrients is represented by a crowding factor, which is simply the fraction of the maximum possible basal area in the 10 × 10 m plot that is actually covered by trees. The function is:

$$gspace = 1.0 - \left[\frac{gareabs}{kareabs_{mx}} \right]$$

gspace = effect of crowding on tree growth (subsumes moisture and nutrients)
gareabs = total basal area on the 10 × 10 m plot covered by trees (cm^2)
kareabs$_{mx}$ = maximum possible plot basal area covered by trees (cm^2)

As the total basal area (the sum of the area at breast height of all the trees) approaches the maximum possible, competition for nutrients and/or water is assumed to increase and the growth factor approaches 0.

5.5.2   Tree mortality

There are two kinds of mortality in this model. One type is related to aging and the other to growth. The aging mortality computes a probability that the tree dies dependent on its age:

$$gmrtage = (1 - kmrtage)^{xage} \qquad (5.10)$$

gmrtage = annual probability of tree mortality due to aging
kmrtage = tree aging mortality parameter
xage = age of tree (yr)

The parameter *kmrtage* is chosen so that on the average only a small fraction of trees survive to their maximum observed age *kage*$_{mx}$, even under optimal growth conditions.

If a tree grows less than 0.01 cm in a year in this model an additional source of mortality *kmrt* is imposed analogous to that in Eq. 5.10 but at a much larger rate than *kmrtage* such that the tree has a much higher chance of dying. The rationale for this function is that if a tree cannot maintain a certain minimum growth rate it will be more susceptible to factors that cause mortality, such as disease.

5.5.3   Tree establishment

Trees are assumed to be established at a size (DBH) of 0.5 cm. The numbers of trees established of each species per year for shade-tolerant species depends on both the prevailing temperature (degree days) and the soil moisture level on the site. Available soil moisture *zevap* is measured as the annual evapotranspiration for the site. To establish, a species must have *zevap* between its maximum and minimum moisture requirements ($kevap_{mx}$ and $kevap_{mn}$, respectively). Similarly, the degree days for that year must be between the maximum and minimum number of degree days for growth for that species. A list is then generated of those species that can establish under the given temperature and soil moisture conditions. A random sample of these species is then made (using a Monte Carlo technique; see Chap. 7) and either 0, 1, or 2 trees of each of the species chosen at random from the list (the number again being chosen by Monte Carlo methods) are established (added to the existing trees on the stand).

Shade-intolerant species may only establish if the total leaf area on a plot is less than a species-specific threshold (related to light). If establishment can occur, a large number (e.g., 60–75) of trees of the proper species are added. In cases where more than one shade-intolerant species may establish, a random number of trees of each species are entered depending on (inversely proportional to) the shadiness of the plot. For example, cherry is the most shade-intolerant species and tends to predominate on some sites in early succession in large numbers, whereas birches are more shade tolerant and only come in under somewhat lower prevailing light levels. Thus, an early successional stage would have only cherry trees. Once the cherry tree's light cutoff level is passed then birchs can establish. The establishment module is more ad-hoc than the other modules in that establishment was tailored to fit the establishment patterns observed in the region.

Both establishment and mortality have a random component to them that is generated using Monte Carlo methods. For all practical purposes this model can be looked at as deterministic, because the average behavior does not have much variability around it from run to run. This is because the aspects of the model governed by stochastic elements (mortality and establishment) are unimportant compared with the growth functions. The growth module, although itself deterministic, can be affected by environmental conditions that can be generated using Monte Carlo methods (but were not in the original model). The importance of having a stochastic versus deterministic environment in the model to forest succession is uncertain to us.

The final module specifies the site parameters. This amounts to computing driving variables from data. Sometimes this required some artistry and a number of assumptions. Air temperature ztp was calculated from long-term estimates of mean January and July temperatures assuming temperature is sinusoidal over the year. This method will be discussed in more detail later. Temperatures were adjusted for elevation differences using data that show an average drop in temperature of 3.6°F per 1000 feet in July and 2.2°F per 1000 feet in January (Sellers, 1965).

Monthly precipitation and temperature are used to calculate annual evapotranspiration using an equation due to Thornthwaite (Sellers, 1965). The precipitation values for the Hubbard Brook application of this model were taken without change from the nearest weather station. Light level $zinsol$ was normalized to 1.0 for the Hubbard Brook succession model.

Although we have presented this model as a differential equation model, it is solved as a difference equation model, or uses the equivalent of an Euler approximation with a time step of 1 year. Because most of the rates in the model are not large, the error resulting from the difference equation approximation (i.e., using flow rates as if they were flow amounts) is small.

5.5.4   Hubbard Brook forest parameters

Table 5.3 gives the parameters for some of the 13 species in the model for the Hubbard Brook ecosystem. The model produces as output the average basal area for each species during secondary succession after clearcut. The basal area (cm^2) is shown in Fig. 5.13A for red spruce, a later succession or *climax* species, and in Fig. 5.13B for white birch, an early successional species. Average total basal area of all species is also given in Fig. 5.13C. Apparently, early succession is dominated by white birch, whereas later succession has a more diverse mix of species. Crown closure (all ground space shaded by foliage) appears to occur around year 40, judging by no further increase in basal area after this time.

Table 5.3   Parameter values for forest succession on a 10 × 10 m forest stand in Hubbard Brook, New Hampshire.

Parameter	Units	Sugar maple	Beech	Pin cherry	Balsam fir	Spruce	White birch
$kgrow_{mx}$	cm/yr	108.3	68.2	81.6	80.0	20.0	288.1
$karealf$	—	1.57	2.2	2.45	2.5	2.5	0.486
$kage_{mx}$	yr	200	300	30	80	350	80
$kdbh_{mx}$	cm	152.5	122	28	50	50	46
kht_{mx}	cm	4011	3660	1126	1830	1830	1830
$kdbhht_1$	cm	137	137	137	137	137	137
$kdbhht_2$	—	50.9	57.8	70.6	67.9	67.9	73.6
$kdbhht_3$	/cm	0.167	0.237	1.26	0.679	0.679	0.8
$kdegdy_{mn}$	deg day/yr	2000	2100	1100	1100	600	1100
$kdegdy_{mx}$	deg day/yr	6300	6000	8000	3700	3700	4000
$kevap_{mn}$	mm/yr	300	300	190	190	190	190
$kevap_{mx}$	mm/yr	—	—	—	—	—	600
$kgrowlt_1$	—	1.0	1.0	2.24	1.0	1.0	2.24
$kgrowlt_2$	—	4.64	4.64	1.14	4.64	4.64	1.14
$kgrowlt_3$	—	0.05	0.05	0.08	0.05	0.05	0.08
type*	—	2	2	1	2	2	1

*Type classifies trees according to shade tolerance. 1 = shade intolerant, 2 = shade tolerant. Parameters are given for selected species using values from Botkin et al. (1972).

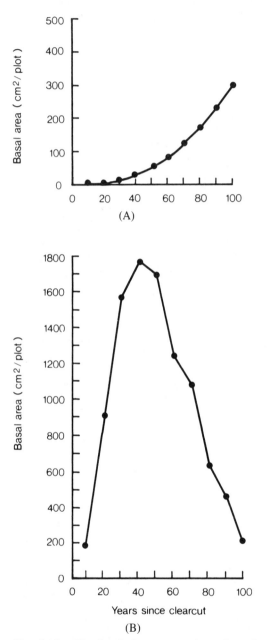

Fig. 5.13.   Simulated change over time, from the forest stand succession model of Botkin, Janak, and Wallis (1972), in total basal area (cm^2/plot) for a 10 × 10 m plot in the Hubbard Brook experimental forest. These are shown for A. red spruce, B. white birch.

130

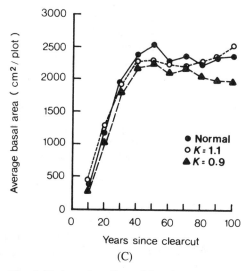

Fig. 5.13 *(cont.).*   C. total basal area for all species. The three curves in C. reflect
normal (K = 1), diminished (K = 0.9), and augmented (K = 1.1) growth rates and
provide a measure of the sensitivity of forest stand growth to growth rates. From Hall,
C. A. S., and J. W. Day, Jr., eds., 1977. Reprinted by permission of John Wiley &
Sons, Inc.

5.6 Summary

We have arrived at last at fully fledged ecological simulation models. We have
given examples that span as wide a range of ecological systems and processes
as possible while remaining simple in structure. Our emphasis in this chapter is
on giving, through example, a good method for presenting models to others so
that they may understand and be able to code them themselves.

Based on a review of over 500 models we have come to appreciate what does
and what does not work in model documents. This is distilled in a section on
model documentation. Because diagrams appear to be an essential part of the
communication of models we have spent some time comparing the most commonly used diagram frameworks in the ecological simulation literature. We
personally prefer the straightforward compartment diagram, but see each of the
methods as useful for certain applications. In fact, the literature contains examples of diagrams that mix and match from the three frameworks presented.
Certainly, anyone reading the literature needs to be familiar with all three.

We have purposely chosen three very different models to use as examples.
This was done in part to indicate the wide range of model types but also to
show how the different objectives and ecosystems led to basically different
structures for the models. The bird model, which represented the population
dynamics and energetics of a breeding bird population, was built as a tool to
indicate which areas for future research might prove most fruitful. With this in
mind only a single species was considered and biomass was separated from

numbers. Only the summer, breeding season was modeled and a weekly time unit was chosen because the model could then be fitted around the weekly sampling intervals used to collect much of the supporting data. Some investigation into the sensitivity of model results to parameters led to a conclusion that respiration was a likely area of study to improve our understanding of lark bunting breeding season energetics.

The San Joaquin model represented a different challenge. The objective was to examine the effects of eutrophication on plankton seasonal dynamics. Because plankton blooms and not species composition was of primary concern, phytoplankton and zooplankton were represented as single trophic levels rather than by species groups or individual species. Because of the rapid nature of plankton dynamics a daily time step was chosen. Nitrogen was assumed to be the major nutrient of concern in this system and was the only nutrient considered. Because the nutrient additions were quite large and the system was eutrophic, refinements of nutrient recycling and the effect of phytoplankton internal nutrients on photosynthesis rate were not included and might not have been needed even though they can be very important under different circumstances (e.g., an oligotrophic system) or different objectives (e.g., understanding phytoplankton growth processes).

The final challenge was to choose a spatial framework for the model. The San Joaquin River Delta was divided into segments, with the area around Mossdale bridge being the one considered in this chapter. Interaction was assumed with the two adjacent segments, with flow from the upriver segment and flow out to the downriver segment. Because river flow conditions are known to have a major effect on river flushing and eutrophication, river flow was chosen as a driving variable in addition to temperature and solar radiation. The latter two were chosen because of their known strong influence on process rates; solar radiation plays a major role in the control of phytoplankton photosynthesis. Model experiments examined the effect of discharge of a source of nitrogen (in addition to inflow from the upriver segment) on plankton dynamics and specifically how this was influenced by river flow. This was possible because the river flow pattern was quite different between the 2 years of the model run.

The emphasis in the San Joaquin model was on structural and functional simplicity. A conscious effort was made to find the minimal model that could represent the observed plankton dynamics, while allowing some study of the effects of nutrient additions. This is apparent with the processes, where simple curve forms were chosen, especially when insufficient data were available to support more complex alternatives.

The Hubbard Brook model was built to replicate long-term forest succession dynamics in a northeastern United States deciduous forest ecosystem. Owing to the heterogeneous nature of the land around Hubbard Brook a fairly small representative plot, 10×10 m, was chosen as the spatial boundary of the ecosystem modeled. It was assumed that forest succession takes place through the interaction of individual trees, not primarily through their actual locations in the plot but through their contribution to canopy foliage and their coverage of the plot. Thus, competition for light and space (moisture and nutrients) is seen as averaged over the entire area, although it affects the tree's growth as individual trees and not as average biomasses of each species, as the San Joaquin type of model would.

Owing to the long-term nature of this model a time step of a year was chosen. Seasonal dynamics of tree growth was seen as unimportant to long-term succession. As such, environmental conditions were annual totals. The environmental conditions in the model were average insolation and an air temperature degree-day index. An evapotranspiration index was also used for tree establishment. Freed from having to deal with the locations of individual trees, the modelers could make simplifying assumptions about the average shape and form of the trees. A series of empirical relationships between tree leaf area and height and the diameter at breast height (DBH) were used to convert volume growth and leaf area growth into diameter growth, thereby making a single variable, the DBH, the primary state variable in the model.

The final singular aspect of the Hubbard Brook model was that it represented forest succession as a process whereby the change in tree species composition affected the environment perceived by the trees. This changing of environmental conditions over time is a striking feature of succession that is included in this model as an explicit part of the model equations.

The representation in this model of individual trees and species allows a detailed account of the size and species composition of trees to be reconstructed. This approach is especially appropriate for forests where data on individual trees are readily obtainable, yet the model does allow us to see the forest through the trees.

One basic difference between the models was in their use of data. The bird model had data collected on the study site about the species of concern. Lack of some data, however, like the age distribution of the birds, resulted in having some part of the newly hatched birds contribute to the breeding population and having newly hatched birds weigh as much as adults. The San Joaquin model was of a much larger part of the ecosystem. It depended for parameter estimates on laboratory experiments on biota that were not the same as those in the San Joaquin system. Data time traces were used, via calibration, to finalize the parameter values.

The Hubbard Brook model, although it represented a single site, was designed to be general over a wide geographic range. As such, species specific, rather than site specific, parameters were estimated from observed tree growth and establishment patterns. Much of the basis for tree form and function within this model was empirical. These three models exemplify the wide difference in data availability and means of using data in ecological models.

In presenting the models in this chapter we have delivered the information needed to understand, code, and run the models without considering in detail what is to be done with the models once they are built. Such modeling activities as model evaluation, model experiments, and using the model to generate various management scenarios will be considered in later chapters. Before we do this we must first develop the computer tools to be able to run models like those presented here. This we will do in Chap. 6.

5.7 Annotated Bibliography

Andersen, K. P., and E. Ursin: A multispecies extension to the Beverton and Holt theory of fishing, with accounts of phosphorus circulation and primary production. *Medd. Dan. Fisk. Havunders,* 7:319–435, 1977.
A complicated multispecies fishery model calibrated to fishery catch data on the North Sea. Innovative in its use of size selective feeding by fish and in its direct treatment of fish early life history.
Brown, M. P.: A mathematical model of PCB bioaccumulation in plankton. *Ecol. Model.* 15:29–47, 1982.
A relatively simple ecological model that can be used for an example of how to document and code an ecological model. The complete documentation is given.
Canale, R. P. ed.: *Modeling Biochemical Processes in Aquatic Ecosystems,* Ann Arbor Science, Ann Arbor, Mich., 1976.
A collection of case studies and models of plankton dominated ecosystems that gives more or less complete documentation for the underlying models.
Coughenour, M. B., W. J. Parton, W. K. Lauenroth, J. L. Dodd, and R. G. Woodmansee: Simulation of a grassland sulfur-cycle. *Ecol. Model,* 9:179–213, 1980.
A good example model giving a complete documentation.
Dale, V. H., T. W. Doyle, and H. H. Shugart: A comparison of tree growth models. *Ecol. Model.,* 29:145–170, 1985.
Reviews a collection of approaches to forest succession and tree growth modeling.
Di Toro, D., D. J. O'Connor, and R. V. Thomann: A dynamic model of the phytoplankton populations in the Sacramento-San Joaquin Delta. *Adv. Chem. Ser.,* 106:131–150, 1971.
The first of a long line of aquatic simulations combining an engineering approach with an ecological perspective. This model was a pioneer in its attempt to give rationale for equations.
Di Toro, D., D. J. O'Connor, R. V. Thomann, and J. L. Mancini: Phytoplankton–zooplankton–nutrient interaction model for western Lake Erie, in *Systems Analysis and Simulation in Ecology, vol. 3,* ed. B. C. Patten, Academic Press, New York, 1975, pp. 423–474.
The original San Joaquin model expanded to include spatial variability and more biota applied to Lake Erie with a surprise "hindcast" as a measure of model validity. The hindcast unfortunately did not work.
Forrester, J. W.: *Industrial Dynamics,* MIT Press, Cambridge, Mass., 1961.
An introduction to Forrester's systems notation now commonly applied in ecological modeling.
Hall, C. A. S., and J. W. Day, Jr., eds.: *Ecosystem Modeling in Theory and Practice: An Introduction with Case Studies,* Wiley, New York, 1977b.
Probably the most wide-ranging collection of models in a single, integrated book. The book attempts to serve as a tutorial to simulation modeling by presenting case studies mixed with more text-like presentations. The quality of the applications is generally high.
Innis, G. S. ed.: *New Directions in the Analysis of Ecological Systems, Parts 1 and 2,*

Simulation Councils Proceedings, 5, Society for Computer Simulation, La Jolla, Calif., 1977.

Proceedings of a workshop on ecological simulation.

Innis, G. S. ed.: *Grassland Simulation Model,* Ecological Studies, 26, Springer-Verlag, New York, 1978.

A collection of papers on the ELM grassland model project. This book treats each trophic level in a chapter and also has an overview to the model and a chapter on sensitivity analysis. Model documentation is complete and thorough.

Innis, G. S., and R. V. O'Neill, eds.: *Systems Analysis of Ecosystems,* Statistical Ecology, 9, International Co-operative Publishing House, Fairland, Maryland, 1979.

This book contains a collection of recent topics in ecological simulation. It also contains Innis' well-written introduction to simulation.

Jorgensen, S. E.: A model of fish growth. *Ecol. Model.,* 2:303–313, 1976.

The documentation of this model is complete. It is a good simulation example.

Jorgensen, S. E. ed.: *Ecological Modelling.*

This journal has been published since 1975 and is the major forum for publication of ecological simulation models for modelers. As such, it is oriented toward presenting complete model documents including equations and parameter values or toward the use of new tools for enhancing simulation models or expediting their development or analysis.

Kickert, R. N.: Names of published computer models in the environmental biological sciences: a partial list and new potential risks. *Simulation,* 43:22–39, 1984.

This review article contains the most comprehensive bibliography on existing ecological simulation models ever assembled. The listing is by the name of the model, which is a more common way of remembering models than the title. The article shows the burgeoning number of simulation application areas. Over 300 models are included.

Kremer, J. N., and S. W. Nixon: *A Coastal Marine Ecosystem,* Ecological Studies, 24, Springer-Verlag, New York, 1978.

This book gives a good account of the process of simulation model development as well as a good discussion of model evaluation and the kinds of experiments that can be done with simulation models. The model documentation is complete.

Krishnamurthy, L: Herbage dynamics of some semi-arid Indian grazing lands. *Ecol. Model.,* 5:77–84, 1978.

Another good example model

Lauenroth, W. K., G. V. Skogerboe, and M. Flug, eds.: *Analysis of Ecological Systems: State-of-the-Art in Ecological Modelling,* Developments in Environmental Modelling, 5, Elsevier, New York, 1983.

A proceedings of a meeting of the International Society of Ecological Modelling. The number of papers and range of applications is mind boggling, but the quality control is lacking and you cannot tell the value of a paper without getting into it.

Lehman, J. T., D. B. Botkin, and G. E. Likens: The assumptions and rationales of a computer model of phytoplankton population dynamics. *Limnol. and Oceanogr.,* 20:343–364, 1975.

This paper gives an excellent discussion of model rationale. It is a consummate example of how a model can be concisely, although incompletely, described.

Middlebrooks, E. J., D. H. Falkenborg, and T. E. Maloney, eds.: *Modeling the Eutrophication Process,* Ann Arbor Science, Ann Arbor, Mich., 1974.

An early collection of simulation models predicting the effects of nutrient loading on lake ecosystems. This book provides a balanced picture of the state of the art in modeling in 1974.

Mitsch, W. J., R. W. Bosserman, and J. M. Klopatek, eds.: *Energy and Ecological Modelling,* Developments in Environmental Modelling, 1, Elsevier, New York, 1981. Proceedings of a meeting of the International Society of Ecological Modelling. The papers are of variable quality and reflect the burgeoning range of applications of simulation modeling, mostly to management.

Odum, H. T.: *Environment, Power and Society,* Wiley, New York, 1971.
A fascinating account of Odum's energy transaction theory of the universe.

Odum, H. T.: *Systems Ecology: An Introduction,* Wiley, New York, 1983.
Energy is the basis of all ecological processes and Odum is its prophet. The energy-based diagrams are commonly used in model presentation. They are starting to look more and more like other diagrams.

Park, R. A., R. V. O'Neill, J. A. Bloomfield, H. H. Shugart, Jr., R. S. Booth, R. A. Goldstein, J. B. Mankin, J. F. Koonce, D. Scavia, M. S. Adams, L. S. Clesceri, E. M. Colon. E. H. Dettmann, J. A. Hoopes, D. A. Huff, S. Katz, J. F. Kitchell, R. C. Kohberger, E. J. LaRow, D. C. McNaught, J. L. Peterson, J. T. Titus, P. R. Weiler, J. W. Wilkinson, and C. S. Zahorcak: A generalized model for simulating lake ecosystems. *Simulation,* 23:33–50, 1974.
A document for a lake ecosystem model that attempts to include the best available information about each process. The model addresses all relevant trophic levels and the discussion of the processes is interesting. The documentation itself is incomplete and more is now known about many of the processes, yet in its time this model was valuable and served as a springboard for a whole series of model refinements.

Riley, G. A.: Factors controlling phytoplankton populations in George's Bank. *J. Mar. Res.,* 6:54–73, 1946.
Perhaps the first nonlinear simulation model reported in the literature. This model was solved by hand calculation.

Shugart, H. H.: *A Theory of Forests Dynamics: An Investigation of the Ecological Implications of Several Computer Models of Forest Succession,* Springer-Verlag, New York, 1984.
An attempt to develop a theory of forest dynamics using forest succession models applied to a wide variety of ecosystems. The models are based on the Hubbard Brook model discussed in this chapter.

Shugart, H. H., and D. C. West: Forest succession models. *Bioscience,* 30:307–314, 1977.
A good review of succession models, many of which were derived from the Hubbard Brook model.

Steele, J. H.: *The Structure of Marine Ecosystems,* Harvard University Press, Cambridge, Mass., 1974.
The first model documentation in book form. Steele's presentation of the model equations and parameter values set a precedent for later models in book form. The book views the model as a tool for helping to better understand plankton dynamics in the North Sea ecosystem. There are mistakes, however, in the model document that make replication of the time traces given in the book difficult.

Stewart, D. J., D. Weininger, D. V. Rottiers, and T. A. Edsall: An energetics model for lake trout. *Can. J. Fish. Aquat. Sci.,* 40:681–698, 1983.
Unusual in that process experiments were performed as part of the model building process.

Stross, R. G., P. A. Nobbs, and S. W. Chisholm: SUNDAY, a simulation model of an arctic *Daphnia* population. *Oikos,* 32:349–362, 1979.
Another good example model for exercise purposes.

Swartzman, G. L.: Simulation modeling of material and energy flow through an eco-system: methods and documentation. *Ecol. Model.*, 7:55–81, 1979a.
Reviews the steps in model building and a documentation using a coniferous forest stand growth model as an example.
Yezzi, D. J. Jr., and A. P. Uzzo, Jr.: Dynamic model of nutrification in Huntington Bay, New York. *Ecol. Model.*, 6:59–75, 1979.
A complete, yet complicated, model description.

5.8 Exercises

Derivations

5.1. Derive Eq. 5.8, the differential equation for diameter used in the Hubbard Brook model using the volume differential Eq. 5.5, Eqs. 5.6 and 5.7 and the chain rule for differentiation with $xdbh^2 \times xht$ substituted for $xvol$.

Units conversion

5.2.

a. Convert the Di Toro model from mgC/L to g/m^3 for phytoplankton and zooplankton. Assume that the carbon to dry weight (biomass) ratio is 0.4. How do the parameter values change? Make a parameter list analogous to Table 5.2 in the new units.

b. Di Toro et al. (1971) give zooplankton respiration in terms of μl O_2 consumed per mg dry weight per hour. Convert the scale to units of /day. Assume that 2 mg dry weight equals 1 mg carbon and that $[CH_2O]$ is being oxidized. Note: One mole (g molecular weight) of a gas at standard pressure and temperature is 22.4 liters.

Diagrams

5.3. Draw Odum and Forrester type diagrams for the Hubbard Brook model. Contrast and compare these with the compartment diagram given in the text.

Model documentation

5.4. Choose a model from the list given at the end of this chapter and develop a complete model document according to the guidelines given in this chapter. Pay particular attention to notation, units, and data sources for the parameters. Is the information given about the model sufficient to completely document it? Do you have any questions about functions, parameter values, driving variables, or rationale that you would like to put to the authors? What improvements in documentation would you suggest to the authors of this model?

Chapter 6

Coding Nonlinear Models

6.1 Introduction

In this chapter we develop the Fortran code for running nonlinear simulation models. Our goal is to develop a simulator that is general enough to be easily adapted to modeling many different systems. The code is described using the San Joaquin estuary model as an example. At the end of the chapter we modify the simulator for running a version of the Hubbard Brook forest succession model. In the exercises at the end of the chapter Steele's (1974) marine ecosystem model is described.

6.2 Nonlinear model code

Many of the features of linear simulation model code are also present in the code for nonlinear models. There are, however, many features that are unique to these models. These include notational, organizational, and documentation features.

Ideally the notation used in these models should be identical to the notation used in the equations in Chap. 5. Doing this, however, would lead to a number of problems with passing variables and parameters between subroutines, especially between the differential equations solvers (which use either the Euler or Runge–Kutta routines) and the subroutines that compute the intermediate variables and flow rates. The problem in a nutshell is that it is more efficient to pass arrays between subroutines than to pass long lists of variables. The k, x, g, z notational scheme used for these models can be easily represented as arrays, as they were with the linear models. To maintain compatibility between the array notation and the mnemonic notation, which we find much preferable for model communication, we provide translations between the two both in the code equations (using comments) and in the input–output routines. Mnemonic variable names with subscripts that appear in the computer code comments have the subscript added to the end of the name (e.g., $knitC_{rt}$ becomes $knitCrt$).

138

The code is organized into the following subprograms:

main
envin
envout
parin
parout
incond
driver
kutta or euler
plot
print

Except for *driver* all the other routines existed in the linear model simulator.

The *main* program serves as the commander and calls the other subprograms in their time of need.

```
c-- main program for nonlinear system simulation
c
c  Solves a system of nonlinear differential equations over time
c  using the 4-th order Runge-Kutta or Euler method.
c
c  Variables:
c     nstate -- integer -- number of states
c     maxst  -- integer -- maximum number of states allowed
c     nisv   -- integer -- number of inter. system variables
c     maxisv -- integer -- maximum number of inter. system variables
c     nparam -- integer -- number of parameters
c     maxpar -- integer -- maximum number of parameters
c     nenv   -- integer -- number of env. variables
c     maxenv -- integer -- maximum number of env. variables
c     parnam -- integer -- array of parameter names
c     nstep  -- integer -- time step accumulator
c     err    -- integer -- error code (0 => no error, 1 => error)
c
c     dt     -- real    -- time step
c     tzero  -- real    -- starting time
c     tend   -- real    -- ending time
c     tprint -- real    -- print interval
c     tplot  -- real    -- plotting interval
c     time   -- real    -- time
c
c     x      -- real    -- state variable array (nstate)
c     g      -- real    -- isv array (nisv)
c     k      -- real    -- parameter array (nparam)
c     z      -- real    -- env. variable array (nenv)
```

```
c
      real teps

      integer maxst,maxisv,maxpar,maxenv
      common /maxsiz/ maxst,maxisv,maxpar,maxenv

      integer IN,DIN,OU,PL
      common /inout/ IN,DIN,OU,PL

      integer nstate,nisv,nparam,nenv,nstep,parnam(25,9),err
      real dt,tzero,tend,tprint,tplot,time,
     +     x(8),g(20),k(25),z(5)

      data teps /1.0e-5/
      maxst = 8
      maxisv = 20
      maxpar = 25
      maxenv = 5

      IN = 5
      DIN = 8
      OU = 6
      PL = 7
      open(unit = 8,file = "driver.d")

c  read in and echo out simulation environ. checking for errors
      call envin(nstate,nisv,nparam,nenv,
     +     dt,tzero,tend,tprint,tplot,err)
      if (err .eq. 1) stop
      call envout(nstate,nisv,nparam,nenv,
     +     dt,tzero,tend,tprint,tplot,err)

c  read in and echo out parameters
      call parin(k,parnam,nparam)
      call parout(k,parnam,nparam)

c  read in and print initial conditions
      call incond(nstate,x)
      call print(nstate,tzero,x)
      call plot(nstate,tzero,x)

c  initialize time
      time = tzero

c  get initial driving variables:
      call driver(z,nenv,time)
```

```
c   time loop
        nstep = 0
    10 if (time .ge. tend) go to 20
            nstep = nstep + 1
            time = tzero+nstep * dt

c   if new day compute driving variables
            if (abs(amod(time,1.0)) .lt. teps) call
            driver(z,nenv,time)

c   update state variables by one dt
            call kutta(x,g,k,z,nstate,dt,time)

c   if time to print
            if (abs(amod(time,tprint)) .lt. teps)
    +           call print(nstate,time,x)

c   if time to plot
            if (abs(amod(time,tplot)) .lt. teps)
    +           call plot(nstate,time,x)

        go to 10

    20 continue


        end
```

The *main* program and the subroutines follow the same organizational pattern that was used in Chap. 3 programs. Following the name and description of the subroutine is a commented list of the variables passed to that subroutine. Arrows indicate if the variable is input ($\rightarrow$), output ($\leftarrow$), or both ($\leftrightarrow$) for the subroutine. This is followed first by type declarations (e.g., integer, real) of the variables and then by any common blocks. Finally, we reach the code for what is done in the subroutine.

The main program also contains statements at the beginning to set the *maxst*, *maxisv*, *maxpar*, and *maxenv* variables equal to the dimension of their corresponding array (x, g, k, z, respectively). These variables are passed in the common block *maxsiz* to subroutines for array bound checking. Also set at the beginning of *main* are the unit numbers for input–output. These are passed to subroutines in the common block *inout*. Input comes from two sources; input files for specifying the simulation environment, parameters and initial conditions (IN) and input files containing the driving variables to be used in the model (DIN). These files are distinguished from each other because the first is read only at the start of the simulation, whereas the driving variables may be input throughout the simulation. Units 5 and 8 are used here for IN and DIN. There are also two output file units, OU for printed output and PL for output to be plotted. These are given units 6 and 7 here. The plot file must be postprocessed through a plotting routine.

Input–output units are handled differently in different implementations of Fortran. In this version the driving variable file must be opened and given a name, here *driver.d*. The *.d* convention is used to identify a file as a data file. Units 5 and 6 (IN and OU) are usually the standard input and output (the terminal), and, therefore, do not need to be opened. The plotfile unit, PL, will be automatically opened when the program is run and the resulting file will have a name such as *fort.7* or *for007.dat*.

The main program first calls the routines *envin* and *envout*.

```
      subroutine envin(nstate,nisv,nparam,nenv,
     +                 dt,tzero,tend,tprint,tplot,err)
c
c   Read in the simulation environment. The input file
c   should contain a one line description before
c   the actual values (the description line is
c   not actually read) An example:
c
c      nstate, nisv, nparam, nenv
c      5 20 18 2
c      dt, tzero, tend
c      0.1 0.0 365.0
c      tprint, tplot
c      10.0 5.0
c
c      nstate <-- number of state variables
c      nisv   <-- number of intermediate system variables
c      nparam <-- number of parameters
c      nenv   <-- number of environmental variables
c      dt     <-- time step
c      tzero  <-- starting time
c      tend   <-- ending time
c      tprint <-- print interval
c      tplot  <-- plotting interval
c      err    <-- integer error switch, 0 => no error, 1 => error
```

```
c
      integer maxst,maxisv,maxpar,maxenv
      common /maxsiz/ maxst,maxisv,maxpar,maxenv

      integer IN,DIN,OU,PL
      common /inout/ IN,DIN,OU,PL

      integer nstate,nisv,nparam,nenv,err
      real dt,tzero,tend,tprint,tplot

      err = 0

      read(IN,*)
      read(IN,*) nstate,nisv,nparam,nenv
      read(IN,*)
      read(IN,*) dt,tzero,tend
      read(IN,*)
      read(IN,*) tprint,tplot

c   check for errors
      if (nstate .gt. maxst) then
          write(OU,110) maxst
          err = 1
      end if
      if (nisv .gt. maxisv) then
          write(OU,120) maxisv
          err = 1
      end if
      if (nparam .gt. maxpar) then
          write(OU,130) maxpar
          err = 1
      end if
      if (nenv .gt. maxenv) then
          write(OU,140) maxenv
          err = 1
      end if
      if (dt .lt. 0.0) then
          write(OU,150)
          err = 1
      end if
      if (tzero .gt. tend) then
          write(OU,160)
          err = 1
      end if

      return
```

```
110 format(1x,'error, too many state variables, max. is ',i3)
120 format(1x,'error, too many inter. state var., max. is ',i3)
130 format(1x,'error, too many parameters, max. is ',i3)
140 format(1x,'error, too many environmental var., max. is ',i3)
150 format(1x,'error, negative dt')
160 format(1x,'error, tzero greater than tend')

    end




    subroutine envout(nstate,nisv,nparam,nenv,
   +                  dt,tzero,tend,tprint,tplot)
c
c  Print out the simulation environment setup.
c
c    nstate --> number of state variables
c    nisv   <-- number of intermediate system variables
c    nparam <-- number of parameters
c    nenv   <-- number of environmental variables
c    dt     --> time step
c    tzero  --> starting time
c    tend   --> ending time
c    tprint --> printing interval
c    tplot  --> plotting interval
c

    integer IN,DIN,OU,PL
    common /inout/ IN,DIN,OU,PL

    integer nstate,nisv,nparam,nenv
    real dt,tzero,tend,tprint,tplot

    write(OU,110) nstate,nisv,nparam,nenv,
   +     dt,tzero,tend,tprint,tplot

    return

110 format(1x,'number of state variables: ',i3,/,
   +       1x,'number of intermediate system variables: ',i3,/,
   +       1x,'number of parameters: ',i3,/,
   +       1x,'number of environmental variables: ',i3,/,
   +       1x,'time step: ',f7.4,/,
   +       1x,'time start: ',f8.4,3x,'end: ',f8.4,/,
   +       1x,'print interval: ',f8.4,1x,'plot interval: ',f8.4)

    end
```

These subprograms read in and print out, for purposes of corroboration, the simulation environment. This includes the structural environment—numbers of state variables, *nstate,* intermediate system variables, *nisv,* parameters, *nparam,* and driving or environmental variables, *nenv*—and the temporal environment— time step, *dt,* starting time, *tzero,* ending time, *tend,* and print and plot intervals, *tprint* and *tplot.* Two read statements are used for each line read in as was done in the linear model *envin* subroutine. The first read does nothing but skip over a line in the input file. The skipped line contains a description of the values that are to be read in on the following line. Editing of the data files to change the simulation environment, parameters, or initial conditions is easier with this file setup because the names of the variables are on the line above their values. The values are all read unformatted.

An error switch *err* is used to indicate, as with the linear model, some illegal data or condition, such as an array overflow (e.g., number of parameters greater than the dimension of the parameter array) or negative time step. If such an error is found an error message is printed and the program stops ($err = 1$).

An example model data file for the San Joaquin model is shown in Fig. 6.1, which illustrates how two lines are used for each line of data. Parameters and

```
nstate, nisv, nparam, nenv
3 16 21 3
dt, tzero, tend
.05 1.0 360
tprint, tplot
30.0 5.0
kphtmx
3.0
kphtltop
300
kCChlrt
50
kphttp
.0333
kphtniths
.025
krsptp
.005
kflowp
.250
knitCrt
.17
kfdmx
.39
kassim
.60
```

Fig. 6.1. Input file for the San Joaquin model setting parameter values for the simulation environment, for model processes, and setting initial conditions.

```
kfdpyhs
3.00
kfdtp1
0.0
kfdtp2
.20
krsptpz
.01
kmrt
.075
kflowz
.125
kdisnit
5.67e9
kflownit
.10
kd
1.2
kextlt
4.0
kvol
2.75e10
xbiop
0.04
xbioz
0.20
xnit
0.55
```
                   Fig. 6.1.   (continued)

initial conditions are read from this same file with subroutines *parin* and *incond,*
and are printed out by subroutines *parout* and *print,* respectively.

```
      subroutine parin(k,parnam,nparam)
c
c  Read in the model parameters, each parameter requires 2 lines
c   the first line has the parameter name (in columns 1-9) and
c   the second line has the parameter value. The parameter name
c   and its value will be printed out if subroutine parout is
c   called.
c
c    k       <-- parameter array
c    parnam <-- parameter name array
c    nparam --> number of parameters
```

```
c

      integer IN,DIN,OU,PL
      common /inout/ IN,DIN,OU,PL

      integer parnam(25,9),nparam,i,j
      real k(1)

      do 10 i = 1, nparam
          read(IN,110) (parnam(i,j),j=1,9)
          read(IN,*) k(i)
   10 continue

      return

  110 format(9a1)

      end

      subroutine parout(k,parnam,nparam)
c
c Output the model parameters and their names, also
c outputs the value of the k array where each parameter
c is stored.
c
c   k      --> parameter array
c   parnam --> parameter name array
c   nparam --> number of parameters
c

      integer IN,DIN,OU,PL
      common /inout/ IN,DIN,OU,PL

      integer parnam(25,9),nparam,i,j
      real k(1)

      write(OU,110)

      do 10 i = 1, nparam
          write(OU,120) (parnam(i,j),j=1,9),i,k(i)
   10 continue

      return

  110 format(/,1x,'Parameters:')
  120 format(1x,9a1,' k(',i2,') ',g9.4)

      end
```

```
      subroutine incond(nstate, x)
c
c Read in and print the initial condtions with variable
c names. Each parameter requires 2 lines in the input file,
c the first line contains the name of the state variable
c (in columns 1-9) and the second line its initial value.
c
c    nstate --> number of state variables
c    x        <-- array of state variables at time tzero
c    of length nstate
c

      integer IN,DIN,OU,PL
      common /inout/ IN,DIN,OU,PL

      integer nstate,stname(10,9),i,j
      real x(10)

      write(OU,210)

      do 10 i = 1, nstate
          read(IN,110) (stname(i,j),j=1,9)
          read(IN,*) x(i)
          write(OU,220) (stname(i,j),j=1,9),i,x(i)
   10 continue

      write(OU,230) (i,i=1,nstate)
      write(OU,*)

      return

  110 format(9a1)
  210 format(/,1x,'Initial Conditions:')
  220 format(1x,9a1,'x(',i2,')   ',g9.3)
  230 format(/,1x,'   Time      State Variables:',/
     +        8x,8(4x,i2,4x))

      end


      subroutine print(nstate,time,x)
c
c Print out the time and state variables.
c
c nstate --> number of compartments
c time   --> current time
c x      --> state variable array
```

```
c
      integer IN,DIN,OU,PL
      common /inout/ IN,DIN,OU,PL

      integer nstate, i
      real x(1)

      write(OU,110) time,(x(i),i=1,nstate)

      return

110 format(1x,f8.3,8(1x,g9.3))

      end
```

As with the *envin* subroutine two lines are read in for each parameter or initial condition, however, this time the parameter names, on the first of the line pairs, are read into the two-dimensional array *parnam(i,j)* where *i* is the parameter array index and *j* is the character index (up to nine characters per name). This allows *parout* to print the parameter's mnemonic name, its array name, and its value. Space is provided for up to 25 parameters (*maxpar* = 25). Although this is far from the most elegant way to do input–output, it does work. Furthermore, it is simple, general, and expandable and will not require reformatting when used with a different model. The initial values for the state variables are read by *incond*, which also reads the mnemonic names and prints a state variable list with the initial conditions using both mnemonic and array names of the variables.

Most of the simulation model computations are embedded within a time loop in the main program. Time starts at *tzero* and is incremented in steps of *dt* until *tend* is reached, as was done in the linear model code. Inside the time loop the main program calls *driver* on a daily basis, and the *plot* and *print* routines at intervals *tplot* and *tprint*. They simply print the time and state variables to output files 6 and 7. The rest of the time loop involves using the *euler* or *kutta* routines to update the state variables.

6.2.1   Driving variables

In Chap. 5 we talked about driving variables for three models. Now we are at the stage of implementing driving variables into a simulation model. We will consider two ways of doing this. The first is to approximate driving variable data by an equation involving time such as a sinusoid, and the second is to read in driving variable data at equal time intervals (e.g., daily). The solar radiation driving variable used by Di Toro et al. (1971) is (in ly/day):

$$zinsol = 450 + 300 \times sin(0.01745 \times time - 1.0472) \qquad (6.1)$$

This sinusoid function is computed daily in subroutine *driver* (*time* in Eq. 6.1 is thus given in Julian days). Temperature and flow rate data for the San Joaquin in 1966–1967, which are also driving variables, were given in Fig. 5.10 and are read in from unit DIN (8) on a daily basis by subroutine *driver*.

```
      subroutine driver(z,nenv,time)
c
c Read in or compute the current days driving variables.
c
c nstate --> number of driving variables
c time    --> current time
c z       <-- current day driving variable array
c
      integer IN,DIN,OU,PL
      common /inout/ IN,DIN,OU,PL

      integer nenv
      real z(1)

c compute solar radiation, zinsol:
      z(1) = 450 + 300 * sin(.01745 * time - 1.0472)

c read in temperature, ztp and flow, zflow
      read(DIN,*) z(2),z(3)

      return
      end
```

The *driver* routine is entirely specific to the San Joaquin model driving variables and would have to be changed for a different model.

6.2.2 Runge–Kutta and Euler methods

Subroutine *kutta* or *euler* does the solving of the model's differential equations. Either one is called by the main program.

```
      subroutine kutta(x,g,k,z,nstate,dt,time)
c
c Use the Runga-Kutta (4-th order) method to solve a system of
c nonlinear differential equations. A subroutine der is needed
c to compute the derivatives of the state variables.
c
c x      <-> state variable array
```

```
c  g         <-- isv
c  k         --> parameter array
c  z         --> environmental variable array
c  nstate  --> number of state variables
c  dt        --> time step
c  time     --> current time
c

      integer nstate, i
      real x(1),g(1),k(1),z(1),dt,time,
     +     dxdt(8),xplus(8),sumdx(8),dto2

      dto2 = dt / 2.0

      call der(x,g,k,z,dxdt,time)

      do 10 i = 1, nstate
          xplus(i) = x(i) + dto2 * dxdt(i)
          sumdx(i) = dxdt(i)
10 continue

      call der(xplus,g,k,z,dxdt,time)

      do 20 i = 1, nstate
          xplus(i) = x(i) + dto2 * dxdt(i)
          sumdx(i) = sumdx(i) + 2.0 * dxdt(i)
20 continue

      call der(xplus,g,k,z,dxdt,time)

      do 30 i = 1, nstate
          xplus(i) = x(i) + dt * dxdt(i)
          sumdx(i) = sumdx(i) + 2.0 * dxdt(i)
30 continue

      call der(xplus,g,k,z,dxdt,time)

      do 40 i = 1, nstate
          sumdx(i) = sumdx(i) + dxdt(i)
          x(i) = x(i) + dt * sumdx(i) / 6.0
40 continue

      return
      end
```

```
        subroutine euler(x,g,k,z,nstate,dt,time)
c
c  Use the Euler method to solve a system of nonlinear
c  differential equations. A subroutine der is needed to
c  compute the derivatives of the state variables.
c
c  x      <-> state variable array
c  k      --> parameter array
c  z      --> environmental variable array
c  nstate --> number of state variables
c  dt     --> time step
c  time   --> current time
c
      integer nstate,i
      real x(1),g(1),k(1),z(1),dt,time,dxdt(8)

      call der(x,g,k,z,dxdt,time)

      do 10 i = 1, nstate
          x(i) = x(i) + dt * dxdt(i)
10 continue

      return
      end
```

Both subroutines need to call a model-specific subroutine, *der* that computes the derivatives of the state variables with respect to time.

```
        subroutine der(x,g,k,z,dxdt,time)
c
c  Compute the derivatives of the state variables for the
c  San Joaquin delta model.
c
      real x(1),g(1),k(1),z(1),dxdt(1),time

c  first compute all the intermediate variables
      call pphs(x,g,k,z)
      call prsp(x,g,k,z)
      call pinfl(x,g,k,z)
      call zgraz(x,g,k,z)
      call zrsp(x,g,k,z)
      call zmrt(x,g,k,z)
      call zinfl(x,g,k,z)
      call nuptk(x,g,k,z)
      call ndisch(x,g,k,z)
      call ninfl(x,g,k,z)
```

```
c       dxbiop/dt = xbiop * (gpht - grsp) - xbioz * gfd + gflowp
        dxdt(1) = x(1) * (g(1) - g(5)) - x(2) * g(7) + g(6)

c       dxbioz/dt = xbioz * (kassim * gfd - grspz - gmrt) + gflowz
        dxdt(2) = x(2) * (k(10) * g(7) - g(10) - g(11)) + g(12)

c       dxnit/dt = -guptnit + gdisnit + gflownit
        dxdt(3) = -g(13) + g(14) + g(15)

        return
        end
```

Subroutine *der* calls routines that calculate the process flow rates for the particular system being modeled. For the San Joaquin model these include:

pphs	phytoplankton photosynthesis
prsp	phytoplankton respiration
pinfl	inflow and outflow of phytoplankton
zgraz	zooplankton grazing rate
zrsp	zooplankton respiration rate
zmrt	zooplankton natural mortality
zinfl	inflow and outflow of zooplankton
nutpk	uptake of nutrients by phytoplankton
ndisch	discharge of nitrogen into the river
ninfl	inflow and outflow of nitrogen

These routines, which form the main body of the model document, are given in Appendix 4. Some of these routines call other routines to compute the subprocesses that comprise a process, e.g., temperature, light, and nutrient effects on photosynthesis are computed in subroutines *ppht, plite,* and *pphn.* The equations for these subroutines are documented in Chap. 5 as part of the description of the San Joaquin model.

For a given time step the Runge–Kutta method calls *der* (which calls the process subroutines) four times compared with the single call of *der* per time step for the Euler method. The first time *der* is called the state at the start of the time step x is included in the argument list for *der*. On returning from *der*, *kutta* temporarily updates the state variables by going half the time step dt at rates *dxdt* (computed by *der*) to compute the temporary states *xplus*, which are used as arguments in the second call to *der*. The derivatives returned from *der* (M_2 in the derivation of the fourth order Runge–Kutta technique in Chap. 2) are again added to initial states x after multiplication by $dt/2$ to give a second set of temporary state variables *xplus*, which are used as arguments in *der* to produce a third set of derivatives *dxdt*. Multiplying these derivatives by dt and adding the result to the initial states x gives a third set of temporary state variables, which are used as arguments in *der* to give a fourth set of derivatives.

The four sets of derivatives are weighted by the proper factors as given in Eq. 2.11 and a composite derivative is computed (denoted by *sumdx*/6 in *kutta* and by Φ in Chap. 2). This is then multiplied by time step *dt* and added to *x* to give the approximate solution to the differential equations for *x* at the next time step.

6.3 San Joaquin Delta model

In this section we finally put all the pieces of the simulator together to run the Sacramento–San Joaquin Delta model. The model will be run for 360 days with a time step of 0.05. Printed output is given every 30 days and output for plotting every 5 days (i.e., *tzero* = 0, *tend* = 360, *dt* = 0.05, *tprint* = 30.0, *tplot* = 5.0). The three state variables, nitrogen concentration (*xnit*), phytoplankton biomass (*xbiop*), and zooplankton biomass (*xbioz*) are labeled $x(1)$, $x(2)$, and $x(3)$, respectively, in the code. The complete input file is shown in Fig. 6.1. The driving variables *zflow* and *ztp* are shown in Fig. 5.10 and are listed in Appendix 5 so that you will be able to run the model.

Example printed output from the model is shown in Fig. 6.2 and plotted

```
number of state variables:    3
number of intermediate system variables:   16
number of parameters:  21
number of environmental variables:    3
time step:   0.0500
time start:    1.0000    end: 360.0000
print interval:   30.0000 plot interval:    5.0000

Parameters:
kphtmx      k( 1) 3.000
kphtltop    k( 2) 300.0
kCChlrt     k( 3) 50.00
kphttp      k( 4) .3330e-01
kphtniths   k( 5) .2500e-01
krsptp      k( 6) .5000e-02
kfloup      k( 7) .2500
knitCrt     k( 8) .1700
kfdmx       k( 9) .3900
kassim      k(10) .6000
kfdpyhs     k(11) 3.000
kfdtp1      k(12)    0.
kfdtp2      k(13) .2000
krsptpz     k(14) .1000e-01
```

Fig. 6.2.   Output file produces from the San Joaquin model using the input file in Fig. 6.1. The Runge–Kutta solution is used here.

```
kmrt        k(15)  .7500e-01
kflowz      k(16)  .1250
kdisnit     k(17)  .5670e+10
kflownit    k(18)  .1000
kd          k(19)  1.200
kextlt      k(20)  4.000
kvol        k(21)  .2750e+11
```

Initial Conditions:
```
xbiop       x( 1)  0.400e-01
xbioz       x( 2)  0.200
xnit        x( 3)  0.550
```

Time	State Variables: 1	2	3
1.000	0.400e-01	0.200	0.550
30.000	0.549	0.978e-01	0.610
60.000	3.17	0.138	0.290
90.000	4.29	0.504	0.196e-01
120.000	2.17	1.57	0.451e-01
150.000	1.66	1.81	0.700e-01
180.000	1.37	1.95	0.106
210.000	1.37	1.84	0.109
240.000	1.39	1.68	0.135
270.000	1.74	1.28	0.166
300.000	2.45	0.682	0.296
330.000	0.680	0.104	0.570
360.000	0.339	0.101	0.498

Fig. 6.2.   (continued)

output in Fig. 6.3. The phytoplankton have an early spring bloom and then a smaller peak in the fall. The zooplankton peak after grazing on the spring phytoplankton bloom and then die off in late summer from lack of food. Their die-off allows the phytoplankton to have the second bloom in the fall.

6.4   Hubbard Brook forest succession model

As another example we use our simulator to run the Hubbard Brook forest succession model described in Chap. 5. This will involve both some simplification of the model presented in Chap. 5 and some modifications to our simulator code.

The state variables are tree diameter (*xdbh*), tree height (*xht*), tree leaf area (*xarealf*), and tree age (*xage*). The original model kept track of these variables

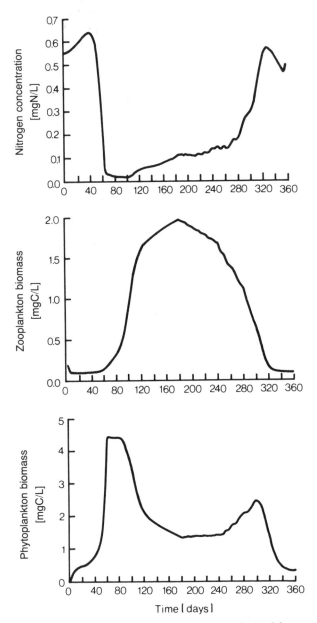

Fig. 6.3.   Graphical output for the San Joaquin model generated by the nonlinear simulation program given here with parameter values set by the file in Fig. 6.1.

156

Table 6.1   Notation and units for the state and intermediate variables used in the implementation of the Hubbard Brook forest succession model. Notation is given both for the documentation and for the Fortran code.

Variable	Fortran name	Description	Units
xdbh	x(1,j)	Tree diameter	cm
xht	x(2,j)	Tree height	cm
xarealf	x(3,j)	Leaf area	cm^2
xn	x(4,j)	Tree numbers	no.
xage	x(5,j)	Tree age	yr
ggrow	g(1,j)	Growth rate	cm/yr
ggrowlt	g(2,j)	Light effect on growth	—
garealf	g(3,j)	Leaf area of all trees above the jth tree	cm^2
ggrowtp	g(4,j)	Temperature effect on growth	—
gspace	g(5,j)	Competition effect on growth	—
gareabs	g(6,j)	Total basal in stand	cm^2
gmratage	g(7,j)	Age mortality	/yr
gmrt	g(8,j)	Slow growth mortality	/yr
glt	g(9,j)	Light at any level in the canopy	—

for each tree of 13 species on a 10 × 10 m plot. This is beyond the capabilities of our present simulator. We will simplify things by only modeling an ''average'' tree of each species and keeping track of how many ''average'' trees we have with a new state variable *xn*. We will also reduce the number of species considered.

To account for the multispecies aspect of the model we will index our state variables by species number *j* (e.g., *xdbh* (*j*) refers to the diameter of the ''average'' tree for species *j*). The intermediate variables and parameters will be similarly indexed. Because we want the computer code to pass arrays, the code will now have doubly indexed arrays for state and intermediate variables and for parameters. For example, $x(2,3)$ refers to the height of the third tree species [*xht*(3) in mnemonic notation] and $k(5,2)$ is the maximum diameter for species 2 [$kdbh_{mx}(2)$]. Table 6.1 shows all the state and intermediate system variable names and their corresponding names in the code. We have to change most of the simulator code to account for double indexing of the arrays, but the changes are simple (mostly adding loops over the species index). New simulation environment variables, *nspec*, denoting the number of species and *maxsp* denoting the maximum *nspec* allowed to prevent overflowing the arrays (the error check is added to subroutine *envin*) are created. Subroutines *parin* and *incond* will now read in *nspec* parameter values and initial conditions on each line.

Model-dependent code includes a subroutine *update* for updating the state variables (because the Runge–Kutta or the Euler techniques are not used with difference equation models such as the Hubbard Brook model).

```
      subroutine update(x,g,k,z,nspec,time)
c
c Update the difference equations calling rate subroutines first.
c

      integer nspec,i
      real x(8,6),g(20,6),k(25,6),z(5),dxdt(6),time

      call growth(x,g,k,z,nspec)
      call gmorta(x,g,k,z,nspec)
      call gmorts(x,g,k,z,nspec)

      do 10 i = 1, nspec

c         compute growth change and add to diameter:
c         dxdbh/dt = ggrow * xarealf *
c                    (1.0 - xdbh * xht / (kdbhmx * khtmx) /
c                    (xdbh * (2.0 * kdbhht1 + 3.0 * kdbhht2 * xdbh -
c                    4.0 * kdbhht3 * xdbh ** 2
          dxdt(i) = g(1,i) * x(3,i) *
     +               (1.0 - x(1,i) * x(2,i) / (k(5,i) * k(6,i))) /
     +               (x(1,i) * (2.0 * k(1,i) + 3.0 * k(2,i) * x(1,i)
     +               4.0 * k(3,i) * x(1,i) ** 2))
c         xdbh = xdbh + dxdbh/dt
          x(1,i) = x(1,i) + dxdt(i)

c         update ht. and leaf area by direct computation:
c         xht = kdbhht1 + kdbhht2 * xdbh - kdbhht3 * xdbh ** 2
          x(2,i) = k(1,i) + k(2,i) * x(1,i) - k(3,i) * x(1,i) ** 2
c         xarealf = karealf * xdbh ** 2
          x(3,i) = k(4,i) * x(1,i) ** 2

c         update tree no, if growth is large no slow growth mort:
c         if (dxdbh/dt > kdbhth) gmrt = 0.0
c         xn = xn * (1.0 - (gmrtage + gmrt))
c         if (xn < 0.0) xn = 0.0
          if (dxdt(i) .gt. k(15,i)) g(8,i) = 0.0
          x(4,i) = x(4,i) * (1.0 - (g(7,i) + g(8,i)))
          if (x(4,i) .lt. 0.0) x(4,i) = 0.0

c         update age:
c         xage = xage + 1
          x(5,i) = x(5,i) + 1

10    continue

      return
      end
```

Subroutine *update* calls the *growth* subroutine (shown in Appendix 4), which computes the growth rate as a function of solar radiation, temperature, and soil moisture. We have subdivided growth into individual subprocesses coded in separate subroutines, which are shown in Appendix 4.

Tree mortality due to both slow growth and old age are computed as fractions of trees dying rather than as probabilities because tree numbers rather than individual trees are being modeled. For example, if the cherry trees are age 9 then their age mortality is computed by:

$$gmrtage = 1.0 - (1.0 - kmrtage)^{xage}$$
$$= 1.0 - (1.0 - 8.38 \times 10^{-3})^9$$
$$= 0.0729$$

and fraction $0.0729 \times xn$ cherry trees die for this year. The *kmrtage* parameter for a species is set so that only 2 percent of the trees will reach the maximum age of the species. The slow growth mortality parameter, *kmrt*, is set so that only 1 percent of the slow growing trees live for 10 years. The mortality fractions are computed in subroutines *gmorta* (age mortality) and *gmorts* (slow growth mortality) shown in Appendix 4.

The tree establishment module of the original Hubbard Brook model cannot be implemented here because we are not modeling individual trees. Instead we look at different tree establishment patterns by varying the initial conditions. Another approach to establishment is to read in new tree numbers every 10 years or so. This is left as a problem at the end of the chapter.

The site specification module specifies the site-specific parameters. The two driving variables, annual degree days above 40°F, *ztp*, and total annual insolation, *zinsol,* have constant values [3000 and 1.0 (normalized), respectively, in our example] to reflect averages from Hubbard Brook data. It is left as a problem to examine the effects of varying driving variables on model behavior. The other site-specific value is $kareabs_{mx}$, the maximum total basal area for the 10×10 m site being modeled. Information about the site being modeled is needed to estimate this parameter.

For our example we have chosen four tree species—two shade-intolerant species, pin cherry (*Prunus pensylvanica*) and white birch (*Betula papyrifera*) and two shade-tolerant species, sugar maple (*Acer saccharum*) and spruce (*Picea rubens*). The model parameters are shown both in Table 6.2 and in the complete input data file given in Fig. 6.4.

For the first run we start with all age 1 trees of 0.5 cm diameter. The initial numbers are 50 cherry, 30 birch, and 5 each of maple and spruce. A 100-year run is illustrated in Fig. 6.5. In this run the cherry reaches its peak basal area after 19 years and is gone from the stand after about 40 years. The birch peaks at around 40 years and then disappears from the stand by 100 years. The shade-tolerant maple is still increasing even after 100 years and the slow growing spruce is gradually increasing in the stand.

Table 6.2   Parameter values used in the Hubbard Brook implementation of the forest succession model including notation, units, values for each species in the version simulated here.

Parameter	Units	Cherry	Birch	Maple	Spruce
$kdbhht_1$	cm	137	137	137	137
$kdbhht_2$	—	70.6	73.6	50.9	67.9
$kdbhht_3$	/cm	1.26	0.8	0.167	0.679
$karealf$	—	2.45	0.486	1.57	2.5
$kdbh_{mx}$	cm	28	46	152.5	50
kht_{mx}	cm	1126	1830	4011	1830
$kgrow_{mx}$	cm/yr	81.6	288.1	108.3	20.0
$kgrowlt_1$	—	2.24	2.24	1.0	1.0
$kgrowlt_2$	—	1.136	1.136	4.64	4.64
$kgrowlt_3$	—	0.08	0.08	0.05	0.05
$kdegdy_{mn}$	deg-day/yr	1100	1100	2000	600
$kdegdy_{mx}$	deg-day/yr	8000	3700	6300	3700
$kmrtage$	—	8.38e-3	1.21e-3	1.95e-4	6.37e-5
$kmrt$	—	0.080	0.080	0.080	0.080
$kdbh_{th}$	cm	0.01	0.01	0.01	0.01
$kextlt$	/cm^2	1.167e-4	1.167e-4	1.167e-4	1.167e-4
$kareabs_{mx}$	cm^2	10000	10000	10000	10000

```
nspec, nstate, nisv, nparam, nenv
4 5 9 17 2
dt, tzero, tend
1.0 0.0 100.0
tprint, tplot
10.0 2.0
kdbhht1
137.0 137.0 137.0 137.0
kdbhht2
70.6 73.6 50.9 67.9
kdbhht3
1.26 0.80 0.167 0.679
karealf
2.45 0.486 1.57 2.50
kdbhmx
28.0 46.0 152.5 50.0
khtmx
1126.0 1830.0 4011.0 1830.0
kgrowmx
81.633 288.07 108.28 20.0
kgrowlt1
2.24 2.24 1.0 1.0
```

Fig. 6.4.   Input file for the Hubbard Brook tree succession model. The file specifies the modeling environment, sets process parameter values, and gives initial conditions for state variables.

```
kgrowlt2
1.136 1.136 4.64 4.64
kgrowlt3
0.08 0.08 0.05 0.05
kdegdymn
1100.0 1100 2000.0 600.0
kdegdymx
8000.0 3700.0 6300.0 3700.0
kmrtage
0.008378 .001207 .000195 .000064
kmrt
0.08 0.08 0.08 0.08
kdbhth
.005 .005 .005 .005
kextlt
.000167 .000167 .000167 .000167
kareabsmx
100000.0 100000.0 100000.0 100000.0
xdbh
0.5 0.5 0.5 0.5
xht
172.0 173.6 162.4 170.8
xarealf
0.61 0.12 0.39 0.63
xn
50 30 5 5
xage
1.0 1.0 1.0 1.0
```

Fig. 6.4.   (continued)

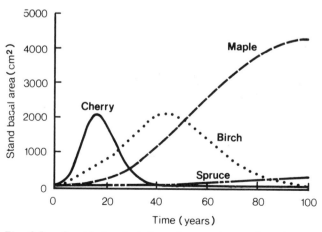

Fig. 6.5.   Graphical output for a four-species version of the Hubbard Brook model showing successional changes in the species composition of stand basal area beginning with a clearcut stand. The results are generated using the input file in Fig. 6.4.

Table 6.3   Initial conditions for run 2 of Hubbard Brook forest model.

State variable	Cherry	Birch	Maple	Spruce
xdbh	20.0	15.0	0.5	0.5
xht	1045.0	1061.0	162.4	170.8
xarealf	980.0	109.4	0.39	0.63
xage	15	15	1	1
xn	10	5	3	3

A more realistic run is made by starting with older and larger shade-intolerant trees and a few young shade-tolerant ones. Under this scenario the spruce and maple should become established as the cherry and birch get shaded out. A run of this type with initial conditions shown in Table 6.3 is plotted in Fig. 6.6. As was predicted the shade-intolerant trees die out and the maple and spruce take over.

As a final example we will try to establish cherry, birch, and maple under a canopy of spruce and maple. To do this we need to have two maple cohorts in our model—the large old trees of the canopy and the young ones we are trying to establish. The easiest way to implement this is to add another ''species'' to our model. No changes are needed in the simulator code, only in the data input file. The number of species, *nspec* is now 5 and we need to replicate the maple species parameters for the second cohort as if it were a second species. The initial conditions for this run are shown in Table 6.4. The results of a 150-year

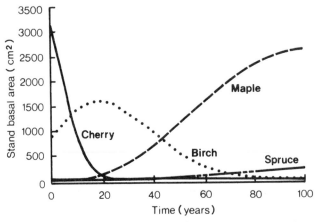

Fig. 6.6.   Graphical output for the four-species Hubbard Brook model showing tree succession beginning with a mixed age stand of trees. The results were generated using the initial conditions in Table 6.3.

Table 6.4   Initial conditions for run 3 of Hubbard Brook forest model.

State variable	Cherry	Birch	Maple 1	Maple 2	Spruce
xdbh	0.5	0.5	40.0	0.5	12.0
xht	172.0	173.6	1905.8	162.4	854.0
xarealf	0.61	0.12	2512.0	0.39	360.0
xage	1	1	80	1	80
xn	20	10	5	10	3

run (shown in Fig. 6.7) show that the shade-intolerant cherry and birch cannot become established under the canopy, whereas the second maple cohort can.

6.5   Summary

Computers are inextricably linked with ecological simulation in this book and, most obviously, in this chapter. We have taken two of the models in Chap. 5 and have developed Fortran coded versions of them. The code was organized into subroutines that were presented using pseudocode. Code that is model dependent was distinguished from code that can apply to any model. We have done this to expedite your developing different models using the same basic code structure presented here. Only the model-dependent code must be changed to do a new model. You will find, of course, that there is no truly general code and that some bending of the rules is necessary to code most new models.

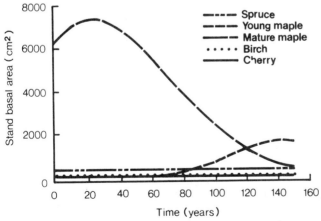

Fig. 6.7.   Graphical output for the five-species Hubbard Brook model showing tree succession beginning with a mature maple stand. The results use the initial conditions in Table 6.4.

One striking feature of coding these models is how much more difficult it is to code the fourth order Runge–Kutta routine than the Euler routine. Because the Runge–Kutta routine requires four updates between each time step, space must be provided for an array of temporary variables. The question with the Runge–Kutta routine is whether the method saves sufficient running time and gives sufficiently improved solution accuracy over the Euler technique that it is worth the extra effort. In studies done on many large models, the Runge–Kutta routine was found to be superior to the Euler routine. In ecological simulation models, which are not especially large as simulation models go, the question has not been studied in detail. We give both approaches here and hope that some future work will shed light on when to use which approach.

The careful reader might notice some minor discrepancies between the model descriptions in Chap. 5 and the code in Chap. 6. These do exist and were made either to simplify some equation or to reflect a more current understanding of a process. The behavior of the coded versions is similar to the model behaviors given by the original models.

6.6   Annotated Bibliography

Landry, M. R.: The structure of marine ecosystems: an alternative. *Mar. Biol. (Berl.)*, 35:1–7, 1976.

Contains a description of Steele's marine ecosystem model and some modifications including multiple cohorts of zooplankton.

Steele, J. H.: *The Structure of Marine Ecosystems,* Harvard University Press, Cambridge, Mass., 1974.

This model is treated in detail as exercises in this chapter.

6.7   Exercises

Modifications to San Joaquin model

6.1.   Try the Runge–Kutta technique with time steps of 0.05, 0.1, and 0.5 days with the original model. How is the solution affected by changing the time step? Which time step do you consider most appropriate for this model?

6.2.   Replace subroutine *kutta* by subroutine *euler* to compute the Euler numerical approximation to the system of differential equations. Compare the Euler and Runge–Kutta methods, by running the Euler at time steps of 0.01, 0.05, 0.1, and 0.5 days. Make a table of solution accuracy that compares the average levels and maximum levels of nitrogen, phytoplankton, and zooplankton for each time step and solution technique combination. Finding the average levels will require either adding a subroutine *average* to the program or writing a program to take averages of the plot file (unit 7) output.

6.3.   Conduct a simulation experiment to see how sensitive model behavior is to initial conditions. Make simulation runs beginning with 0.1, 1.0, and 10.0 times the initial conditions of the standard run for nitrogen, phytoplankton, and zooplankton in all possible combinations ($3^3 = 27$ runs total). Use the average and maximum phytoplankton, zooplankton, and nitrogen as measures of model performance. Are results sensitive to changes in initial conditions?

6.4.   Excretion of nitrogen by zooplankton was not treated in the San Joaquin model. Add nitrogen excretion to the model code. Represent nitrogen excretion rate *gexcnit* by the equation

$$gexcnit = knitC_{rt} \times (1 - kassim) \times gfd$$

In this formulation nitrogen excretion is the unassimilated phytoplankton multiplied by their N/C ratio. Nitrogen excretion by zooplankton is an input to the nitrogen pool and must be multipled by zooplankton biomass *xbioz* in the differential equation updating nitrogen. How important is nitrogen excretion to overall nitrogen dynamics?

6.5.   The San Joaquin model has a single, herbivorous zooplankton compartment. Many lakes, however, have substantial numbers of carnivorous zooplankton. In this exercise you will add a carnivorous zooplankton compartment to the model. The differential equation governing these zooplankton is the same as for herbivorous zooplankton in Eq. 5.3. The two groups are identical except that the prey of the carnivores are the herbivores. If we let $xbioz_1$ denote herbivores and $xbioz_2$ denote carnivores, then the prey density effect on carnivore grazing should be:

$$gfdpy_2 = \frac{xbioz_1}{xbioz_1 + kfdpy_{hs_2}}$$

All parameters for the two species are the same except that $kfdpy_{hs_2} = 0.5$ mgC/L, the inflow rate $kflowz_2 = 0.05$ mgC/L/day, and the initial condition $xbioz_2 = 0.05$ mgC/L for the carnivorous zooplankton. Simulate this expanded system and compare its behavior with the baseline system. How important are carnivorous zooplankton in this modified system?

6.6.   There is no consideration in this model of multiple phytoplankton species. Often the species composition in an aquatic system is of interest, because dominance by some group, such as blue-greens, can be highly undesirable and can foreshadow or reflect highly eutrophic conditions with depleted oxygen levels and a depauperate (only a few species) biotic community. In this exercise a second phytoplankton species, mimicking a blue-green algal group is added. Parameters for blue-green algae are as follows:

$kpht_{mx}$	$kphtlt_{op}$	$kCChl_{rt}$	$kphttp$	$kphtnit_{hs}$	$krsptp$	$kflowp$	$knitC_{rt}$
2.5	200.0	50.0	0.04	0.0075	0.005	0.10	0.17

a. Simulate the San Joaquin system with blue-green algae added with an initial concentration of 0.03 mgC/L. Use the original initial conditions for the other state variables. Assume that the equations governing blue-greens are the same as for the other algae but that there is no grazing on blue-green algae.

b. Increase the temperature average by 5°C to investigate the hypothesis that increased temperature can lead to dominance by blue-green algal communities.

6.7. The San Joaquin model has been applied to a several other sites with fundamentally the same model structure but with different parameters to represent the different biota on the new sites. The table below shows the parameters for a Lake Erie version of the model (Di Toro et al., 1975).

Lake Erie parameters

parameter	value
$kpht_{mx}$	1.5
$kphtlt_{op}$	350.0
$kphttp$	0.06
$kphtnit_{hs}$	0.025
$krsptp$	0.004
$knitC_{rt}$	0.14
kfd_{mx}	0.34
$kassim$	0.65
$kfdtp_1$	0.12
$kfdtp_2$	0.21
$krsptpz$	0.02
$kmrt$	0.015

In cases where parameter values are not given and for initial conditions, use the same values as in the San Joaquin model (see Fig. 6.1). For solar radiation and temperature it is permissible to use the same driving variables as for the San Joaquin model. The Lake Erie model also differs from the San Joaquin model in having spatial segmentation and in including phosphorous flow in addition to nitrogen. For purposes of this exercise we will ignore these differences and instead investigate how Lake Erie biota would respond to San Joaquin conditions. Thus, it is also acceptable to use the flow conditions from the San Joaquin model as well as the other driving variables.

Modifications to Hubbard Brook model

6.8.   How sensitive is the Hubbard Brook model behavior to the environmental conditions? Try simulation runs with temperature and solar radiation both increased and reduced by 20 percent, first starting with all
small trees and then with the mixed age initial conditions. Changes of
this sort can be looked at as changes in elevation or latitude and could
give some indication of how the succession sequence might change over
the range of the various species considered.

6.9.   How does the model respond to changing driving variables over time?
The driving variables used in the Hubbard Brook model presented in
this chapter were constants with *zinsol* = 1.0 and *ztp* = 3000. Let us
assume that both solar radiation and temperature vary as sinusoids with
a period of 10 years, an amplitude of 0.5 for solar radiation and of 500
for temperature, and an average level the same as in the original model.
Try a 100-year simulation with the mixed age initial conditions using a
sinusoid equation like that given below:

$$z = \left(\frac{z_{mx} - z_{mn}}{2}\right) \times \sin\left(\frac{2\pi t}{R}\right) + \left(\frac{z_{mx} + z_{mn}}{2}\right)$$

where R is the period (in years) and z_{mx} and z_{mn} are the maximum and
minimum values of the sinusoid.

6.10.   We have already made runs of the Hubbard Brook model with different
initial conditions and have seen how changing the initial state can markedly alter the succession sequence. Examine how the model behaves
when more shade tolerant trees are around at the start of the simulation.
   a. Try a simulation with initial conditions as in Table 6.3 except have
      10 trees each of spruce and maple.
   b. Try another with the 20 shade-tolerant trees but without the shade-
      intolerant trees. This simulation might give some idea about growth
      pattern differences between plantations and naturally established
      stands.
   c. Now try a simulation without the shade-tolerant trees using the initial
      conditions for birch and cherry from Table 6.3.
   d. From these simulations what do you think is the major factor con-
      trolling the decline in volume of the shade-intolerant trees?

6.11.   We have simulated a succession sequence with only four tree species,
whereas the original Hubbard Brook model had up to 13 species. Simulate succession with the addition to our model of beech, a later succession species, and red maple, another shade tolerant, but not so long-
lived, species. Start with five 1-year-old trees each of the new species
with *xdbh* = 0.5 cm plus the original initial conditions for the other

trees (50 cherry, 30 birch, 5 maple, and 5 spruce all age 1 and diameter 0.5 cm). Parameters for the new species are given below

Hubbard Brook model parameters

parameter	species	
	Beech	Red maple
$kdbhht_1$	137	137
$kdbhht_2$	57.8	46.3
$kdbhht_3$	0.237	0.152
$karealf$	2.2	1.75
$kdbh_{mx}$	122.0	152.5
kht_{mx}	3660	3660
$kgrow_{mx}$	68.2	137.1
$kgrowlt_1$	1.0	1.0
$kgrowlt_2$	4.64	4.64
$kgrowlt_3$	0.05	0.05
$kdegdy_{mn}$	2100	2000
$kdegdy_{mx}$	6000	12400
$kmrtage$	8.67e-05	3.45e-04
$kmrt$	0.080	0.080
$kdbh_{th}$	0.01	0.01

6.12. The increase in atmospheric CO_2 resulting from fossil fuel burning has been linked to the potential heating of the earth's environment, resulting in the so called *greenhouse effect* of permanently elevated temperatures. An average daily temperature increase of 5°C could add as much as 1000 degree-days to *ztp*. Simulate the effect of this on the succession sequence in Hubbard Brook working with only the four original species.

6.13. The Hubbard Brook model contains no explicit competition for moisture between the trees, nor any effect of drought on growth. Moisture stress has been considered in a several applications of this model to other sites. One such measure of moisture stress is due to Solomon and Shugart (1983) and is given by:

$$gmois = 1.0 - zdry/(kgrowdy \times kdryprp)$$

zdry = the number of drought (low soil and moisture) days in the season
kgrowdy = the number of growing season days
kdryprp = the proportion of the growing season a species can withstand soil moisture below the wilting point

Enter this function into the Hubbard Brook model code as a multiplicative effect on the growth rate *ggrow* and assume a growing season of 150 days with 15 drought days per year. Assume that *kdryprp* has values of 0.03, 0.02, 0.01, and 0.05 for birch, cherry, maple, and spruce, respectively. Simulate succession over a 100-year period. How does moisture stress affect the successional sequence in this model?

Steele marine ecosystem model

Steele (1974) developed a trophic dynamic model of the North Sea. That model is described here, complete with equations and parameter values as a prelude to a problem set based on it.

The model is made up of nitrate, phytoplankton, and herbivorous zooplankton compartments. Carnivorous zooplankton and fish are implicitly included by having predation on the zooplankton. The state variables (Table 6.5) are phytoplankton biomass (*xbiop*), zooplankton numbers (*xn*), average individual weight (*xwt*), age (*xage*), egg biomass (*xbioegg*), and nitrate concentration (*xnit*). For convenience we have added zooplankton biomass (*xbioz*) as another variable. The zooplankton are calanoid copepods (*Pseudocalanus spp.*), which dominate in the North Sea. All units of weight (biomass) are in micrograms carbon (μgC). Nitrate is in units of micrograms carbon equivalent (μgC eq.). The model time unit is a day.

The zooplankton process equations in this model are more complicated than those we have seen in the San Joaquin model. This model is a *cohort* model in that the zooplankton are presumed to be all of the same age based on the time they were spawned. This is in contrast with a *dynamic pool* model, such as the San Joaquin model, where all zooplankton are in the same pool regardless of age. Both zooplankton numbers and weight are modeled by differential equations and biomass is determined by their product. The numbers are decreased by natural mortality and predation. The zooplankton gain weight until they reach adult size and then any further weight gain is converted into eggs with an egg conversion efficiency of *keggeff*. The adult zooplankton have a maximum lifespan $kage_{mx}$ and on reaching it all zooplankton die and a new cohort based on the number of eggs produced is spawned.

Table 6.5   State variables with initial conditions for the marine ecosystem model.

Variable	Description	Units	Initial Condition
xbiop	Phytoplankton biomass	μgC/L	75.0
xn	Zooplankton numbers	no./L	20.0
xwt	Zooplankton average weight	μgC	0.20
xbioz	Zooplankton biomass	μgC/L	0.40
xage	Adult zooplankton age	days	0.0
xbioegg	Zooplankton egg biomass	μgC/L	0.0
xnit	Nitrate concentration	μgC eq./L	50.0

The differential equations are:

$$\dot{xbiop} = gpht \times xbiop - gfd \times xbioz - gmixp$$

$$\dot{xn} = -gmrt \times xn$$

$$\dot{xwt} = \begin{cases} (kassim \times gfd - grspz) \times xwt & \text{if } xwt < kwt_{mx} \\ 0 & \text{otherwise} \end{cases}$$

$$\dot{xbioegg} = \begin{cases} (kassim \times gfd - grspz) \times xbioz \times keggeff & \text{if } xwt > kwt_{mx} \\ 0 & \text{otherwise} \end{cases}$$

$$\dot{xnit} = gmixnit + gexcnit - guptnit$$

$gpht$ = photosynthesis rate

gfd = zooplankton feeding rate

$gmixp$ = phytoplankton mixing

$gmrt$ = density dependent predation and natural mortality rate

$kassim$ = assimilation rate

kwt_{mx} = maximum individual zooplankton weight

gfd = zooplankton grazing rate

$grspz$ = zooplankton respiration rate

$gmixnit$ = nitrogen mixing

$gexcnit$ = zooplankton excretion

$guptnit$ = phytoplankton uptake

Phytoplankton processes

The photosynthesis rate depends on the nitrogen concentration through a Michelis–Menton function:

$$gpht = kpht_{mx} \times gphtnit$$

$kpht_{mx}$ = maximum photosynthes rate

$gphtnit$ = nitrogen effect on photosynthesis

$$= \frac{xnit}{xnit + kphtnit_{hs}}$$

$kphtnit_{hs}$ = nitrogen uptake half saturation

The mixing term represents phytoplankton lost from the epilimnion (the segment modeled) to the hypolimnion (the deeper waters). Mixing in the turbulent North

Sea is assumed to occur year round. The process is proportional to the phyto-plankton present:

$$gmixp = kmix \times xbiop$$

$$kmix = \text{mixing coefficient}$$

Zooplankton processes

The zooplankton mortality has two components, a constant natural mortality *kmrt* and a density dependent zooplankton predation mortality, representing fish and carnivorous zooplankton predation on the copepods.

$$gmrt = kmrt + gpred$$

$$gpred = kpred_{mx} \times gpredpy$$

$$kpred_{mx} = \text{maximum predation rate}$$

$$gpredpy = \text{density dependent predation function}$$

$$= \begin{cases} \dfrac{(xn - kpredn_{th})(xwt - kpredwt_{th})}{(kpredpy_{hs} + xbioz)} & \text{for } xn > kpredn_{th}, \\ & xwt > kpredwt_{th} \\ 0 & \text{otherwise} \end{cases}$$

$$kpredn_{th} = \text{threshold numbers for predation}$$

$$kpredwt_{th} = \text{threshold weight for predation}$$

$$kpredpy_{hs} = \text{half saturation for predation}$$

The zooplankton grazing rate is given by the product of two subprocesses: a weight effect and a phytoplankton density effect:

$$gfd = kfd_{mx} \times gfdwt \times gfdpy$$

$$kfd_{mx} = \text{maximum grazing rate}$$

$$gfdwt = \text{weight effect on grazing}$$

$$= xwt^{kfdwt}$$

$$kfdwt = \text{weight effect exponent}$$

$$gfdpy = \text{prey density effect on grazing}$$

$$= \frac{(xbiop - kbiop_{th})}{(xbiop - kbiop_{th} + kfdpy_{hs})}$$

$$kbiop_{th} = \text{phytoplankton threshold for grazing}$$

$$kfdpy_{hs} = \text{half saturation for phytoplankton density effect on grazing}$$

The zooplankton respiration rate, like the zooplankton grazing rate, depends on the zooplankton weight. There is both a feeding independent and a feeding dependent component (called specific dynamic action).

$$grspz = (krspsda \times gfdpy + krsp) \times grspwt$$

$$krspsda = \text{ingestion dependent respiration rate}$$

$$krsp = \text{maximum respiration rate}$$

$$grspwt = \text{weight effect on respiration}$$

$$= xwt^{krspwt}$$

$$krspwt = \text{weight effect on respiration exponent}$$

Nitrogen processes

The nitrogen dynamics are governed by three processes, nitrogen uptake by phytoplankton, mixing in the water column, and zooplankton excretion of nitrogen.

Uptake by phytoplankton is modeled with the same Michelis–Menton function (*gphtnit*) used for phytoplankton photosynthesis:

$$guptnit = kuptnit_{mx} \times gphtnit \times xbiop$$

$$kuptnit_{mx} = \text{maximum uptake rate}$$

Mixing is a proportional to the amount of nitrogen above or below the initial concentration. This assumes that the hypolimnion has a constant nutrient concentration that is equal to the epilimnion concentration in the winter when the model run begins and that this concentration affects the model through mixing with the epilimnion throughout the year.

$$gmixnit = kmix \times (knit - xnit)$$

$$kmix = \text{mixing rate}$$

$$knit = \text{initial (and average hypolimnion) nitrogen concentration}$$

Excretion of nitrogen by zooplankton is assumed to be proportional to the zooplankton respiration rate:

$$gexcnit = kexcnit \times grspz \times xbioz$$

$$kexcnit = \text{zooplankton nitrogen excretion per unit of respiration}$$

Coding the Steele model

Most of the model equations will fit right into the simulator that was developed for the San Joaquin model. The unique part, not yet described, is the termination of the existing zooplankton cohort and the spawning of a new cohort. This occurs at discrete times whereas all the other processes occur continuously. Although we have not considered *events* in simulation models many models do have events. These include external factors such as the turning on or off a power plant. To accommodate birth and aging, an aging function is needed. It will update the state variable *xage* by one *dt* every time step if $xwt > kwt_{mx}$, i.e., when the zooplankton are adults. A cohort spawning subroutine is needed to remove all the adult zooplankton and spawn the new cohort. This subroutine will be called when $xage > kage_{mx}$. The total biomass of eggs, *xbioegg*, is converted into juvenile zooplankton each with weight, kwt_{mn}. The equations needed for the event are:

$$xn = xbioegg/kwt_{mn}$$

$$xwt = kwt_{mn}$$

$$xage = 0.0$$

$$xbioegg = 0.0$$

Table 6.6   Parameters for the marine ecosystem model.

Parameter	Description	Units	Value
$kpht_{mx}$	Maximum photosynthesis rate	/day	0.2
$kphtnit_{hs}$	Nitrogen uptake half saturation	μgC eq./L	96.0
$kmrt$	Zooplankton mortality rate	/day	0.0
$kpred_{mx}$	Maximum zooplankton predator mortality	/day	0.08
$kpredn_{th}$	Numbers threshold for zooplankton predator	no./L	0.0
$kpredwt_{th}$	Weight threshold for zooplankton predator	μgC	0.0
$kpredpy_{hs}$	Zooplankton predator half saturation	μgC/L	1.0
kwt_{mn}	Minimum zooplankton weight	μgC	0.2
kwt_{mx}	Maximum zooplankton weight	μgC	100.0
$kassim$	Zooplankton assimilation efficiency	—	0.7
kfd_{mx}	Maximum grazing rate	μgC^{-kfdwt}/day	1.6
$kfdwt$	Exponential weight effect on grazing	—	−0.3
$kbiop_{th}$	Phytoplankton threshold for zooplankton grazing	μgC/L	0.0
$kfdpy_{hs}$	Grazing half saturation	μgC/L	175.0
$krspsda$	Maximum ingestion dependent respiratory rate	$\mu gC^{-krspwt}$/day	0.3
$krsp$	Maximum ingestion independent respiratory rate	$\mu gC^{-krspwt}$/day	0.2
$krspwt$	Exponential weight effect on respiration	—	−0.3
$keggeff$	Egg conversion efficiency	—	0.1
$kage_{mx}$	Maximum adult zooplankton life span	day	12.0
$kuptnit_{mx}$	Maximum uptake rate	μgC eq./μgC/day	0.2
$knit$	Initial nitrogen concentration	μgC eq./L	756.0
$kexcnit$	Zooplankton exer. per unit respiration	μgC eq./μgC	0.4
$kmix$	Mixing rate between epilimnion and hypolimnion	/day	0.01

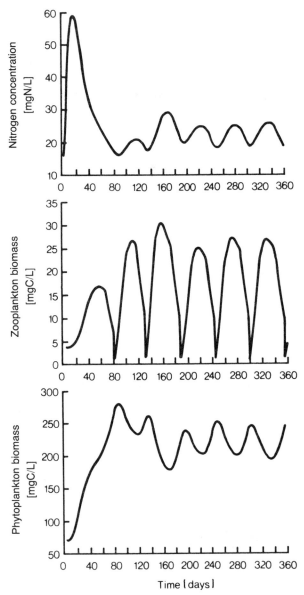

Fig. 6.8. Graphical output for the Steele marine ecosystem model showing phytoplankton and zooplankton biomass and nitrogen concentration over a 1-year simulation using parameter values and initial conditions given in Table 6.6 and Table 6.5.

The *der* subroutine will be a little different from what has been considered before. The derivative for zooplankton weight will be zero when the zooplankton weight is at the adult weight and their weight will be converted to zooplankton egg biomass. Similarly, when zooplankton are less than adult weight the zooplankton egg biomass derivative will be zero.

6.14. Code the Steele model using the Euler differential equation solver and run it with $dt = 0.1$ day using the initial conditions from Table 6.5 and with the parameter values shown in Table 6.6. A run of the model according to the specifications above should give a solution like that graphed in Fig. 6.8.

Modifications to the Steele model

6.15. The natural mortality rate *kmrt* was set to 0.0 in the Steele model. Set it first at 0.01 and then at 0.1 to see what effect it might have on the model behavior.

6.16. The thresholds included in the model equations represent levels of phytoplankton biomass density (for grazing) and of zooplankton numbers and weight (for predation on zooplankton) below which the prey are protected. These were all set to 0.0 in the parameter list Table 6.6. Include these thresholds one at a time. Raise $kbiop_{th}$ to 50.0 μgC/L, $kpredn_{th}$ to 5.0/L, and $kpredwt_{th}$ to 0.25 μgC. How important are these three factors?

Chapter 7

Stochastic Methods and Models

7.1 Introduction

The world of models that we have been examining so far is a deterministic world (except for Markov chains). Behind it all is the realization, however, that much of the information and data leading to our understanding of the ecosystems modeled is highly uncertain. It is a physical impossibility to have enough observing and data gathering capability and computer capacity to replicate the cause and effect relationships in an ecosystem. Therefore, we are left with having to allow for uncertainty involved (1) in our understanding of the controlling processes, (2) in our collapsing of a spatially heterogeneous system into a spatially homogeneous average, and (3) in the data we collect both to estimate the parameters and to compare with model behavior.

Not all these sources of uncertainty are amenable to analysis for discovering their effect on model behavior. The effect of parameter uncertainty can, however, be investigated. This uncertainty can be included either directly in the model through treating model variables and parameters as *random variables* or indirectly by using estimates of parameter variability to investigate the effects of this variability on model behavior. In either case, one method most commonly used to conduct the investigation is the *Monte Carlo* method. This method consists of generating *random numbers* that have some *probability distribution* (e.g., uniform, normal, or exponential) and using these in some way to simulate the variability in some aspect of the model. Random variables and probability distributions are described in Appendix 3.

To give you the general picture about how this method works, we first discuss how uniform random numbers are generated on the computer and then how these random numbers can be transformed into a particular probability distribution. Following this more technical discussion, we give some examples of where Monte Carlo methods are used in ecological simulation. Further examples using Monte Carlo for sensitivity analysis are treated in the next chapter on model evaluation.

Stochastic models can also work with the actual distribution of the random variable. A good example in ecological modeling is to apply the stochastic models of *queuing theory* to problems of fish feeding. These methods are discussed at the end of this chapter.

7.2 Random number generators

Almost all computers have some random number generator. These generators usually produce numbers approximately *uniformly distributed* between 0 and 1. A uniform distribution, denoted U(a,b), means that any value between a and b is equally probable. We will restrict our use of *uniformly distributed* to U(0,1) unless otherwise stated. If X is a uniformly distributed *random variable,* we will write $X \sim U(0,1)$ where the $\sim$ denotes "distributed as" and means that the random variable on the left has the probability distribution on the right side of the $\sim$. The *cumulative distribution function* for a random variable is the probability that the variable has a value less than x (for all x). For a uniformly distributed random variable, this is given by

$$F(x) = Pr(X < x) = \int_0^x dx = x \qquad \text{for } x \in [0,1][1]$$

Because we often use the random number generators already available on the computer, we only briefly review the most commonly used method for generating random numbers. All random numbers are generated by an algorithm (a numerical procedure) that is deterministic and, therefore, they are not truly random. Because the resultant numbers appear random, they are called *pseudorandom numbers*.

The most popular method for producing pseudorandom numbers is the *congruential* random number generator. The algorithm begins with an initial integer called the *seed,* X_0, and three parameters; a multiplier a, an increment c, and a modulus m. The algorithm is:

$$X_{n+1} = modulus \ (aX_n + c, m)$$

$$U_{n+1} = \frac{X_{n+1}}{m}$$

(7.1)

If $c = 0$ the generator is called a *multiplicative congruential generator* and for $c \neq 0$ a *mixed congruential generator*.

The constants a, c, and m are chosen with several criteria in mind. The algorithm should produce values that have the properties of uniform random numbers and, because thousands or even millions of numbers are generated in a single application program, the computations must be efficient. The integer

---

[1] See note on p. 204.

modulus function is most easily performed on a computer if $m = 2^{k-1}$ where k is the computer word size.

The *period* of a random number generator is the number of unique values that occur before the cycle repeats. For a congruential generator it is easy to see that the maximum possible period is m. Such a generator is said to have full period. Although true random numbers do not repeat in a cycle, a full period pseudo-random generator with $m = 2^{k-1}$, $k = 32$ produces 2,147,483,648 different numbers before repeating, which is adequate for most applications. The values of a, c, and X_0 will determine if the generator has full period. A multiplicative generator, although it involves fewer computations, can never have full period. If $m = 2^k$ then the maximum period is 2^{k-2}. There are several theorems available for the selection of a and c to insure a full period. We refer the reader to Kennedy and Gentle (1980), Knuth (1981), or Ripley (1983).

Having a large period does not ensure a good random number generator. For truly random numbers, successive values should be independent of each other. Looking at Eq. 7.1, independence does not seem likely. The correlation between successive values in a congruential generator depends on the values of a, c, and m. A great deal of research has been done on the selection of combinations of these parameters that minimize dependencies in the resulting sequences yet meet the criteria for a full (or large) period. We refer the reader to the references listed above.

Congruential generators are not the only method used to generate random numbers. Some algorithms modify the congruential generator output to overcome the problems mentioned above. A *shuffle* generator keeps a list of values from a congruential generator and returns values at random from the list, thus shuffling the output and overcoming the dependency problem. Other generators combine the output from several congruential generators (see Wichmann and Hill, 1982).

Most computer systems have random number generators, usually implemented in Fortran as a function such as *ran(x)*. On the first call to *ran* the argument x contains a positive integer, the seed, randomly chosen by the user, to start the algorithm. The function returns as its value a pseudorandom U(0,1) random number and in the argument x a new random integer to be used as the seed at the next call to *ran*. Not all of these generators are good ones and we advise the user to investigate the algorithm and parameters used and consider using an alternative if necessary.

We have not supplied a program for a generator because good efficient code for one is machine dependent. Schrage (1979) presents Fortran code for a generator that will run on any computer that can represent all the integers in the interval $[-2^{31} + 1, 2^{31} - 1]$. Wichmann and Hill (1982) have a generator that uses three mixed congruential generators (and hence three seeds) that will run on virtually any computer with a Fortran compiler. This generator has a huge period (6.95×10^{12}) but is rather slow.

Figure 7.1 shows 500 pairs of uniformly distributed random numbers (x,y) generated by a mixed congruential algorithm.

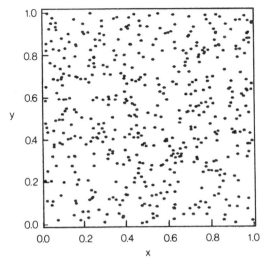

Fig. 7.1.   Five hundred pairs of uniformly distributed $U(0,1)$ random numbers graphed in the x–y plane.

7.3   Probability distributions

Although the uniform distribution is the easiest for generating large numbers of random numbers, often it is not the desired probability distribution for the application. Random numbers Y_n that are uniformly distributed over the interval (a,b) instead of $(0,1)$ can be obtained from U_n by:

$$Y_n = (b - a) \times U_n + a$$

Random integers, K_n in the interval [I,J] can be computed in a similar fashion:

$$K_n = int((J - I + 1) \times U_n) + I$$

The function $int(x)$ takes the integer part of the argument x.

In statistical applications, the assumption of the *normal distribution* function is often made. For spatial processes, the *Poisson distribution* is commonly used, whereas for events in time the *exponential distribution* is common. We will discuss methods for transforming uniformly distributed randon numbers into random numbers from each of these distributions.

7.3.1   Exponential distribution

The probability density function for an exponentially distributed random variable X is:

$$f(x) = \lambda e^{-\lambda x} \qquad \text{for } x \geq 0, \lambda > 0$$

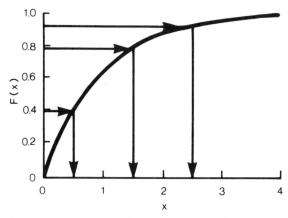

Fig. 7.2. The exponential cumulative distribution function $F(x) = 1 - e^{-x}$ showing how random numbers can be generated from an exponential probability distribution by projecting values of $F(x)$ that are generated from a uniform $U(0,1)$ distribution onto values of x that are exponentially distributed. This method works for every continuous probability distribution for which a closed form cumulative distribution function exists.

and the cumulative distribution function is

$$F(x) = \int_0^x \lambda e^{-\lambda x}\, dx = 1 - e^{-\lambda x} \qquad \text{for } x \geqq 0$$

$F(x)$ is a *monotonic* function of x, which means that for every value of $F(x)$ there is one and only one value of x. This can be seen in Fig. 7.2, which shows the exponential cumulative distribution function. The values of $F(x)$ are uniformly distributed between 0 and 1, i.e., $F(x) \sim U(0,1)$. The selection of a uniformly distributed random number U corresponds to a particular value of $F(x)$. Thus, for an exponential probability distribution:

$$U = F(x) = 1 - e^{-\lambda x}$$

$$e^{-\lambda x} = 1 - U$$

$$x = -\frac{1}{\lambda}\log(1 - U) = F^{-1}(U)$$

Because $1 - U$ is also uniformly distributed on $[0,1]$, we can also say that

$$x = -\frac{1}{\lambda}\log U \qquad\qquad (7.2)$$

is also exponentially distributed. Thus, to generate exponentially distributed random numbers, generate uniform random numbers and use Eq. 7.2 to transform them to exponential random numbers.

7.3.2 Normal distribution

The standard normal probability density function [with mean 0 and variance 1, denoted N(0,1)] is given by:

$$f(x) = \frac{1}{\sqrt{2\pi}} e^{-\frac{x^2}{2}} \qquad -\infty < x < \infty$$

The problem with generating this distribution directly from a uniformly distributed U using $F^{-1}(U)$ is that $F(x)$ [and hence $F^{-1}(U)$] for a normal distribution does not exist in closed form as it does for the exponential distribution.

A simple method to generate normally distributed pseudorandom variates, but one inefficient in use of computer time, is the Box–Muller algorithm. The algorithm generates two uniformly distributed random numbers U_1 and U_2 and converts them into two standard normally distributed random numbers X_1 and X_2 as follows:

1. *Generate U_1 and $U_2 \sim U(0,1)$*
   Set $V_1 = 2U_1 - 1$
   $V_2 = 2U_2 - 1$
   $S = V_1^2 + V_2^2$
2. *If $S \geqq 1$ go to 1*
3. *Set $W = \left(\dfrac{-2\log S}{S} \right)^{\frac{1}{2}}$*
   Return $X_1 = WV_1$
   $X_2 = WV_2$

The method is an exact method in that X_1 and X_2 are truly (not approximately) normally distributed. The time consuming part on the computer is in taking the square root and logarithm. Another disadvantage is that two random variates are produced at every call. More efficient but also more complicated methods are available; see Kennedy and Gentle (1980). A normal variable Y with mean μ and standard deviation σ [$Y \sim N(\mu,\sigma^2)$] may be obtained from the standard normal variable X by the linear transformation $Y = \sigma X + \mu$.

7.3.3 Poisson distribution

The Poisson distribution is a *discrete* probability distribution, unlike the normal and exponential distributions that are *continuous*. Furthermore, it takes on only integer values. If X is a Poisson random variable, then its probability function is:

$$Pr(X = n) = p(n) = \frac{e^{-\lambda}\lambda^n}{n!} \qquad n = 0,1,2,\ldots \lambda > 0$$

Because the Poisson distribution is discrete its cumulative distribution function is not uniformly distributed so $F^{-1}(U)$ cannot be used to generate Poisson random numbers. An algorithm for generating Poisson distributed variates with parameter λ (Knuth, 1981) works as follows,

1. *Set $P = e^{-\lambda}$*
   $S = 1$
   $X = -1$
2. *Generate $U \sim U(0,1)$*
   Set $S = S \times U$
   $X = X + 1$
3. *If $S < P$ return X*
   Else go to 2

Here a series of uniformly distributed random numbers are generated until a condition is satisfied. The resultant value X will be a random number from a Poisson distribution.

7.4   Random walk predation model

Predator–prey dynamics have been treated so far in our models as spatially averaged. In reality, encounters between predators and their prey occur over space as well as time and are a series of events of encounter followed by pursuit and subsequent capture or escape. *Stochastic processes,* which are a collection of random variables related over time or space, suggest themselves as a model for the random encounters between predator and prey over time and space. We develop here a *random walk* stochastic process model for predator movement and prey encounter. It captures one essential aspect of the predator–prey process missing in deterministic models—search, encounter, pursuit, and capture at variable (and uncertain) time intervals.

We are interested in how ignoring space and time variability of these events, as was done in many ecological simulation models such as the San Joaquin model (which assumed spatial homogeneity and average grazing rates) affects model predictions. Unfortunately, there are two major stumbling blocks to answering this question. First, we do not know in detail many of the factors necessary to accurately represent predation in time and space as a series of capture events (for example, we do not know predators' decision rules for search or pursuit). Second, even if we did have all these details, we may not have tractable mathematical methods for including them. We are left with a compromise. We first assume simple rules for search and capture success based on the best of our observations and on simplifying assumptions. We then use a stochastic model to explore the implications of these simple encounter rules on average capture rates and population dynamics.

We include the uncertain nature of a predator's search for and capture of its prey by using Monte Carlo techniques to generate a series of simulations of predators walking over a grid at random (the random walk) and feeding when encountering prey. Results of the simulations can be summarized to produce average feeding rates as well as a measure of variability around this average from simulation to simulation. Notice we use simulation here differently than when solving differential equations numerically. Here we generate random numbers that control the predator's movement on a grid containing prey.

The advantages of this simulation approach over more analytical approaches are (1) it requires less technical knowledge of stochastic processes and (2) it allows search and capture assumptions to be unconstrained by analytical tractability (i.e., ability to solve the equations analytically). The disadvantages are (1) general results are difficult to obtain and (2) it can foster the bad habit of moving ahead to simulation before the problem is well thought out.

For a simple initial example consider a ''predator'' moving at random on an $n \times n$ grid. The predator's location on the grid is denoted by (i,j) $i = 1, \ldots,$ $n, j = 1, \ldots, n$. Assume that the predator moves at each time step (time is assumed to be discrete with time step of 1). The direction of movement is random with the probability of moving up being *pup*, right *pright*, down *pdown*, and left *pleft* $= 1 - pup - pright - pdown$. The probabilities sum to 1 for all possible directions. This random walk model is an example of a Markov chain, with each state being a grid location of which there are n^2 states. The transition matrix is $n^2 \times n^2$.

Movement can be modeled as follows:

> *Generate U $\sim$ U(0,1)*
> *if (U < pup)*
>   *i = i − 1*
> *else if (U < pup + pright)*
>   *j = j + 1*
> *else if (U < pup + pright + pdown)*
>   *i = i − 1*
> *else*
>   *j = j − 1*

Because we are interested in predator–prey encounters, we need some way of including the prey in this framework. For the present case let us assume that the prey does not move over the feeding period. Field data show that a Poisson distribution describes the spatial distribution of many organisms. The average density λ can be used to generate numbers of prey in each cell of the grid. Furthermore, let us assume that the predator can take at most $prey_{mx}$ prey items from a grid location in a single time step.

We can now combine predator movement with prey distribution using the following steps:

1. *Initialize prey over grid with prey $(i,j) \sim Poisson(\lambda)$*
2. *Initialize predator location (i,j) with*
   $i = int(n \times U_1) + 1$
   $j = int(n \times U_2) + 1$
   where $U_k \sim U(0,1)$ k = 1,2

3. *Move predator and consume prey as follows:*
   initialize time t = 0
   while (t < t_{mx})
      move predator according to probability rule given above
      consumption = min(prey(i,j), $prey_{mx}$)
      prey(i,j) = max(0, prey(i,j) − consumption)
      total consumption = total consumption + consumption
      t = t + 1
   end while
4. *Compute summary statistics (e.g., average prey consumed)*

Some rule must be devised for movement at the boundary of the grid. With a *reflecting boundary,* movement out of the grid becomes movement in the other direction. Thus, if we are at (20,3) on a 20 × 20 grid, and we are supposed to move down to (21,3) (which does not exist), we instead move up to (19,3). This has the effect of biasing results toward the center. In *wrap-around boundaries* movement down from (20,3) would result in entering the grid at the top at (1,3). This boundary rule does not have the central tendency bias of the reflecting algorithm. The reflecting boundary could be likened to an animal feeding over a home range, whereas the wrap-around boundary would be more like an animal foraging over a wide range with movement outside the boundary being similar to entering a new field.

We now present a program for the predator–prey random walk model. The program has been divided up into modules, each with a specific task:

ranwlk	main program that controls the other modules
setup	read in simulation parameters
inprey	initialize the prey on the grid
inpred	place predator on the grid and initialize summary values
mvpred	move predator
feed	allow the predator to feed
sumary	compute summary statistics
poiss	generate a Poisson pseudorandom number

Also needed but not listed is a function *ran(seed)* that returns a uniform random number given an integer seed. The function should also return a new integer value in seed.

```
c   Program ranwlk to simulate a random walk predator-prey model
c
c   Variables:
c      maxrow      -- integer -- max. no. of rows allowed in grid
c      maxcol      -- integer -- max. no. of columns allowed in grid
c      nrow        -- integer -- number of rows in grid
c      ncol        -- integer -- number of columns in grid
c      time        -- integer -- time
c      tmax        -- integer -- maximum feeding time
```

```
c      seed      -- integer -- starting number for random no. generator
c      err       -- integer -- error flag, 0 => no error
c      pup       -- real    -- probability predator moves up on the grid
c      pright    -- real    -- probability predator moves right on grid
c      pdown     -- real    -- prob. predator moves down on the grid
c      lambda    -- real    -- mean number of prey per grid point
c      preymx    -- real    -- maximum number of prey taken per encounter
c      encsum    -- real    -- sum of the prey encountered
c      encssq    -- real    -- sum of squares of the prey encountered
c      prysum    -- real    -- sum of the prey consumed
c      pryssq    -- real    -- sum of squares of the prey consumed
c      prey(i,j) -- real    -- no. of prey at location i,j in the grid
c                             (nrow by ncol)
c
c

       integer maxrow,maxcol
       common /maxsiz/ maxrow,maxcol

       integer i,j,nrow,ncol,time,tmax,seed,err
       real pup,pright,pdown,lambda,preymx,
      +      encsum,encssq,prysum,pryssq

c maximum grid size is 20x20
       real prey(20,20)
       maxrow = 20
       maxcol = 20

c Setup simulation environment
       call setup(nrow,ncol,tmax,seed,pup,pright,pdown,
      +            lambda,preymx,err)
       if (err .ne. 0) stop

c Initialize prey on grid
       call inprey(nrow,ncol,prey,lambda,seed)

c Initialize predator
       call inpred(nrow,ncol,i,j,
      +            encsum,encssq,prysum,pryssq,seed)

c Movement and feeding loop:
       do 10 time = 1,tmax

c          Move predator
           call mvpred(nrow,ncol,i,j,pup,pright,pdown,seed)
c          Write location and prey available
           write(*,110) time,i,j,prey(i,j)
```

```
c          Allow predator to feed
           call feed(i,j,prey,preymx,encsum,encssq,prysum,pryssq)


   10 continue

c  Compute and print summary statistics
       call sumary(encsum,encssq,prysum,pryssq,tmax)

   110 format(4x,i4,3x,i2,3x,i2,3x,f6.2)
       end



       subroutine setup(nrow,ncol,tmax,seed,
      +                 pup,pright,pdown,lambda,preymx,err)
c
c  Read in and print out simulation setup.
c  The input file should contain a one line description
c  before the actual values as follows: (the description
c  line is not actually read) An example:
c
c    nrow,ncol,tmax,seed
c    20 20 50 98751
c    pup,pright,pdown
c    0.25 0.25 0.25
c    lambda,preymx
c    10.0 5.0
c
c    nrow   <-- number of rows on the grid
c    ncol   <-- number of columns on the grid
c    tmax   <-- number of time steps for the simulation
c    seed   <-- integer seed to start the random number generator
c    pup    <-- probability of moving up
c    pright <-- probability of moving right
c    pdown  <-- probability of moving down
c    lambda <-- mean of prey density
c    preymx <-- max. prey consumption allowed per time step
c    err     --> error switch (err = 0 => no error,err = 1 => error
c

       integer maxrow,maxcol
       common /maxsiz/ maxrow,maxcol

       integer nrow,ncol,tmax,seed,err
       real pup,pright,pdown,pleft,lambda,preymx,psum

       err = 0
```

```
c  Read simulation parameters
      read(*,*)
      read(*,*)  nrow,ncol,tmax,seed
      read(*,*)
      read(*,*)  pup,pright,pdown
      read(*,*)
      read(*,*)  lambda,preymx

c  Check for errors
      if (nrow .gt. maxrow .or. ncol .gt. maxcol) then
         write(*,110) maxrow,maxcol
         err = 1
      end if
      psum = pup + pright + pdown
      if (psum .gt. 1.0) then
         write(*,120)
         err = 1
      end if

c  Output simulation parameters
      pleft = 1.0 - psum
      write(*,130) nrow,ncol,tmax,seed
      write(*,140) pup,pright,pdown,pleft,lambda,preymx

c  Write header
      write(*,150)

      return

  110 format(1x,'Error,nrow or ncol too large,maximums are:',2i5)
  120 format(1x,'Error,probabilities greater than 1')
  130 format(1x,'nrow = 'i2,', ncol = 'i2,', tmax = 'i4,', seed = 'i7)
  140 format(1x,'probability of movement: up = ', f4.3,', right = ',
     +          f4.3,' down = ', f4.3,', left = ', f4.3,/,1x,
     +          'mean prey density = ', f7.3,4x,
     +          'max. consumption allowed = ', f7.3)
  150 format(/,5x,'time   i    j   prey(i,j)')

      end


      subroutine inprey(nrow,ncol,prey,lambda,seed)
c
c  Place prey on grid using a poisson dist. with mean lambda.
c
c    nrow    --> number of rows in the grid
c    ncol    --> number of columns in the grid
```

```
c     prey   <-- prey matrix (nrow by ncol)
c     lambda --> mean prey density
c     seed   <-> random number generator seed
c

      integer nrow,ncol,seed,i,j,poiss
      real prey(20,20),lambda

      do 20 i = 1,nrow
          do 10 j = 1,ncol
              prey(i,j) = poiss(lambda,seed)
  10      continue
  20 continue

      return
      end



      subroutine inpred(nrow,ncol,i,j,
     +                    encsum,encssq,prysum,pryssq,seed)
c
c place predator at initial random location (i,j) on grid
c and initialize accumulators for summary statistics.
c
c     nrow   --> number of rows in the grid
c     ncol   --> number of columns in the grid
c     i      <-- initial row location for the predator
c     j      <-- initial column location for the predator
c     encsum <-- accumulator for prey encountered
c     encssq <-- accumulator for prey**2 encountered
c     prysum <-- accumulator for prey consumed
c     pryssq <-- accumulator for prey**2 consumed
c     seed   <-> seed for random number generator
c

      integer nrow,ncol,i,j,seed
      real encsum,encssq,prysum,pryssq,ran

      i = ran(seed) * nrow + 1
      j = ran(seed) * ncol + 1
      encsum = 0.0
      encssq = 0.0
      prysum = 0.0
      pryssq = 0.0

      return
      end
```

```fortran
      subroutine mvpred(nrow,ncol,i,j,
     +                         pup,pright,pdown,seed)
c
c  Move predator on the grid using wrap around boundaries
c
c     nrow    --> number of rows in the grid
c     ncol    --> number of columns in the grid
c     i       <-> row location for the predator
c     j       <-> column location for the predator
c     pup     <-- probability of moving up
c     pright  <-- probability of moving right
c     pdown   <-- probability of moving down
c     seed    <-> seed for the random number generator
c

      integer i,j,nrow,ncol,seed
      real pup,pright,pdown,prob,ran

      prob = ran(seed)

      if (prob .lt. pup) then
c         move up
          i = i - 1
c         wrap around top to bottom
          if (i .lt. 1) i = nrow
      else if (prob .lt. (pup + pright)) then
c         move right
          j = j + 1
c         wrap around edge right to left
          if (j .gt. ncol) j = 1
      else if (prob .lt. (pup + pright + pdown)) then
c         move down
          i = i + 1
c         wrap around edge bottom to top
          if (i .gt. nrow) i = 1
      else
c         move left
          j = j - 1
c         wrap around edge left to right
          if (j .lt. 1) j = ncol
      end if

      return
      end
```

```
      subroutine feed(i,j,prey,preymx,
     +                 encsum,encssq,prysum,pryssq)
c
c Feeding by the predator including accumulation of
c prey encountered and consumed for summary statistics.
c
c    i       --> row location for the predator
c    j       --> column location for the predator
c    encsum <-> accumulator for prey encountered
c    encssq <-> accumulator for prey**2 encountered
c    prysum <-> accumulator for prey consumed
c    pryssq <-> accumulator for prey**2 consumed
c    seed   <-> seed for random number generator
c

      integer i,j
      real prey(20,20),preymx,encsum,encssq,prysum,pryssq

      encsum = encsum + prey(i,j)
      encssq = encssq + prey(i,j) * prey(i,j)

      if (prey(i,j) .lt. preymx)  then
c         all prey taken at location,set prey to zero
          prysum = prysum + prey(i,j)
          pryssq = pryssq + prey(i,j) * prey(i,j)
          prey(i,j) = 0.0
      else
c         only take preymx
          prysum = prysum + preymx
          pryssq = pryssq + preymx * preymx
          prey(i,j) = prey(i,j) - preymx
      end if

      return
      end


      subroutine sumary(encsum,encssq,prysum,pryssq,time)
c
c Compute feeding summaries.
c
c    encsum --> sum of all prey encounter
c    encsum --> sum of squares of all prey encounter
c    prysum --> sum of prey consumed
c    pryssq --> sum of squares of prey consumed
c    time   --> number of time steps that have occured
```

```
c

      integer time
      real encsum,encssq,prysum,pryssq,
     +       encave,encsd,pryave,prysd

      encave = encsum / time
      pryave = prysum / time
      encsd = sqrt((time * encssq -  encsum * encsum) /
     +                (time * (time - 1.0)))
      prysd = sqrt((time * pryssq -  prysum * prysum) /
     +                (time * (time - 1.0)))

      write(*,110) time,encave,encsd,pryave,prysd
  110 format(/,1x,'Time = ',i4,
     +              ', Prey encounted: ave =',f7.3,', sd = ',f7.3,/,13x,
     +              ' Prey consumed:  ave = ',f7.3,', sd = ',f7.3)

      end



      function poiss(lambda,seed)
c
c Function to return a Poisson distributed random number with
c mean lambda.
c
c    lambda --> mean of the Poisson distribution
c    seed   <-> seed for the random number generator
c

      integer seed,poiss
      real lambda,ran,explam,p

      explam = exp(-lambda)
      p = 1
      poiss = -1

   10 continue
          p = p * ran(seed)
          poiss = poiss + 1
      if(p .ge. explam) go to 10

      return
      end
```

The program uses the same style and techniques as in Chaps. 3 and 6. In subroutine setup we do simple error checking to prevent overwriting the prey array and to avoid impossible probability values. All reading and writing is done to standard input and output and input is kept free-format. Again we use input lines in pairs with the first line used for a description of the input values that appear on the next line. The function *poiss* illustrates how a Poisson distributed random number generator is coded and how the uniform distributed random numbers are used in the Poisson generator. The parameter *seed* is used to start the random number generator and is changed by the generator for use at the next call to the generator. The *sumary* subroutine is general enough that it can be called at any time during the simulation.

What can be done with a model like this? First, we will discuss simulation model runs to be made and next generalizations to the existing program to make the model more realistic.

7.4.1 Model analysis

A run of this model involves choosing the three movement probabilities, the average prey density λ, the maximum consumption allowed ($prey_{mx}$), and the random number seed and running the feeding model for t_{max} feeding time steps. This simulates the movement of a single predator feeding by moving over a Poisson distributed prey field and sampling *without replacement,* with future movement being independent of prior feeding. There is no learning by the predator. The input file for a simulation run with symmetric movement probabilities (i.e., *pup* = *pright* = *pdown* = 0.25) is given in Fig. 7.3. Table 7.1 shows summaries of several model runs using this input file. Here we see the average number of prey encountered per time step and the average number of prey taken per time step for feeding periods of 10, 20, 30, and 50 time steps. The results are the average of runs with 10 different predators distinguished by having different initial seeds for the random number generator, which gives them a different prey distribution and movement pattern (although according to the same probability distribution and movement rules).

Table 7.1 also shows the average standard deviation in the number of prey encountered and captured. Notice that the average number of prey taken drops with increased feeding time owing to the nonreplacement of "sampled" prey, resulting in a sparser prey density with time over the feeding period. Predators returning at random to a previously visited grid point will find fewer (or no) prey there than on the previous visit. Having a maximum consumption rate

```
nrow ncol tmax seed
20 20 50 23637
pup pright pdown
.25 .25 .25
lambda preymx
10 5
```

Fig. 7.3. Input file for a simulation run of the random walk predator–prey model. The file sets up the grid size, feeding time, seed for Monte Carlo simulation, movement probabilities, and parameters for prey abundance and maximum feeding.

Table 7.1   Results of Monte Carlo simulations of a random walk predator–prey model in which the prey counts on a 20 × 20 grid are generated from a Poisson distribution with mean $\lambda = 10$. A randomly placed predator moves at random, capturing at most five prey items per time step.

Time steps	Prey encountered per time step		Prey taken per time step	
	mean	standard deviation	mean	standard deviation
10	7.32	4.20	4.10	1.64
20	7.12	4.45	3.96	1.72
30	6.90	4.50	3.86	1.81
50	6.39	4.75	3.56	2.05

*Values are the average from 10 model runs.

prevents most grid points from being stripped of prey at the first predator visit. Also, the variability in prey encountered and taken increases with increasing run time because more empty grid points are encountered.

If the predator moves with a bias in any direction the average number of prey encountered and taken increases, but the variance in both these variables decreases. For example, when there is a probability of 0.4 of moving to the right the number of prey encountered and taken for a 50-time step feeding period are 8.0 and 4.3, respectively, and the respective standard deviations are 4.2 and 1.5 (these are averages from 10 runs of the model). This is because having a bias in movement gives the predator a greater propensity for covering new territory, thereby taking the maximum number of prey more frequently than in the other cases in Table 7.1.

This model assumes that the predator moves at random, without learning, over a Poisson distributed prey grid. Encounter of prey at a grid point results in feeding up to the maximum allowable per encounter, or in taking all the prey available at that grid point, whichever is less. There is no pursuit, capture, or digestion in this model. Animals with full guts move just like those with empty guts. Time is discrete and events occur at discrete intervals. As such, aspects of feeding, such as pursuit time, capture time, and capture success probability, are not considered.

7.4.2   Model extensions

Many extensions can be made to the random walk model to improve its realism. Some of these extensions will be discussed in the exercise section. We discuss additional extensions here.

The random predator movement can be altered in a number of ways:

1. The probability of movement in any direction can depend on the distance from the "home territory" of the predator. For example, the probability of movement away from the home territory can be inversely proportional to the

distance from it. This forces a bias in movement toward forays from and returns to the home base.

2. The probability of movement to adjacent grid points may depend on the prey density. Predators could be expected to move with a higher probability toward areas of high prey density. This could be done by giving the predator a memory where the probability of movement in the same direction as the present direction is greater if no prey are encountered and of reversing directions or staying in one place is higher if the prey densities encountered are higher than average. Another approach is to give the predator sensing ability for adjacent grid points and to move with probabilities in proportion to the relative density of the present and adjacent grid points.

3. The predator can choose not to feed on a group of prey but instead move to a new grid location. Then searching and feeding time can be separated.

4. The predator's capture efficiency can be simulated by having the fraction of available prey taken be a random variable with distribution depending on the number of prey encountered.

5. The effect of feeding satiation can be simulated by having the probability of movement away from the present grid location affected by the number of prey in the gut, with prey leaving the gut at a "digestion rate." The more prey in the gut the less the chance of movement. Nonmoving predators are assumed to not feed.

6. In the present model, predators tend to deplete prey in an area. This would be typical of a "home range" predator. Many predators, like raptors (e.g., hawks) or gulls, cover large areas and can move rapidly if feeding conditions in one area are not good. This could be simulated by having the predator "jump" to another random location when, let us say, no prey are encountered in two adjacent locations in a row.

Let us devise some extensions to how prey are treated in this model:

1. The Poisson distribution is neither a regular nor a clumped distribution. More regular or more clumped probability distributions can be used.

2. The present probability distribution for the prey is trend free. The average prey density λ is the same over the entire grid. This could be altered by having λ be a function of space. A linear or other function of location can be used for λ. Such trended distributions are common in ecosystems and reflect gradients in environmental variables, such as moisture or temperature, that affect plant abundance and species distribution.

3. We deal here with a single prey type. Multiple prey types with different densities λ_1, λ_2, etc., can be included. In this way an average diet as well as number of prey taken can be computed.

In this model no direct spatial or temporal interpretation is given to the system. The grid points can be spatial locations or feeding "patches." The time scale can be one feeding foray or a long period with many feedings. Our analogs of up, down, right, and left are just conveniences for purposes of programming and to simplify the model context. A model like this can be adapted to many real feeding scenarios including foraging or burrowing herbivores, such as voles, where the spatial analog holds, raptorial search where feeding is over a wide area and over much of the day and is often not tied to a home territory, woodpecker foraging from tree to tree where each tree is a feeding patch, or fish feeding on zooplankton patches.

Relaxing the underlying assumption of spatial homogeneity of predators and prey can lead to evaluating how the homogeneity assumption affects predictions of feeding and predation rates. The models can also be used to show which assumptions about feeding strategies and prey distribution lead to the most accurate representation of the observed feeding rate and diet. As such, they are a reality-testing scenario for studying predator–prey interactions.

We have included the code and algorithms primarily to give the fundamentals of how to build stochastic models for those who want to embark on this type of modeling (it has, until recently, been overlooked in ecological modeling), and also to familiarize you with the most common approach to the problem of modeling processes affected by spatially heterogeneous environments. We will return to this last topic, with some example models, in Chap. 10.

7.5 Monte Carlo driving variables

In Chap. 6 we discussed how to generate driving variables in a deterministic fashion using either sinusoids or abiotic data sets to drive models. However, what can be done if only partial weather data is available but it needs to be extended to a longer time period? Monte Carlo techniques can be used to generate weather patterns that are more realistic representation than available from a deterministic sinusoid approximation.

Weather is one thing we still talk about with interest in the world today, some of its appeal lies in our uncertainty about how it will change. Figure 7.4A, for example, shows daytime average air temperature (°C) and, Fig. 7.4B precipitation (cm) for 1973 at the W. H. Thompson experimental forest in western Washington, United States. Although there is a definite pattern over the year for each of these variables, there is variability around this pattern. What we desire in generating driving variable data based on data sets like these is to capture both the pattern of the mean and the variability around it. We show here, for example, how Monte Carlo methods are used to generate both temperature and precipitation data in this fashion.

7.5.1 Temperature data generation

First we approximate the average air temperature, as in Chap. 6 as a sinusoid:

$$ztp_{av} = \left(\frac{ztp_{mx} - ztp_{mn}}{2} \right) \sin \left(\frac{2\pi(t - t_0)}{R} \right) + \left(\frac{ztp_{mx} + ztp_{mn}}{2} \right) \quad (7.3)$$

where t_0 is the Julian day on which the average temperature occurs and R is the period in days of the sinusoid (here $R = 365$).

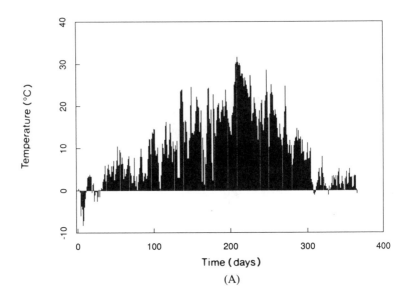

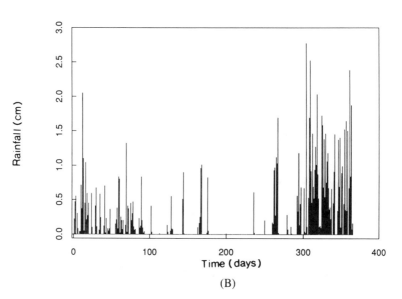

Fig. 7.4. Charts showing the distribution of A. daily average air temperature (°C) and B. daily precipitation (cm) for 1973 at the W. H. Thompson experimental forest near Seattle, Washington.

196

For simulating variability in temperature we assume that temperature is distributed about the daily average temperature according to a normal distribution, i.e., $ztp = ztp_{av} + Z$ where $Z \sim N(0,\sigma^2)$. This assumption can be checked with Quantile–Quantile plots of the residuals $(ztp - ztp_{av})$ as described in Appendix 3. The variance of the residuals can be used to estimate σ^2, which is used in the following data generation algorithm:

> *initialize time t = 1*
> *while (t ≤ t_{max})*
>   *compute ztp_{av} as above*
>   *generate $Z \sim N(0,\sigma^2)$ by one method discussed earlier*
>   *$ztp = ztp_{av} + Z$*
>   *$t = t + 1$*
> *end while*

7.5.2 Precipitation data generation

One way to generate precipitation data is to assume, as we did in Chap. 2, that the occurrence of rainfall on any day can be modeled as a Markov chain (see section 2.8). The empirical distribution of rainfall for all rainy days can then be used to generate a rainfall amount given that it has rained, using Monte Carlo methods. Although this model does not capture some features of weather patterns, such as fronts moving in for several days, it does include enough of the association of weather patterns from one day to the next to be a significant improvement over models ignoring these patterns from one day to the next.

Suppose we wish to generate daily rainfall for a model in western Washington for the period of October to December. By counting the incidence of rainy days following rainy days, and dry days following dry days we develop the following probability transition matrix

$$\mathbf{T} = \begin{bmatrix} P_{dd} & P_{wd} \\ P_{dw} & P_{ww} \end{bmatrix} = \begin{bmatrix} 0.4 & 0.2 \\ 0.6 & 0.8 \end{bmatrix}$$

We already know that this matrix produces a stationary probability vector (see section 2.8):

$$\mathbf{s} = \begin{bmatrix} 0.25 \\ 0.75 \end{bmatrix}$$

Figure 7.5 gives a daily rainfall histogram for October–December, 1973, for the Thompson experimental forest using the data in Table 7.2.

This histogram and the Markov chain model can be combined to generate pseudorainfall data over a desired time period (here October–December). Denote the histogram boundaries by $h_1, h_2, \ldots, h_k$ where $h_0 = 0$ and their associated probabilities by $q_1, q_2, \ldots, q_k$, i.e., $\Pr(h_{i-1} < rain < h_i) = q_i$. The algorithm, which starts with the equilibrium probabilities for rain, is as follows:

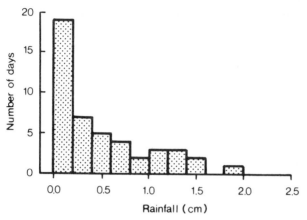

Fig. 7.5.   Histogram for the distribution of daily rainfall for days that it rained at the W. H. Thompson experimental forest from October–December, 1973.

1. *initialize time t* = *1*

   set $\mathbf{p}(t) = \begin{bmatrix} p_0(t) \\ p_1(t) \end{bmatrix} = \mathbf{s}$

2. *while* $(t < t_{mx})$

   generate $X_t \sim U(0,1)$

   if $(X_t < p_0(t))$

   it is dry:

   set $\mathbf{u}(t) = \begin{bmatrix} 1 \\ 0 \end{bmatrix}$

   else $(X_t > p_0(t))$

   it is raining:

   set $\mathbf{u}(t) = \begin{bmatrix} 0 \\ 1 \end{bmatrix}$

   set $Y_t = \dfrac{X_t - p_0(t)}{p_1(t)}$, *(note* $Y_t \sim U(0,1)$)

   initialize sumq = 0, $i = 0$

   while $(sumq < Y_t)$

   $i = i + 1$

   $sumq = sumq + q_i$

   end while

   set rain = $\dfrac{h_{i-1} + h_i}{2}$

   set $\mathbf{p}(t + 1) = \mathbf{Tu}(t)$

   $t = t + 1$

   end while

Table 7.2   Rainfall amounts (cm) for the Thompson forest site in western Washington for 1973. Any day not listed had no rainfall.

Day	Rain	Day	Rain	Day	Rain	Day	Rain	Day	Rain
1	0.22	55	0.15	113	0.01	284	0.11	327	1.38
2	0.47	56	0.21	123	0.13	285	0.01	328	0.54
3	0.55	57	0.38	124	0.03	292	0.56	329	1.46
5	0.30	58	0.08	127	0.05	293	0.33	330	0.75
6	0.08	59	0.83	128	0.55	294	1.18	331	1.06
8	0.04	60	0.80	129	0.08	295	0.13	332	1.18
9	0.04	61	0.02	130	0.07	296	0.44	333	0.67
10	0.71	62	0.25	143	0.51	297	0.68	334	0.36
11	0.37	63	0.20	144	0.90	301	0.67	335	0.39
12	2.05	64	0.07	145	0.01	302	0.04	336	0.22
13	1.10	65	0.20	163	0.10	303	0.02	337	0.66
14	0.05	68	0.13	165	0.16	304	2.77	339	0.45
15	0.45	69	1.32	166	0.25	305	0.11	340	0.91
16	1.04	70	0.03	167	0.96	307	0.53	341	1.45
17	0.21	71	0.41	168	1.01	308	1.69	345	0.34
18	0.59	72	0.37	175	0.02	309	2.52	346	1.37
19	0.27	75	0.45	176	0.82	310	0.92	347	0.69
20	0.45	76	0.20	177	0.05	311	0.46	348	1.40
24	0.59	77	0.30	200	0.01	312	1.46	349	0.10
25	0.10	78	0.47	236	0.61	313	0.69	350	0.89
29	0.41	79	0.11	237	0.02	314	0.52	351	0.99
30	0.67	80	0.06	250	0.20	315	0.94	353	0.44
31	0.11	81	0.07	260	0.17	316	1.27	354	1.52
34	0.06	86	0.23	261	0.15	317	1.01	355	0.35
35	0.58	87	0.09	262	0.93	318	2.03	356	1.64
36	0.24	88	0.12	263	0.97	319	0.87	357	0.34
40	0.12	89	0.83	264	0.27	320	0.52	358	1.50
41	0.70	90	0.20	265	1.12	321	0.11	360	0.67
42	0.02	91	0.09	266	1.03	322	0.12	361	2.38
43	0.23	92	0.02	267	1.69	323	0.09	362	0.84
45	0.08	93	0.04	268	0.02	324	1.72	363	1.87
46	0.05	102	0.41	279	0.28	325	1.58	364	0.07
47	0.08	103	0.03	280	0.07	326	0.68	365	0.16
48	0.36								

If a finer selection of rainfall values is desired we can randomly choose a value within the histogram bin rather than taking the bin average as follows: compute $Z_t = \dfrac{Y_t - h_{i-1}}{h_i - h_{i-1}}$ (Note that Z_t is also $\sim U(0,1)$) and set rain = $h_i +$ $Z_t(h_i - h_{i-1})$. The transformation of X_t to Y_t and Y_t to Z_t is done to allow the same random number to generate both whether it rains and how much rain if it does.

This algorithm would work for longer time periods, but it ignores the strong seasonal precipitation pattern observed, for example, in Fig. 7.4B. What is usually done here is to estimate separate Markov chains and empirical rainfall distributions over different seasons, the seasons chosen based on an examination of the seasonal rainfall pattern. For the northwest United States, precipitation data tend to be divided most conveniently into a rainy winter period (mid-October–February), a less rainy spring (March–June), and a dry summer (July–mid-October).

7.6   Queueing and feeding

We introduce some of the basics of queueing theory in the context of an inconsistency raised by Beyer (1976) in some experiments on larval fish feeding. In experiments on *Trachurus* larvae in the Black Sea, Ivlev (1965) noticed that although the larvae would search a volume of water per day, which, on the average, contained less than five food organisms, the larvae were calculated from stomach content analysis to consume an average of 75 food items daily.

This apparent discrepancy was examined by Beyer using queueing theory, and he showed that the difference could be explained by larvae dying when they had empty guts, with the remaining living larvae having higher than average numbers of encounters with food items. We will show how Beyer generated his results after a brief description of queueing theory.

7.6.1   Queueing theory introduction

Queueing theory is a technique for analyzing systems where arrivals occur at random to some *server* who serves these arrivals, with the service time also being a random variable. If the server is busy when an arrival occurs, that arrival joins a queue (line) where it awaits service, usually on a first come first served basis. The theory consists of methods for calculating the expected waiting time of the arrivals, the average length of the queue, and the probability the server will be idle (i.e., the queue is empty). In the simplest case the number of arrivals per unit time is assumed to be Poisson distributed. Under this assumption the times between arrivals are exponentially distributed. Service times are also assumed to be exponentially distributed. This type of queue (exponential arrivals and service) with a single server is called an $M/M/1$ queue in the queueing literature. The first M refers to the independence and exponential distribution of the arrivals, the second M means the same for the service and the 1 denotes a single server. If the arrival rate is larger than the service rate, then the queue will grow without bound. Thus, only the case where arrival rate is less than service rate is of interest for long-term analysis.

7.6.2   Analogy to fish feeding

Beyer envisioned the feeding process of fish larvae as a queueing system with an $M/M/1$ queue. This is shown pictorially in Fig. 7.6 with the stomach being the queue and the gut the server. In the $M/M/1$ queue we assume that the larva

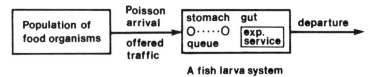

A fish larva system

Fig. 7.6. The feeding process of an individual fish represented as a single server M/M/1 queue. Food organisms arrive at random and wait in the stomach queue to be digested. Service times (digestion times) are exponentially distributed to reflect the prey size distribution being weighted toward small food organisms.

dies if the server becomes idle (empty gut). Using formulas from queueing theory Beyer computed both the probability that the stomach is empty at time t given that there were k food items in the gut at time 0 and the probability that there are i ($i = 1, \ldots, \infty$) food items in the stomach at time t given k items at time 0 and given that the organism has not died by time t. These formulas are given in Beyer (1976) and are too complicated to include here.

By assuming that the initial number of food items in the gut is less than 10, approximations were obtained for the probability that the larva is alive at time t (has a nonempty gut), and the expected number of food items (denoted by $xgut_t$) in the stomach at time t given that the larva is still alive. Let $p_{ki}(t)$ be the probability that there are i items in the stomach at time t, given k items at time 0 (the start of the feeding experiment). Then the probability that the larva is alive at time t is $1 - p_{k0}(t)$.

In the following expressions the ratio of arrival (prey encounter) rate to service (gut processing or digestion) rate, denoted by $a = \dfrac{kenctr}{kdig}$, is termed the *traffic offered*. The relevant expressions are:

$$1 - p_{k0}(t) = \frac{k \times a^{-\frac{k}{2}-\frac{1}{4}}}{(1 - \sqrt{a})^2 \times 2\sqrt{\pi}} \times (t \times kdig^{-\frac{3}{2}}e^{-(1-\sqrt{a})^2 \times t \times kdig})$$

$$E(xgut_t) = \frac{1 + \sqrt{a}}{1 - \sqrt{a}}$$

$p_{k0}(t)$ = probability gut is empty at time t (larva is dead)
a = traffic offered (ratio of encounter to service rates)
$kdig$ = digestion rate
$xgut_t$ = length of queue at time t (number of food items in stomach)

These expressions are accurate to at least one significant digit for times greater than $\dfrac{10}{kdig(1 - \sqrt{a})^2}$. This implies that the closer the traffic offered (a) is to 1.0 the longer the time before the approximations become valid. Table 7.3 shows the relevant expected numbers of food items in the fish stomach at the earliest time that the approximation equations are valid. For this table $k = 2$ (i.e., two

Table 7.3   Average gut contents after a sufficiently long time in a queueing theory
model mimicking Ivlev's feeding experiment.*

Traffic offered	Mean gut contents
a	*xgut*
0.10	1.9
0.20	2.6
0.30	3.4
0.40	4.4
0.50	5.8
0.60	7.9
0.70	11.0
0.80	18.0
0.90	38.0
0.95	78.0

*Results are shown under a number of different levels of the traffic offered (ratio of prey encounter
rate to gut processing rate).
From Beyer (1976).

food items in the stomach at the start of the experiment). The closer *a* is to 1.0
the larger the average number of food items in the larval stomachs. Notice that
the expression for the expected value of *xgut*, depends only on *a*. The numbers
are not in excellent agreement with Ivlev's data, which suggests that transient,
rather than long-term, behavior is most important in his experiment. The results,
however, do show that Ivlev's observing larger ratios than expected could be
due to mortality of fish with empty guts. Those that died saw less than average
numbers of prey, whereas those that survived saw greater than average prey
densities.

7.6.3   Prey density effect on feeding

A different application of queueing theory to ecological modeling was made by
Sjoberg (1980). He envisioned the feeding process much like Beyer did in Fig.
7.6 with encounters with prey being Poisson, having rate *genctr* and digestion
or gut processing being exponential with rate *kdig*. One difference here is that
the encounter rate is not constant but can depend on the state of the system.
Sjoberg considered three cases, all of which have prey encounter rate *genctr*
proportional to the product of the search rate *gsrch* and the prey density *xprey*.
The three cases are:

1. Search rate *gsrch* = $ksrch_{mx}$, a maximum search rate, when the gut is not
   full and 0 when the gut is full. Thus, feeding ceases when the gut is full.
   Also, the maximum gut content is assumed to be a single prey item. This
   might apply to animals feeding on large prey items.
2. This case is the same as above except that the maximum gut content, $kgut_{mx}$,
   is assumed to be arbitrarily large.

3. Here the search rate decreases with increasing gut content *xgut* (gut contents are assumed to occur in integer units). Thus:

$$gsrch = \frac{ksrch_{mx}}{xgut + 1}$$

For each case Sjoberg computes the ingestion rate per unit gut content after a long feeding period. This is called the *stationary* ingestion rate because after a long time the feeding rate equals the digestion rate and the system is in a *stationary state* (does not change its average behavior). The results use the fact that in its stationary state the feeding rates and digestion rates are the same. The actual rate of passage of food through the gut equals the product of the digestion rate *kdig* and the probability that the gut is not empty, $1 - p_0$. This probability can be computed for the M/M/1 queue by $\frac{S - 1}{S}$ where

$$S = 1 + \sum_{k=0}^{\infty} \left[\prod_{xgut=0}^{k-1} \frac{genctr_i}{kdig^k} \right]$$

The complexity of this term derives from the fact that the arrival rate depends on the gut contents. Therefore, the product of the ratio of encounter rates over the digestion rate to the *k*th power for all possible gut contents *k* replaces the simpler traffic intensity term *a* of Beyer's model.

For each of the three cases the feeding rates *gfd* per unit gut content *xgut* become:

$$\frac{gfd}{xgut} = \begin{cases} kdig \times \dfrac{xprey}{\dfrac{kdig}{ksrch_{mx}} + xprey} & \text{case 1} \\[3em] ksrch_{mx} \times \dfrac{xprey}{kgut_{mx}} & \text{case 2 for } xprey < \dfrac{kdig}{ksrch_{mx}} \\[2em] \dfrac{kdig}{kgut_{mx}} & \text{case 2 for } xprey \geq \dfrac{kdig}{ksrch_{mx}} \\[2em] kdig\left(1 - e^{-\frac{ksrch_{mx} \times xprey}{kdig}}\right) & \text{case 3} \end{cases}$$

Case 1 gives a Michaelis or Holling equation, much like presented in Chap. 4 for the effect of prey density on fish feeding. In case 2 a bilinear feeding function results, whereas in case 3 an exponential distribution like that suggested by Ivlev (1960). The striking aspect of Sjoberg's work is that a stochastic model can, under limiting assumptions, generate feeding relationships that have been used in deterministic simulation models. The equations also suggest somewhat different interpretations for the parameters in the process equations. For example, the half-saturation parameter, previously defined only in terms of prey density for half-maximum ration $kfdpy_{hs}$, now is interpreted as the ratio of digestion to prey encounter rates ($kdig/ksrch_{mx}$). Thus, an animal with rapid digestion would, according to this interpretation, have a larger half-saturation level and would be more sensitive to low prey levels than one with slower digestion rates.

7.7 Summary

Much emphasis in the current ecological simulation literature has been given to including variability. Our treatment of this topic is introductory, although we do give enough of the basics to make the more advanced applications comprehensible. Some of these applications will be given in the annotated bibliography of this chapter.

In this chapter we have shown methods for generating random numbers on computers, not so much to expect you to have to do it yourself, because most computer systems contain some random number generator, but mostly to give an introduction to those who, like ourselves, are fascinated with how things work. We showed how to transform uniformly distributed random numbers into other probability distributions, which are more essential to applications.

Our examples are primarily focused on giving the tools to do Monte Carlo simulations, which will also be used in Chap. 8 for sensitivity analysis. The applications in this chapter use random number generators to simulate weather data and to model predator–prey dynamics with a random walk model. We have touched on this latter topic because we believe it will eventually help to greatly improve our understanding of the role spatial processes play in ecosystem dynamics. We provided the code for a simple spatial predator–prey simulator and gave a flavor of the kinds of refinements that would provide greater realism than the basic model. Some of these refinements are implemented in the exercise section.

Our examples from queueing theory showed how random variables can be treated in a direct fashion in a model. The applications to fish feeding showed how both experimental observations and empirical relationships can be explained using a stochastic model and how the models provide insights into interpreting the parameters.

Note

1. The notation $\in$ means "is an element of." Thus $x \in [0,1]$ means that x is some number in the interval between 0 and 1. The brackets [0,1] mean that the boundary numbers (0 and 1) are included (a closed interval), whereas (0,1) means that they are excluded (an open interval). Semiopen intervals, (0,1] or [0,1), are also possible.

7.8 Annotated Bibliography

Bailey, N. T. J.: *The Elements of Stochastic Processes with Applications to the Natural Sciences,* Wiley, New York, 1964.
    Contains a chapter on queueing theory as well as a good review of Markov chains at a level of mathematical sophistication beyond the scope of this book.

Beyer, J. E.: *Survival of Fish Larvae—A Single Server Queue Approach,* Institute of Mathematical Statistics and Operation Research, Technical University of Denmark, Copenhagen, 1976.
Reviews some queueing theory and places fish feeding into this theory.

Cox, D. R., and W. L. Smith: *Queues,* Methuen, London, 1961.
This is a classic reference on queueing theory. It is written at a mathematically sophisticated level and will require careful reading.

DeAngelis, D. L., and G. T. Yeh: An Introduction to modeling migratory behavior of fishes, in *Mechanisms of Migration in Fishes,* ed. J. D. McCleave, G. P. Arnold, J. J. Dodson, and W. H. Neill, Plenum, New York, 1984, pp. 445–469.
Reviews random walk and diffusion approaches to modeling movement.

Kennedy, W. J. Jr., and J. E. Gentle: *Statistical Computing,* Marcel Dekker, New York, 1980.
Gives a review of various uniform random number generators, how to test them for randomness, and shows how to generate many other probability distributions from uniform random numbers.

Morgan, B. J. T.: *Elements of Simulation,* Chapman and Hall, London, 1984.
Reviews methods for generating uniform random numbers, transforming them into other probability distributions, testing them for randomness, and using them in Monte Carlo simulations.

Morrison, K. A., N. Therien, and B. Coupal: Simulating fish redistribution in the LG-2 reservoir after flooding. *Ecol. Model.,* 47, 1985.
An application of a random walk model in which movement is inversely proportional to the density at a point, which causes a spreading phenomenon.

Ripley, B. D.: Computer generation of random variables: A tutorial. *Int. Stat. Rev.,* 51:301–319, 1983.
A review of random number generation on computers.

Schrage, L.: A more portable Fortran random number generator. *ACM Trans. Math. Software,* 5:132–138, 1979.
Contains Fortran code for a machine independent random number generator for any computer that can represent all integers in the interval $[-2^{31} + 1, 2^{31} - 1]$.

Sjoberg, S.: Zooplankton feeding and queuing theory. *Ecol. Model.,* 10:215–225, 1980.
This paper provides a connection between queueing theory and the representation of feeding in deterministic simulation models. It gives a new interpretation to the parameters.

Wichmann, B. A., and I. D. Hill: An efficient and portable pseudo-random number generator. *Applied Stat.,* 33:188–190, 1982.
Presents Fortran code for a portable uniform random number generator with an extremely long period (requires three seeds).

7.9   Exercises

Generating driving variables

7.1. Figure 5.10 shows the driving variables used for the San Joaquin model and the data used to fit the driving variable curves. For both temperature and solar radiation, a sinusoid function, like that given in Eq. 7.3, appears to give an adequate fit. For the temperature driving variables a good approximation was obtained using a sinusoid with $t_0 = 0$, $ztp_{mx} = 24$, and $ztp_{mn} = 8°C$. As the data show, however, there is a significant amount of variability around the driving variable curves. This variability can be included by generating deviations from the sinusoid using Monte Carlo techniques.

a. Use the Box–Muller algorithm given in this chapter to generate normally distributed deviations with mean 0 and $\sigma^2 = 2.25$ to add on to the sinusoid temperature average. Run the San Joaquin model five times with this temperature data generated from five different seeds. How do the results compare with the original model?

b. Run the model five times with temperature data generated as above but having $\sigma^2 = 25.0$. How does having more variability in temperature affect model behavior?

7.2. Table 7.2 shows the rainfall amounts at the Thompson forest site in western Washington for 1973.

a. Using these data estimate the Markov chain for rain–no rain and the rainfall distribution as was done in this chapter (use 0.1-cm wide histogram bins). Simulate the rainfall pattern over a year and compare it to the reported data.

b. Estimate different Markov chains and rainfall histograms for the three seasons suggested in this chapter (mid-October–February, March–June, July–mid-October). Use these to simulate rainfall pattern and compare this with the pattern generated in (a) and the actual data.

Simulating tree establishment

7.3. One major difference between the Hubbard Brook model reported in Chap. 5 and the version run in Chap. 6 is that the latter was deterministic whereas the former had stochastic elements, specifically in tree mortality and establishment. In this problem tree establishment will be simulated using a Monte Carlo method. The computing requirement to do this as in the Hubbard Brook model by establishing individual trees would be too great. Instead the following scenarios can be simulated:

a. Simulate tree establishment at the beginning of the model run by generating numbers of shade-tolerant and shade-intolerant trees. From the four species considered generate numbers of cherry and birch by generating an integer uniform random number between 0 and 100 for the cherry and 0 and 60 for the birch. For the shade-tolerant species, generate numbers from a Poisson distribution with a mean $\lambda = 5.0$. The means of these distributions are the same as the initial numbers used in the first run of the model in Chap. 6. Make five 100-year runs with initial conditions established with five different seeds for the random number generator.

b. Simulate establishment every 20 years in the model run by examining the total stand basal area and seeing whether shade-intolerant trees can establish under these conditions. Species-specific thresholds for basal area are linked to the light conditions resulting from this basal area. Assume that cherry has a threshold of 2000 cm^2 and birch has a threshold of 3000 cm^2. Above these thresholds no establishment of these species can occur. For cherry below its threshold use an integer uniform random number between 0 and 50 for establishment and for birch use an integer uniform random number between 0 and 30. For maple and spruce the number of trees established is assumed to depend on the basal area as well, but here it is a measure of crowding rather than light. As such, assume that the number of trees to be established is a Poisson random variable with mean λ equal to 2.0 ×

$\left(1.0 - \dfrac{gareabs}{6000}\right)$. This relationship will assure that λ will be close

to 2 at the start of the simulation and λ will be less than 2 later in the simulation. The easiest approach to this problem is to run the Hubbard Brook model in 20-year stretches and to separately run the establishment module after each 20-year run, using the established trees to generate a new initial condition for the next 20-year run. Because you may have more than six cohorts at one time you will have to adjust the maximum array sizes in the Hubbard Brook program. Do five 100-year runs using different initial seeds to generate establishment.

Random walk modifications

In the next set of exercises the basic random walk model will be refined to include more realism both for the predator and the prey. If not specified use λ = 10, *prey*$_{mx}$ = 5, *pup* = *pright* = *pdown* = 0.25, and 50 times steps. Average the results from five simulations.

7.4. Instead of using a Poisson distribution for the number of prey at each grid point use a uniform distribution. To have the mean prey density be λ transform the U(0,1) random number into a U(0,2λ) random number by multiplying by 2λ.

7.5. Instead of a constant λ try having a gradient by making λ$_{ij}$ = $2\lambda\left(\dfrac{i^2 + j^2}{2n^2}\right)$, where *n* is the (assumed square) grid size and *i* and *j* are the grid locations. This equation makes a gradient of average density that increases diagonally from the upper left to lower right in the grid, with a maximum of 2λ at the lower right corner of the grid.

7.6. (continuation) With the previous problem, try changing the predator's search probabilities so that it has a bias for search toward the right and down. Try *pup* = 0.1, *pright* = 0.4, and *pdown* = 0.4. How does this affect the average prey encountered and taken? How much variability is there in this model relative to the previous model and to the original model?

7.7. Modify the probabilities of moving in each direction according to the density in each adjacent segment. As a simple algorithm we add up all the densities for adjacent grid points and move with probability equal to the density in each direction over the total density. If all four adjacent grid points are empty move with equal probability in any direction (i.e., revert to the original algorithm).

7.8. Predators do not leave patches if feeding within them is good. In the original algorithm there is movement to an adjacent grid point no matter how good the present grid point is for feeding. As an initial simulation of this phenomena have the predator remain at the same grid point if the patch density *prey(i,j)* is larger than the maximum ration possible in each feeding time step, *preymx*. Compare the average prey taken per time step with this model to the original model.

7.9. In the present simulation only a single predator is foraging. We might expect that as more predators forage over the same area there is increased interference, resulting in reduced feeding rates. Simulate two predators foraging simultaneously over the same 20×20 grid. Compute the average number of prey taken by each predator and compare it with the average for a single predator.

Chapter 8

Simulation Model Evaluation

8.1 Introduction

With the steady increase in the use of simulation models for management, impact assessment, and as aids to understanding ecosystems, the question of how good these models are arises frequently. Are there standard procedures for model evaluation? In this chapter we introduce some means for evaluating simulation models, while showing how it is impossible to *validate* (assess the truth of) an ecological simulation model. The term validation has crept into the literature and, as far as we are concerned, is a misnomer. A more accurate term to describe the process of model evaluation is model *corroboration*. This implies the process of increasing confidence that the model meets its objectives.

Evaluating how good a model is depends on the model objectives. Caswell (1976) distinguished two major types of model objectives. These are (1) to provide accurate predictions of the behavior of the system and (2) to gain insight into how the system operates (ecosystem theory). The first type of model can be corroborated by comparing model output with field data, because it is meant to be predictive only over the range for which it is to be used. The second is expected to hold over a wide range of conditions. Therefore, it can only be shown to hold or not hold on a case-by-case basis. If the model continues to pass various empirical tests, our confidence in it increases and the model is further *corroborated*. Comparison with data is seen as a part of model corroboration.

Most ecological simulation models do not fall entirely into either the predictive or the theoretical category. We would feel uneasy with a model that appeared to predict some observed phenomenon well but was not based on hypotheses about how the system worked. For example, many models are built to examine the effects of perturbations (changes), either man-induced or natural, on system behavior. Thus, they are concerned with system behavior outside the range of data used to build or corroborate the model. On the other hand, most

209

ecological simulation applications have prediction as a tacit objective. A model that cannot come close to observed system behavior is regarded with suspicion.

We believe that comparison of a model to field data has some value, although the question of model corroboration is a wider one than agreement with data. We view model evaluation as a set of tools for ascertaining (1) how accurate the assumptions made in building and running the model are, (2) how realistic the behavior resulting from those assumptions is, and (3) how sensitive model behavior is to changes in these assumptions. These assumptions include (for the most common set of ecological simulation models dealt with in this book):

1. underlying model assumptions such as spatial homogeneity, determinism, and the boundaries of the system modeled.
2. the particular process equations chosen.
3. the parameter values used, the data brought to bear to estimate these values, and how this estimation process is done.

We can never be sure that some other set of underlying assumptions and equations will not fit the data equally well. We have picked choices from a limited menu of alternatives. Of course the same problem exists with individual hypotheses, except there we are usually dealing with a single process, whereas with models we consider a suite of hypotheses. As such, the opportunity for missing a feasible alternative is increased.

An additional set of model evaluation criteria relate to how accurately the model reported is reflected in the code, and is termed model *verification*. A model can also be evaluated in terms of the accuracy and completeness of its documentation. We emphasize the tools and techniques of model evaluation in this chapter including methods for comparing model output with field data and for *sensitivity analysis*.

Richard Levins (1966) described three attributes of models: generality, precision, and realism. He claimed that no ecological model can have all three qualities. These terms are interesting in comparison with Caswell's two sets of model objectives. A simple predictive model would be precise but would lack realism and generality. A model representing a single system might be precise and realistic but not general. Alternatively, models used to represent a wide range of ecosystems over a long time scale as, for example the JABOWA (Hubbard Brook) model (see Chap. 5) sacrifices precision (e.g., trees establish at 0.05 cm DBH; growth is an annual total) for generality.

Levins' three model descriptors are useful when considered from the standpoint of model evaluation. Of the evaluation methods we consider in this book some speak primarily to model realism, others to model generality, and still others to model precision. Figure 8.1 shows several methods for model evaluation, along with model objectives that they relate to. A number of objectives have expanded on the theme of Caswell's original prediction and ecosystem theory (understanding). These include model objectives of (1) generalizing beyond a single site, (2) identifying areas for future research, where the field experiments and models are designed to work hand in hand with each other,

METHODS   **OBJECTIVES**

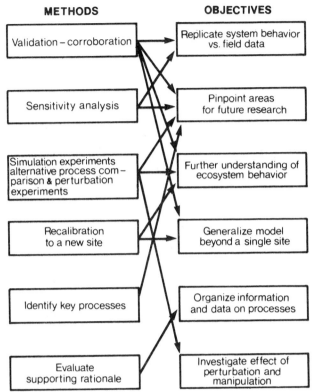

Fig. 8.1.   Connections between the objectives for ecological simulation models and the methods used for model evaluation.

(3) investigating the effects of perturbation and manipulation (using the model as a scenario for investigating effects not practical for direct experimentation on the system), and (4) organizing data and information about processes into a whole ecosystem framework. Four of the methods in Fig. 8.1 will be dealt with in this chapter—comparison with field data (corroboration), sensitivity analysis, perturbation experiments, and detailed examination of a key process. Methods for examining the supporting rationale for model processes were considered somewhat in Chap. 4 and are also discussed in this chapter. Methods for designed simulation experiments are also discussed here.

8.2   Model–data comparison

Methods include using tables, graphs, and statistics (or pseudostatistics) for comparing a series of model state variable outputs with data collected in the field. For rigorous corroboration these data should be independent of data used to build the model, especially data used in model calibration. Such data are difficult to obtain because often all the data that are available are used either explicitly, through calibration, or implicitly, through designing the model equations to "fit" phenomena known to occur in the data set. The temptation to look at all data is usually too much for modelers, who want to make the model speak as well as possible to the information carried by data. Who wants to build a model to fail, despite knowing that we may learn more from our failures than our successes?

8.2.1 Grapical comparison

The most straightforward comparison between model and data is a graphical comparison. Because our eyes and minds are geared to the recognition of pattern in graphical displays, graphing model output versus field data appears to be an acceptable means of model corroboration, especially when the differences between model output and field data can be traced to specific processes or omissions in the model. When several different models are to be compared to a data set, however, graphical comparison is difficult and tables are often more tractable. Also, graphical comparison suffers from lack of standardization of scale. Closeness or distance between model and data can depend on the scale chosen for presentation. Finally, it has recently been emphasized (Cleveland and McGill, 1985) that although our minds are geared to pattern recognition they cannot as easily distinguish differences between rapidly changing curves than differences between curves with smaller slopes. This can mislead us into over-emphasizing differences around peaks and troughs of curves.

Graphical comparison can be done in a variety of ways, especially in the presentation of the field data. Three alternatives are presented here. Figure 8.2 shows zooplankton and phytoplankton model output and field data presented for a model of Lake Ontario by Scavia et al. (1976). Here the data over a 5-year period are presented as *replicates* of system behavior (note these are not true replicates because environmental conditions were different for each of the years). Besides giving the lake wide mean data values for several of the years this graph also uses a shaded area to denote the range of one standard deviation on either side of the mean over all 5 years. Comparison of the model to the data shows that the model captures the pattern of the data for both phytoplankton and zooplankton. Both zooplankton and phytoplankton, however, fall outside the one standard deviation zone at different times of the year. This is an indication that model behavior during these times may not agree with the field data.

Figure 8.3 compares phytoplankton and zooplankton field data from Narragansett Bay to output of a model by Kremer and Nixon (1978). The field data is for a single year and shows the commonly observed phenomenon, when the fit is compared to that in Fig. 8.2, that models tend to fit better with data averaged over several years than they do with data for a single year. The model is aberrant in pattern and, to a lesser extent, in the average magnitude of the biota. The field data in Fig. 8.3 had no replicates, so sampling variability is not available. We are, therefore, at a loss to know to what extent differences in data values from time to time reflect sampling variability. Another feature of this

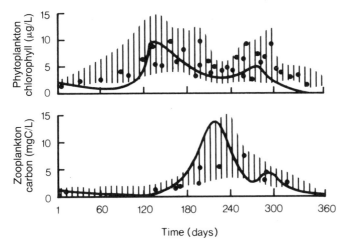

Fig. 8.2.   Graphical evaluation of an ecological simulation model for plankton dynamics on Lake Ontario showing time traces for the phytoplankton and zooplankton state variables compared with data points for individual years and a shaded area representing the range of values within one standard deviation of the mean data values over a 5-year period. From Scavia et al. (1976).

figure is that it compares model output for one spatial segment or subarea for Narragansett Bay, whereas the Lake Ontario model represents an average for the entire lake. This suggests another phenomenon, rigorously demonstrated for Lake Ontario by Simons (1976), that models tend to fit data better when they are spatially aggregated. The average of results from model output for a number of segments tend to fit the average of the field data from these segments better than the individual segment fits. Simons (1976) maintains that this is caused by hydrodynamic features that play a major role in species distribution on the small scale but that average out on the large scale. The same might be said in terrestrial

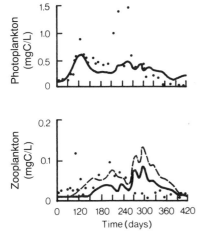

Fig. 8.3.   Graphical evaluation from a model for Narragansett Bay showing phytoplankton and zooplankton state variable time traces for a single year compared with data values collected over the year at the Greenwich sampling station on Narragansett Bay. The dashed line for zooplankton is for the case when they are assumed to consume detritus. From Kremer, J. N., and S. W. Nixon (1978), A marine ecosystem, *Ecol. Studies* 24. Used by permission.

systems where microtopography may play a major role on a specific site but where results may average out over a larger area.

In all these examples, some parameter adjustments were done by the modelers to make the model output more closely fit data traces, despite our earlier statement that model corroboration through data comparison should be done with a data set independent from that used for calibration. This calibration procedure somewhat reduces the value of direct comparison of model output to field data as, given enough parameter adjustment, it would seem possible to fit any data trace. This argument, however, is an overstatement of the case against calibration for several reasons. First, in calibration only a small fraction of the parameters are usually changed. Second, parameter values are usually constrained to be within a range of values reported in the literature. Third, many of the parameters are initially estimated from data other than time-trace data and an attempt is made in calibration not to diverge too widely from these initial estimates. Fourth, in most models there are many state variable data traces to compare with field data and fitting entirely by calibration would be intractable. Finally, many of the parameters are interactive with each other in their effect on model output, and thus changing them during calibration is more an art than a science; i.e., it is not that easy to fit many time traces with the highly constrained equations and parameters of most ecological simulation models. Automatic or optimization based calibration techniques have been sometimes used, but appear impractical for large models and may not be necessary for small models because these are easy to calibrate.

Even after calibration the model often does not accurately fit the field data. What is to be done in these cases? One possibility is to interpret differences as indicating the need for model revision. Certainly such revision may be required, although without an accompanying increase in the realism of the model equations there appears to be little justification for revision, because the independent data set will no longer be independent (little as it might have been when originally used). In other words, model revision to improve fits to data is only justifiable where some ecologically significant change is made. Another interpretation of discrepancies between model and data is that the differences can indicate areas for future research or a need for more detail on individual processes. This is possible only if the model differences from field data can be traced to individual processes. For example, in Kremer and Nixon's (1978) model of Narragansett Bay, the pre-spring bloom (winter) biomass of phytoplankton in the model (and the data) was insufficient to support the observed zooplankton standing stock in the winter (Fig. 8.3). This suggested that some alternative food source was being used by the zooplankton, which indicated the need for further research into the seasonal changes of diet of the dominant zooplankton in Narragansett Bay.

8.2.2   Statistical comparison

In most of our applications simulation model output is looked at as deterministic. As such, any statistical comparisons of model output and data traces can examine whether the model output fits the data trace based only on variability of the data. As pointed out earlier, changes from year to year are often used to estimate the variance in the data. This is not a proper variance estimate for statistical comparison to model output, because the output represents a single year's simulation. That is why we use the term pseudostatistical for the statistical methods used.

There is another, more serious, problem in statistical comparisons of model and data time traces. This is that goodness of fit tests (see Appendix 3) used to compare a series of observations to each other assume independence between the observations. We know, however, that the observations are not independent, because they are linked by the model, which generates future behavior from the present state of the system and future environmental conditions. It is possible to get around the problem of nonindependence over time, as was done by Garratt (1975) and Steinhorst (1979), by including an estimate of the covariance matrix between data points at different times in the statistical comparison of model and data.

When statistically comparing model output and field data it is often assumed that the variance of the model output is the same as that of the data. Certainly, model output does have variance and covariances, even though the model itself is deterministic. This is because the model can be viewed as a transformation, albeit a highly complex and nonlinear one, between the driving variables, which can be viewed as random variables having some underlying probability distribution (see Chap. 7 discussion on generation of driving variables as random variables via Monte Carlo techniques), and model output. At the same time the model is also transforming the uncertainty in parameter values into variance in model output. Thus, the variance and covariance of model output depends on the variance and covariance of the model parameters, the initial conditions and the driving variables as well as on the transformation (model equations) that convert these random variables into model output. Variance estimates for model output are sometimes generated using Monte Carlo methods analogous to those discussed later in the chapter under the topic of sensitivity analysis (Gardner et al., 1981). We will not give any specifics about statistical model corroboration because of some of the difficulties with assumptions in comparing model output with field data. We give annotated references to some of the literature at the end of this chapter for those who are interested in this field.

8.2.3   Tabular comparison

Table 8.1 compares output taken from a model of Andersen and Ursin (1977) with fishery yields over an 11-year period in the North Sea for 11 commercially important fish species. The annual totals are calculated both from the model and from catch statistics. This table condenses the vast amount of model output into a small amount of information. For most models the limited amount of data for corroboration make this type of comparison necessary. However, annual totals are not always the best way to compare model output and data. Often, indices based in ecological theory provide a more meaningful comparison. The choice

Table 8.1   Annual yield (1000 tons) of North Sea fisheries for 11 species compared with results from a simulation model of Andersen and Ursin (1977). The first line for each species is the actual catch and the second line is the model result.

Species	Year										
	1960	1961	1962	1963	1964	1965	1966	1967	1968	1969	1970
Plaice	87	86	88	108	110	97	100	109	111	121	130
	94	86	100	111	116	103	102	138	101	97	118
Dab*	11	11	11	11	11	11	11	11	11	11	11
	9	9	10	11	11	12	12	12	11	11	11
Long rough dab*	1	1	1	1	1	1	1	1	1	1	1
	0	1	1	1	1	1	1	1	0	1	1
Saithe	29	31	22	28	55	69	87	73	97	106	170
	31	29	28	25	42	52	75	73	77	90	151
Cod	105	108	91	110	125	182	229	250	285	199	229
	108	103	89	100	130	152	184	173	218	208	220
Haddock	67	69	53	60	202	225	272	167	140	640	675
	63	57	43	61	164	214	307	162	195	607	627
Whiting	55	85	64	99	88	110	158	91	145	199	183
	35	51	50	77	62	82	109	59	95	130	148
Norway pout	27	26	137	110	44	36	47	178	437	103	217
	12	21	48	53	17	5	6	44	224	581	592
Mackerel	73	86	66	73	115	208	530	930	822	739	322
	74	75	76	76	120	206	523	917	825	689	291
Herring	696	697	628	716	871	1169	896	696	718	547	563
	618	762	778	601	970	1418	980	865	896	564	578
Sandeels	121	178	109	162	128	131	142	189	194	113	191
	138	117	104	91	71	50	42	66	142	272	345
11 species total	1272	1378	1270	1478	1750	2239	2473	2695	2961	2779	2692
	1182	1311	1327	1207	1704	2295	2341	2510	2784	3250	3082

*Accurate information missing but small figures anyway.
From: Andersen and Ursin, 1977. Used by permission.

of these indices often depend on the particular biota modeled. A good example is in modeling of lake plankton dynamics. Important indices here are the timing and magnitude of the spring phytoplankton bloom, because they indicate system productivity for that year and also serve as possible indicators of noxious phytoplankton blooms. Table 8.2 compares model output and data for a model of Lake Ontario plankton (Dale and Swartzman, 1984) using a number of indices including total production and annual averages of phytoplankton, zooplankton, and detritus. The data values are ranges over 5 years of data collection on Lake Ontario. Table 8.2 is an encapsulation of model and data output using indices that speak most to plankton biologists.

Table 8.2   Comparison of selected indexes of ecosystem performance between a model of Lake Ontario and a range of data values taken from 5 years of environmental monitoring on the lake.

Trophic level or nutrient	Total production		Average level	
	Model	Data range	Model	Data range
Phytoplankton	105.1	36.0–142.0	0.288	0.10–0.39
Zooplankton	4.70	4.1–17.3	0.013	0.007–0.031
Ammonia	—	—	0.0275	0.0225–0.0405
Nitrate	—	—	0.127	0.10–0.15
Phosphate	—	—	0.0098	0.0071–0.014

8.3   Sensitivity analysis basics

Sensitivity analysis is the label used for a collection of methods for evaluating how sensitive model output is to changes in parameter values. The most straight-forward method changes parameter values, either singly or in various combinations, each by the same percentage (e.g., 10 or 20%), and observes changes in model output. This gives an idea of the relative sensitivity of model output to each of the parameters and, for some methods, to combinations of parameters.

Sensitivity analysis has been applied as a tool either for model corroboration, for guiding future research by highlighting the most important processes or for parameter estimation by showing which combinations of parameters lead to realistic model behavior. In the first case the sensitivity of model output to parameter changes is used as a test of the *robustness* of model performance to changes in parameter estimates. Under this criterion a model is interpreted as good if it is insensitive to small ($\pm 10\%$) changes in model parameters. Most applications in this chapter focus on this interpretation.

Sensitivity analysis can also be a tool for guiding future research, by indicating which parameters most sensitively affect model output, thereby suggesting where refinement of parameter estimation would most improve model predictions (or the realism of model behavior). An example of this approach from Kitchell et al. (1977) is given later in this chapter. The third approach, championed by Hornberger and Spear (1981), uses sensitivity analysis to estimate variance and covariance between model parameters from those parameter combinations leading to realistic model behavior.

Classical sensitivity analysis (Tomovic, 1963) uses the partial derivative of model output to parameter change $\left(\dfrac{\partial x_i}{\partial k_j} \right)$ as a measure of model sensitivity. It is possible, for a small and simple set of differential equations, to determine the above partial derivatives, even if the differential equations cannot be solved analytically. Steinhorst et al. (1978) show, however, that the Tomovic method is impractical for all but the simplest ecological simulation models, and Gardner et al. (1981) state that several assumptions (e.g., no discontinuities in the process equations) in the Tomovic method are violated in most ecological simulation models.

Alternatives to the classical methods use incremental (discrete) changes in parameters rather than derivatives and are potentially applicable to large simulation models. An example is the case where parameters are changed individually by a constant percentage, and average values of each state variable are compared.

8.3.1 Problems with sensitivity analysis

Four major problems exist in conducting and interpreting sensitivity analyses. These are:

1. The analysis method involves many runs (sometimes upward of 100) of the simulation model, which can prove expensive, time consuming, and unwieldy.
2. There is the possibility of *interactions* between parameters in their effect on model output. Interactions occur when the response of the model to a change in some parameter value depends on the values chosen for other parameters. This aspect of model sensitivity is totally ignored by methods that change parameters one at a time because they assume that interactions are nonexistent or negligible when compared with single parameter effects.
3. Different parameters have different inherent amounts of variability. This is because not all processes are equally well understood and not all parameters are measurable to the same precision. Furthermore, some aspects of a process are more variable from individual to individual or area to area in a model and have parameters that are more variable than others. This observation raises the question about how widely to vary the parameters. Do we wish to use a small consistent deviation (e.g., $\pm$ 10%), a large consistent deviation (e.g., $\pm$ 50%), or a deviation either large or small depending on the (expected or known) variability of the parameter estimate (e.g., $\pm$ 2 standard deviation from the expected or best parameter estimate). The choice among these alternatives strongly affects the interpretation of sensitivity analysis results.
4. There are many potential output variables to use as measures of model behavior. Measures can include any state variable at any time, the average, maximum, minimum, or integrated (over time) values of any state variable, the timing of the maxima or minima, average or totals over some time window, and indexes involving ratios or combinations of state variables (e.g., root to shoot ratio, or diversity index). Sensitivity conclusions may (and usually will) change depending on the choice of output variables for sensitivity analysis.

8.3.2 Qualitative sensitivity analysis

Kitchell et al. (1977) used sensitivity analysis to direct future research using a fish growth and energetics model. In the analysis, parameters were changed singly by plus and minus 10 percent. A qualitative analysis was used, with system response to each parameter classified as either high (sensitive), moderate, or low (insensitive). A single output criterion, fish growth over a year, was used. The model, a yellow perch energetics and growth model, suggested this single output criterion. All of the model parameters were included in the analysis.

Parameter variability and ease of measurement were included in a qualitative fashion by classifying the availability of an estimate for each parameter as either high (easily measured and available for the species under consideration), moderate (either more difficult to measure, measured in the laboratory and somewhat variable in the field, or available only for a related species), or low (either not yet measured, extrapolated from a related species, or measured in the laboratory but highly variable in the field). Priorities for future research were set based on both the parameter sensitivity and the data availability, with emphasis placed on those parameters that had a strong influence on fish growth but were not yet measured or were highly variable in their estimates.

Results of the analysis are shown in Table 8.3. Kitchell found the parameters

Table 8.3 Summary of a sensitivity analysis of model parameters and implications of the need for further research. Qualitative evaluations are given for the availability and sensitivity of model output to change in a parameter value and these are used to suggest further research.

Parameter	Sensitivity	Availability	Research priority
Consumption			
Effect of weight on ration	H	L	H
Exponent of weight on ration	H	M	H
Assimilation	H	L	H
Optimal feeding temperature	M	H	L
Upper feeding temperature	L	L	M
Q_{10} for feeding	M	M	M
Respiration			
Effect of weight on respiration	H	M	H
Exponent of weight on respiration	M	M	M
Activity effect	H	L	H
Optimal respiration temperature	M	H	L
Upper respiration temperature	L	H	L
Q_{10} for respiration	M	H	L
Specific dynamic action respiration	M	M	M
Waste losses			
Egestion loss coefficient	L	L	M
Temperature egestion effect	L	L	M
Feeding level egestion effect	L	L	M
Excretion coefficient	M	L	H
Temperature excretion effect	M	L	H
Feeding level excretion effect	L	L	M

H stands for high, M for moderate, and L for low.
Adapted from Kitchell et al. (1977). Used by permission of the *Journal of the Fisheries Research Board of Canada*.

most in need of future research were the weight effect on ration and respiration, the exponent in the effect of weight on ration, an activity parameter for respiration, an assimilation parameter for consumption, a parameter for the fraction of assimilated food ingested, and a parameter for the temperature effect on the egestion rate.

The benefit of this qualitative sensitivity analysis is that it avoids problems inherent in interpreting differences between numbers. Another advantage is that nonquantifiable factors, such as the ease of estimating parameters, can be combined with other factors such as the variability and even expense of estimates to arrive at those parameters most in need of improved estimates.

8.4   Sensitivity analysis methods

To consider each of the problems with sensitivity analysis discussed above Rose (1981) examined several methods using the following criteria for a successful method:

1. The method should be clearly defined, straightforward, and should specify the number of model runs required. Results must be consistent from one analysis to the next and not change if more than the required number of simulation runs are made.
2. The effects of interactions between parameters must be distinguishable from single parameter effects.
3. The method must include information on the variability associated with parameter estimates.
4. The analysis must allow interpretation for several output variables.

We describe two alternatives garnered from Rose's analysis. One is a *systematic sampling* and the other a *random sampling* (Monte Carlo) method. The methods are each described as two-step procedures; the first step being generating or sampling parameter values and making the model sensitivity runs and the second being the analysis of model output to judge relative parameter sensitivity. The analyses suggested are derived from statistical methods, although they are organizational and advisory tools rather than hypothesis tests.

By systematic sampling we mean varying parameters by prechosen amounts according to some statistical method. Random sampling involves generating parameter values from some prechosen probability distribution using a random number generator. In this sense random sampling schemes are not repeatable (except by using the same seed in the random number generator). The methods should be judged on a model specific basis according to (1) the clarity of design, (2) the number of runs required, (3) the ease of interpretation of results, and (4) the ability to examine interactions between parameters.

8.4.1   Systematic sampling

In the systematic sampling method, a *fractional factorial design* is used to vary parameters and model output is analyzed by an *analysis of variance (ANOVA)*

(Steinhorst et al., 1978). Appendix 3 contains a review of the statistics needed to understand this method. In a *complete factorial design* runs are made such that all *levels* of each parameter are combined with all levels of each other parameter. The simplest case (which we discuss here) has two levels for each parameter. These may be the *nominal* value (the best estimate) and a value 10 percent (or two standard deviations) above the nominal value, or the nominal value and 10 percent (or two standard deviations) below this value. A complete factorial design requires 2^n runs for n parameters. If m levels are used m^n runs would be required. For more than eight parameters, even with $m = 2$, this is a cumbersome number of runs.

To reduce the number of runs needed for a complete factorial design a *fractional factorial design* is used, which involves 2^{n-c} runs, where c is chosen by the design. The loss in doing a fractional instead of a complete design is in the inability to distinguish effects due to some combinations of parameters from others. This indistinguishability between some parameter combinations is termed *aliasing* or *confounding*. This concept is illustrated by an example in Table 8.4, where a two-level, three-parameter design is shown both as a complete factorial and as a fractional factorial design. The three parameters will be denoted by A, B, and C. There are $2^3 = 8$ model runs that will be denoted in the standard design of experiments notation as: a, b, b, c, ac, bc, abc, and (1). The lower case letter denotes the perturbed value of each parameter.

The nominal value is represented by the digit 1, which is handled in combination with letters using the usual algebraic conventions. For example, ab represents ab(1), the run with parameter combinations of the perturbed values of A and B and the nominal value of C. The symbol (1) alone represents the run where all parameters are at their nominal level. In this design seven effects can be distinguished. These are the individual effects of parameters A, B, and C, joint effects, (two-way interactions) between A and B, B and C, and A and C, and the joint effect (three-way interaction) between A, B, and C.

The effect of an individual parameter on the model output variable (denoted by the same symbol as the model run) is the average difference between values

Table 8.4   An example of a three-parameter, two-level factorial design.

Model run number	Level of factor			Value of output variable
	A	B	C	
1	−	−	−	(1)
2*	+	−	−	a
3*	−	+	−	b
4	+	+	−	ab
5*	−	−	+	c
6	+	−	+	ac
7	−	+	+	bc
8*	+	+	+	abc

Starred (*) runs are those included in the fractional design.

of that output variable in runs that have the parameter's level changed ($-$ to $+$) and those where it is not. For example, the effect caused by parameter A (denoted A) is:

$$A = \frac{(a + ab + ac + abc)}{4} - \frac{((1) + b + c + bc)}{4} \qquad (8.1)$$

Effects caused by B and C are computed in an analogous fashion. The interaction effect between two parameters (e.g., BC) is the difference between the effect of B at the perturbed and the nominal values of C. The effect of B when C is at the nominal level is $(ab + b - a - (1))/4$ and when C is at its perturbed level the B effect is $(abc + bc - ac - c)/4$. The interaction is the difference between these two effects:

$$BC = \frac{(abc + bc - ac - c)}{4} - \frac{(ab + b - a - (1))}{4}$$
$$= \frac{(abc + bc + a + (1))}{4} - \frac{(ac + ab + b + c)}{4} \qquad (8.2)$$

Suppose a fractional factorial design with only one-half of the eight runs in the full factorial design are made—those starred in Table 8.4. This is referred to as a 2^{3-1} or a half replicate of a 2^3 design. In that case only four output variables are available, a, b, c, abc. The effect of A is then:

$$A = \frac{(a + abc)}{2} - \frac{(b + c)}{2}$$

whereas that of BC is:

$$BC = \frac{(a + abc)}{2} - \frac{(b + c)}{2}$$

We can see that these effects are confounded or aliased because they are identical. Similarly, B is aliased with AC and C is aliased with AB. By having fewer runs we give up the ability to distinguish some parameter effects from others. By choosing which runs to do (and how many) we can choose which parameter combinations are distinguishable from others.

Fractional factorial designs can be extended to any number of parameters and other fractions besides one-half. Judicious selection of which runs to make enables individual and lower order interactions (two-way interactions) to be aliased with only higher order interactions. The higher order effects can then be assumed negligible, any significance being attached only to the lower order parameter effects. Designs for fractional factorial experiments with different numbers of parameters and different aliasing features are presented by a number of authors including Cochran and Cox (1957) and Box et al. (1978).

Analysis of the results of the fractional factorial design can proceed using analysis of variance (ANOVA) methods as was originally proposed by Steinhorst et al. (1978). We suggest forgetting about the ANOVA and looking at the effects themselves for several reasons. First, the ANOVA suggests statistical tests and significance values that cannot be done because there is no random component in the deterministic models being considered here. Second, the sums of squares, mean squares, and F statistics for the effects in an ANOVA table are all monotone functions of the effects computed above so the effects themselves contain the same information. Last, the effects are in the original units of the output variable being considered, which can make interpretations easier.

8.4.2 Random sampling

In random sampling procedures a probability distribution is assumed for each parameter and for each model sensitivity run a value is "sampled" (using Monte Carlo methods) from this distribution for each parameter. O'Neill et al. (1980) recommended the triangular probability distribution if a most likely or average value and a range for the parameter is known. Tiwari and Hobbie (1976) suggest a uniform distribution if only a range of possible values is known. The parameter values are assumed to be uncorrelated.

If the chosen distributions have a mode (e.g., normal distribution, triangular distribution) then generated parameter values tend to be clustered around this mode and values in the tails of the distribution are infrequent. This results in considerable redundancy in the information obtained from the sensitivity analysis because the region near the parameter modes is too well explored whereas tail regions are often unexplored, especially if the number of simulation runs is limited. This redundancy can be eliminated and a more balanced exploration of the parameter space assured by using *stratified sampling*. This method divides the parameter space into sections, sampling from each section with a certain probability. A further extension is *Latin hypercube sampling*, which divides each parameter distribution into n strata of equal probability $(1/n)$ and samples once from each strata (Iman and Conover, 1980).

Figure 8.4 shows the steps in constructing a Latin hypercube sample with six equal probability ($\frac{1}{6}$) regions for normal, uniform, and triangular probability distributions. For every run each parameter is assigned randomly to one of its six regions, but the sampling is without replacement such that each region is represented only once in six runs. The random sampling then takes place for each run with the probability interval specified in advance for that run. One possible random assignment of the probability intervals among the six runs is shown in Fig. 8.4. Within each interval the values are generated using uniformly distributed random variables. Spread of the sensitivity analysis runs over the entire range of parameter values in the parameter space is assured by using a Latin hypercube sampling design.

The output from a large number of randomly sampled parameter changes can be analyzed using *correlation, partial correlation,* and *partial rank correlation*. See Appendix 3 for definitions of these statistics. The premise is that the greater the correlation between an output variable and a parameter, the more influence the parameter has on that variable. Partial correlations are used to remove any

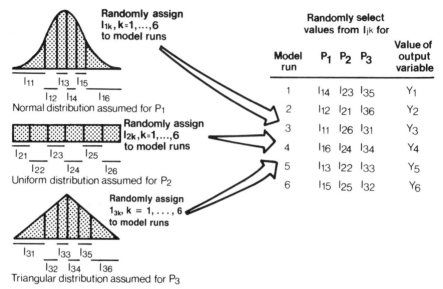

Fig. 8.4. Diagram illustrating a sensitivity analysis for three parameters according to the Latin hypercube sampling method. The probability distribution for each parameter, normal, uniform, and triangular, is divided into equal area (probability) segments to assure that the entire range of each parameter is sampled.

association the output variable may have with other parameters involved in the analysis. The rank correlation is suggested to allow for monotonic yet nonlinear association. None of these correlation methods allow for a nonmonotonic response (e.g., hyperbolic) of the output variable to the parameter changes.

Remembering again our criteria for sensitivity analysis that the method be able to indicate the minimum number of runs needed and that the results be consistent for higher number of runs, we discuss results of an experiment by Rose (1981) who used a model of phytoplankton growth to compare Latin hypercube sampling with simple random sampling using each of the correlation measures above. The parameters were ranked for each of the statistics for two trials of simple random sampling and results were compared after 100, 200, 300, up to 1000 runs to see which correlation measure gave the most consistent results. Rose (1981) found that the partial rank correlation required 400 runs for consistency of ranking of parameters between the two trials, whereas partial correlation required 500 and correlation did not give consistent results even at 1000 runs. Using partial rank correlation, the most economical sampling measure, simple random sampling was compared with Latin hypercube sampling. Results for 400 runs with simple random sampling were compared with 64 and 200 runs using Latin hypercube sampling. The 200 run Latin hypercube sample gave similar ranking to the 400-run simple random sample whereas the 64-run Latin hypercube sample did not. Based on this analysis Rose recommended using a Latin hypercube sampling design with 200 runs using partial rank correlation for interpretation of results.

8.4.3   Comparison of methods

The number of runs in the fractional factorial design depends directly on the number of parameters, whereas for the Latin hypercube sampling the relationship between the number of parameters and number of runs required is not certain. Although the number of runs appears to depend only on the number of probability intervals chosen for the parameters, it could be that having more parameters could require more runs before partial rank correlation results become consistent.

As mentioned earlier it is desirable that the methods can incorporate information about the variability of parameter estimates. Both methods reviewed have this capability; Latin hypercube sampling by changing the spread of the assumed probability distributions and the fractional factorial designs by defining the parameter levels based on nominal values plus or minus some multiple of a standard deviation rather than a constant percentage. Only the fractional factorial design explicitly includes interaction effects, although the LHS, by covering several parameter intervals does represent the changes in parameter values while randomizing over other parameter effects. These methods have the disadvantage of requiring significantly more runs than the single parameter at a time method, although we have paid attention to how to keep the number of runs to a minimum.

8.5   San Joaquin model sensitivity analysis

To demonstrate the fractional factorial design we will do a sensitivity analysis with the San Joaquin model. The analysis involves five parameters, which were chosen to represent what we thought to be the most important controlling parameters for model behavior. These parameters are $kpht_{mx}$, the maximum photosynthesis rate, $kphtlt_{op}$, the light intensity for optimum photosynthesis, $kphtnit_{hs}$, the nitrogen level at which photosynthesis is reduced to half its maximum rate, kfd_{mx}, the maximum zooplankton feeding rate, and $kfdpy_{hs}$, the prey density at which the feeding rate is reduced to half its maximum. Two analyses were done. In the first the altered level is either plus or minus 20 percent of the nominal value. The second analysis has the altered level at plus or minus a standard deviation from the nominal level, the standard deviation for each parameter being based on estimates for that parameter value in the literature. In the first sensitivity analysis the direction of change for the altered value ($+$ or $-$) was chosen at random. In the second set of sensitivity analysis runs, the parameters were altered in the opposite direction from what was chosen for the first analysis. Table 8.5 shows the parameters, their nominal values, and the altered values in the sensitivity analysis. The five parameters are labeled A through E for the analysis. The design chosen was a half factorial design, which means that half the number of runs in a full factorial design are made. The full design has 2^5 runs. Therefore, the half factorial design has 16 runs. Table 8.6 shows the runs made in this design, with $+$ denoting the altered value and $-$

Table 8.5   Parameter values chosen for the San Joaquin model fractional factorial design sensitivity analysis. The nominal value as well as the 20 percent and the one standard deviation altered values are shown.

Parameter	Label	Nominal Value	Altered ±20% Value	Altered ±S.D. Value
$kpht_{mx}$	A	3.00	3.60	2.16
$kphtlt_{op}$	B	300	360	264
$kphtnit_{hs}$	C	0.025	0.020	0.040
kfd_{mx}	D	0.39	0.47	0.19
$kfdpy_{hs}$	E	3.00	2.40	5.00

Table 8.6   Fractional factorial design used for the sensitivity analysis of the San Joaquin model.*

Run	A	B	C	D	E
(1)	−	−	−	−	−
de	−	−	−	+	+
cd	−	−	+	+	−
ce	−	−	+	−	+
bd	−	+	−	+	−
be	−	+	−	−	+
bc	−	+	+	−	−
bcde	−	+	+	+	+
ad	+	−	−	+	−
ae	+	−	−	−	+
ac	+	−	+	−	−
acde	+	−	+	+	+
ab	+	+	−	−	−
abde	+	+	−	+	+
abcd	+	+	+	+	−
abce	+	+	+	−	+

*A − denotes the nominal level of the parameter (factor) and a + the altered value.

Table 8.7   The confounding pattern for the fractional factorial design for the sensitivity analysis of the San Joaquin model.*

A = BCDE	E = ABCD	AE = BCD	CD = ABE
B = ACDE	AB = CDE	BC = ADE	CE = ABD
C = ABDE	AC = BDE	BD = ACE	DE = ABD
D = ABCE	AD = BCE	BE = ACD	I = ABCDE

The I effect is the overall mean.

denoting the nominal value. Confounding of the effects in this design is shown in Table 8.7. Notice that there is no confounding of second order interactions with each other and that main effects are confounded only with fourth order interactions.

8.5.1   Results

A number of model output variables are used as measures of model behavior for the sensitivity analysis. These include maxima and timing of maxima for phytoplankton and zooplankton biomass and the annual averages for phytoplankton, zooplankton, and nitrogen. Phytoplankton annual production is also used as a measure of model behavior. Table 8.8 gives the factorial effects for each of these output variables for the 20 percent design and Table 8.9 shows the results for the one standard deviation design. The top five effects for each output variable are shown in parentheses. The effects numbers are calculated as linear combinations of the output variables as described earlier in this chapter. To label the effects as in Tables 8.8 and 8.9 it is assumed that three-way and higher order interactions are negligible. For example, the A effect is confounded with the BCDE effect (see Table 8.7), but by assuming BCDE small any significance can be attributed to the main effect A and not to the interaction BCDE.

Relative sensitivities differ depending on the output variable chosen. Results also differ in the 20 percent and standard deviation runs. Overall, the most sensitive parameters appear to be the zooplankton grazing rate and prey half saturation and, usually, their interaction. Maximum photosynthesis rate is important in controlling phytoplankton output variables. Scrutiny of Tables 8.8 and 8.9 shows that the effects are generally larger in the standard deviation runs than in the 20 percent runs. This is because the standard deviations are larger than 20 percent of the parameter values so that the altered values differ more in these runs from the nominal values. Because the altered parameter values were in opposite directions in these two sensitivity analyses we would expect the differences owing to the main effects to be in the opposite directions in the two analyses, and this is most often the case. It can happen, however, that either raising or lowering a parameter value can push model output in the same direction. This was the case with the effect of maximum photosynthesis rate on peak phytoplankton biomass.

8.5.2   Analysis methods

A good way to examine the relative sensitivity of the parameters for any output variable is with *quantile plots* (see Appendix 3). These plot the effects in Tables 8.8 and 8.9 against quantiles from a normal distribution. If the output variables all come from the same normal distribution then the quantile plot should be a straight line. Values that greatly deviate from the line are assumed to arise from some other distribution. Although we do not assume normality for the model output these plots are still useful for looking at the sensitivity analysis results.

Table 8.8   The effects for the 20 percent design for the San Joaquin model sensitivity analysis. The five largest (in absolute value) effects are denoted in parentheses.

Effect	Maximum phytoplankton biomass	Time maximum phytoplankton	Phytoplankton production	Average phytoplankton biomass	Maximum zooplankton biomass	Time maximum zooplankton	Average zooplankton biomass	Average nitrogen concentration
A	0.422	−65.7(1)	283.0(1)	0.854(3)	1.40(3)	16.4(2)	0.658(2)	−1.48(4)
B	−0.006	32.8(2)	−147.0(4)	−0.524(4)	−0.389	−7.79(5)	−0.279(4)	0.616(2)
C	0.468(4)	−4.10(5)	25.3	0.054	0.159	0.333	0.070	−0.169
D	−3.03(1)	6.68(3)	−188.0(2)	−2.40(1)	2.16(1)	−27.7(1)	1.08	1.84(1)
E	−1.70(2)	5.50(4)	−154.0(3)	−1.77(2)	1.80(2)	−17.5(3)	0.826(1)	1.52(3)
AB	0.271	−0.280	3.54	−0.032	0.107	1.89	0.041	−0.102
AC	0.014	0.260	−2.804	0.005	0.016	−1.15	0.004	0.024
AD	0.633(3)	−0.520	47.1	−0.063	0.744(4)	1.79	0.287(3)	−0.484
AE	0.376	−0.460	36.1	−0.036	0.581	0.85	0.221(5)	−0.368
BC	−0.103	−0.480	2.25	−0.005	−0.074	9.97(4)	−0.035	−0.019
BD	−0.280	0.140	−12.5	0.046	−0.196	2.67	−0.100	0.141
BE	−0.173	0.200	−10.7	0.028	−0.166	−2.95	−0.080	0.118
CD	−0.001	0.800	2.50	−0.006	0.025	2.87	0.015	−0.020
CE	−0.032	0.580	2.48	−0.005	0.021	2.09	0.013	−0.020
DE	−0.460(5)	0.440	−70.2(5)	0.151(5)	−0.712(5)	−2.73	−0.214	0.630(5)

Table 8.9 The effects for the one standard deviation design for the San Joaquin model sensitivity analysis. The five largest (in absolute value) effects are denoted in parentheses.

Effect	Maximum phytoplankton biomass	Time maximum phytoplankton	Phytoplankton production	Average phytoplankton biomass	Maximum zooplankton biomass	Time maximum zooplankton	Average zooplankton biomass	Average nitrogen concentration
A	0.173	66.7(4)	−297.0(2)	−1.93(3)	−0.437(4)	−11.0(4)	−0.234(4)	1.06(2)
B	−0.100	−12.0	59.6(5)	0.412(4)	0.041	1.10	0.036	−0.178
C	−0.573	8.90	−41.1	−0.246(5)	−0.069	0.800	−0.040	0.186
D	18.2(1)	683.0(1)	−372.0(1)	16.3(1)	−7.00(1)	714.0(1)	−2.84(1)	−1.07(1)
E	7.01(2)	188.0(3)	−56.7	5.23(2)	−2.64(2)	627.0(2)	−1.08(2)	−0.767(3)
AB	0.087	2.80	2.85	−0.016	0.055	1.18	0.028	−0.070
AC	−0.141	−2.64	−4.66	0.025	−0.044	0.720	−0.027	0.068
AD	−0.204	−61.9(5)	80.5(4)	−0.107	0.435(5)	11.0(5)	0.218(5)	−0.576(5)
AE	0.606(4)	−3.60	53.5	−0.064	0.411	9.38	0.176	−0.458
BC	0.582(5)	0.920	49.6	−0.062	0.411	9.34	0.173	−0.446
BD	0.097	7.62	−12.1	0.025	−0.041	−1.10	−0.033	0.076
BE	−0.135	1.68	−7.326	0.026	−0.043	0.720	−0.027	0.063
CD	0.540	−4.06	9.79	−0.064	0.067	−0.800	0.037	−0.091
CE	0.112	−1.72	6.58	−0.027	0.056	1.18	0.028	−0.068
DE	3.56(3)	198.0(2)	−120.0(3)	−0.125	1.61(3)	613.0(3)	0.709(3)	0.722(4)

Effects that deviate from the line made by the points around the origin can be considered significant and need further analysis. We give two examples: maximum zooplankton biomass and phytoplankton production.

Figure 8.5 shows quantile plots for maximum zooplankton biomass for both the 20 percent and standard deviation sensitivity analyses. Here the two curves look like flipped versions of each other. The relative sensitivities appear to be about the same, only in opposite directions, as expected from the symmetry of the designs for the two analyses. The zooplankton parameters (D and E) and their interaction deviate most from the straight line made by the middle points as does the maximum phytoplankton photosynthesis rate (A). Difference in scale

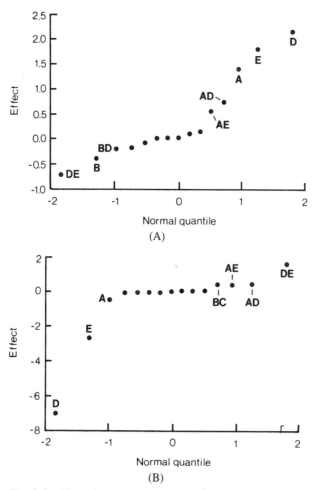

Fig. 8.5. Normal quantile plots of the effects for zooplankton maximum biomass from the sensitivity analysis of the San Joaquin model. A. 20 percent design. B. one standard deviation design.

is due to the larger differences in the parameter values from the nominal values in the standard deviation runs.

Quantile plots for total phytoplankton production are shown in Fig. 8.6. Here results between the two analyses are different from each other. Although both analyses have the same eight parameters and parameter interactions as the most sensitive (as shown by the quantile plots), their order and the direction of change are related in an odd fashion. For example, although both analyses mark maximum photosynthesis (A), maximum grazing (D), and their interaction as important, the D effect is negative in both designs, the A effect is negative in the standard deviation design and positive in the 20 percent design, and the AD interactions are positive in both designs.

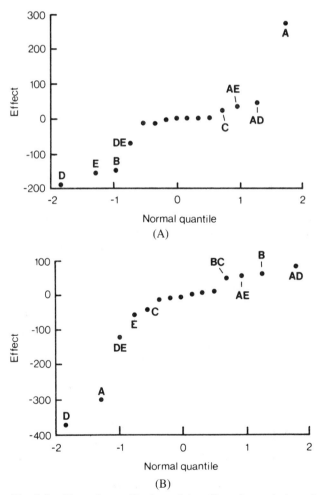

Fig. 8.6.   Normal quantile plots of the effects for total phytoplankton production from the sensitivity analysis of the San Joaquin model. A. 20 percent design. B. 1 standard deviation design.

These relationships can be explored further using *interaction diagrams*. The four means of the output variables, one for each combination of the two factors involved in the interaction are plotted versus the first factors levels. Lines are then drawn between the means with the same level of the second factor. Letting N denote the nominal level of a factor and P denote the perturbed value, the four combinations are NN, NP, PN, and PP. For example, the means used for the AD interaction diagram are:

$$m_{NN} = \frac{((1) + ce + be + bc)}{4}$$

$$m_{NP} = \frac{(de + cd + bd + bcde)}{4}$$

$$m_{PN} = \frac{(ae + ac + ab + abce)}{4}$$

$$m_{NP} = \frac{(ad + acde + abde + abcd)}{4}$$

On the diagram, m_{NN} and m_{NP} are plotted above the N on the x-axis and m_{PN} and m_{PP} are plotted above the P. Lines are drawn between m_{NN} and m_{PN} and between m_{NP} and m_{PP}.

Figure 8.7 shows total phytoplankton production AD and DE interaction diagrams for both sensitivity analyses. Because the lines are not parallel in Fig. 8.7 the interactions are significant. The AD interaction, for example, is the difference between the values of high and low levels of D at the high level of A minus the same difference at the low level of A. This interaction is positive for AD in both analyses (the difference is more negative at the low than at the high end of A). The interaction diagram also shows that the D effect is consistent between the 20 percent and one standard deviation analysis in that the D effect is larger at the lower value of A than at the higher value (the distance between the endpoints of the two lines is larger at the nominal A end in both diagrams). The results are inconsistent for A in that the effect of A is greatest at the high level of D in the 20 percent case and at the low D level in the one standard deviation case (the P–P line in the 20 percent diagram has the larger slope whereas the N–N line has the greater slope in the one standard deviation diagram). This explains the unusual orientation of A, D, and AD effects in the quantile plots.

For the DE interaction the one standard deviation design illustrates a crossed-over interaction diagram in which the effect of increasing E (the prey density half saturation for grazing rate) can either raise or lower phytoplankton productivity depending on whether A is at its high or low level. This crossover does not occur in the 20 percent design. Despite this crossover the D, E, and DE effects appear in similar locations on the quantile plots with the D, E, and DE effects all being negative in both phytoplankton productivity quantile plots.

In comparing the 20 percent and standard deviation analyses bear in mind

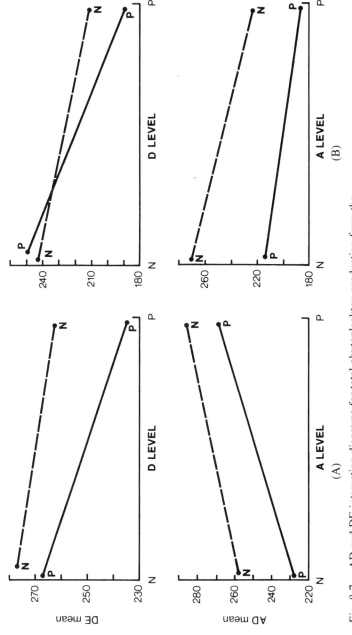

Fig. 8.7. AD and DE interaction diagrams for total phytoplankton production from the San Joaquin model sensitivity analysis. A. 20 percent design. B. 1 standard deviation design.

that the plus or minus one standard deviation analysis considers how sensitive the model is to parameter changes over their expected range of values, whereas the 20 percent analysis gives a more model-oriented result irrespective of the relative variability of parameter estimates.

8.6 Individual process evaluation

Laboratory controlled microcosms offer an unusual opportunity for evaluating models built for more complex ecosystems and for uncovering processes that require further examination. Most "nuts and bolts" work in ecological modeling is at the process level because an understanding of most biota is not possible within the ecosystem framework. Models can be used to aid in the study of individual processes by bringing test animals from the field and both modeling and observing them under controlled conditions. An excellent example is provided by Bailey and Batty (1983), who studied predation by the medusa *Aurelia aurita* on the larvae of the herring *Clupea harengus*. In their experiments single medusae were left for an hour in 5-L glass jars each with from 5 to 30 herring larvae/L and the number of larvae eaten or killed (apparently by the medusa) were recorded. Four replicates were made of each experiment. The medusae were observed through closed circuit TV to record their swimming speed (umbrella beats per 30 seconds times average distance swum per beat).

The feeding process was modeled as consisting of the separate processes of search, encounter, and capture. Search was assumed to be at random at a swimming speed that increased linearly with medusa size:

$$gswim = 4.12 + 0.49 \times xradenctr$$

$xradenctr$ = the encouter radius for the medusa
$gswim$ = swim speed (mm/sec)

The predation rate *gpred* is the product of the encounter rate *genctr* and the capture efficiency *gcapeff*. The encounter rate was taken from the literature for the case where the prey are mobile (herring are assumed to swim at an average swim speed of 3.68 mm/sec). This equation is:

$$genctr = \pi \times xradenctr \times xprey \times \frac{gswim^2 + 3(3.68)^2}{3(3.68)}$$

$genctr$ = encounter rate between medusae and herring larvae
$xprey$ = herring larval density

Here the encounter radius is the sum of the umbrella radius of the medusa and of the herring larva treated as a sphere of half its length, *xlen* projected onto a plane normal to the path of the larva relative to the medusa. This radius can be shown by integration to equal $\dfrac{2 \times xlen}{\pi^2}$. Capture efficiency was observed to increase with medusa size according to the following formula:

$$gcapeff = -0.24 + 0.16 \times xradenctr$$

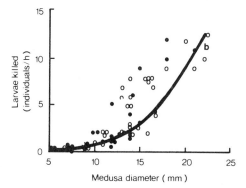

Fig. 8.8. Predation rates by the medusa *Aurelia aurita* on herring larvae as a function of the medusa size. The solid line is predicted by the predation model. Closed circles show data from laboratory experiments conducted in the dark and open circles are from experiments in the light. From Bailey and Batty (1983). Used by permission of Springer-Verlag and the author.

These three equations are sufficient to predict the feeding response of medusae of different sizes to various larval densities. Figure 8.8 shows the results of a series of experiments with different sized medusae feeding on larvae at an initial concentration of 15 larvae/L. Numbers of larvae killed are somewhat under-estimated by the model for large medusae. Figure 8.9 shows results of experiments where 12- to 14-mm medusae were fed on different densities of larvae. The tendency again is to slightly underestimate larval mortality, although results are within the range of data values. Bailey and Batty attributed the discrepancy to the possibility of feeding satiation, especially for the larger medusae. Swimming speeds of medusae declined as the numbers of larvae captured increased. For example, medusae average swim speed was over 9 mm/sec with three or fewer larvae attached, but was only 6 mm/sec for medusae with six larvae attached. This difference was enough to account for the discrepancy between the model and experimental results.

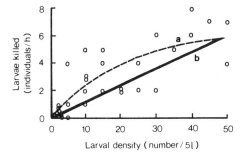

Fig. 8.9. Predation rates by the medusa *Aurelia aurita* on herring larvae related to larval density. Results of laboratory experiments run for 1 hour with 12 to 14 mm medusae are shown with open circles. Model results from the predation model are shown with a solid line. The dotted line is the Ivlev curve that gives the best fit to the data. From Bailey and Batty (1983). Used by permission of Springer-Verlag and the author.

Here is an example of where model–data discrepancies led to further investigation that uncovered some previously overlooked fact about a process. In this case the information could be incorporated into the model to improve model predictions. Notice again that the model led to a certain level of agreement with the data. This gave confidence that the model was indeed addressing the major processes involved with medusa feeding. Having observed a tendency of the model to underpredict for larger medusae led the investigator to further examine the feeding process with an eye to what might be causing the discrepancy. Satiation and reduced swim speeds after satiation was a hypothesis that indeed was observed and was able to explain the observed discrepancy. This illustrates a fundamental principal of modeling: that a model itself can only go as far as to suggest areas for further research through its uncovering of discrepancies and of tendencies within the discrepancy. Further progress requires additional experimentation or reexamination of existing data.

8.7   Model prediction

The question of prediction has plagued ecological modeling because the systems modeled are generally so large, the sampling intervals so long, and the systems so spatially patchy that many sets of hypotheses may give equally plausible fits to field data. What is needed is a data set that considers a system of enough complexity that the results could be said to be indicative of a natural ecosystem, but yet controllable enough and sampled frequently enough that the behavior of the system can be monitored. Such data sets are few and far between.

8.7.1   Microcosm model

We consider here a model of one such data set from experimental ecosystems in jars called *aquatic microcosms*. The data consist of measurements on the quantities of various biota, amounts of nutrients, and other measures of ecosystem function such as *pH, dissolved oxygen,* and chlorophyll *a* at frequent intervals (semiweekly). Each microcosm was set up with 10 species of phytoplankton grown in laboratory culture and inoculated at equal densities into a 10-L jar with a standard balanced nutrient medium and identical sediment. The jars were placed in a room at constant temperature under light banks set on a 12-hour on–off cycle. The experiments was run for 63 days. On day 4 zooplankton were added in identical amounts to each jar. These included the cladoceran *Daphnia magna*, rotifers, ostracods (small mussel-like zooplankton that are free swimming), amphipods (freshwater shrimp), and protozoans. The treatments consist of toxicants that were added on day 7 of the experiment (there were six jars for each treatment). Nontoxicant-treated microcosms were included as controls for the toxicant-dosed microcosms. The experiment (Taub and Crow, 1980) was repeated with different toxicants including streptomycin, an algicide, malathion, that kills zooplankton, and copper, a biocide that induces mortality on all biota in the microcosm.

These experiments were originally designed to investigate the practicality of using multispecies microcosms as bioassays to examine the effect of various chemicals on ecosystems. At present, bioassays to test the safety of prospective new chemicals are standardly conducted on single "indicator" species. The advantage of using multispecies microcosms for this purpose is that they include trophic interactions and are, therefore, more realistic than single species bioassays. Another powerful argument for using multispecies bioassays is that, like single species bioassays, the results are repeatable. Figure 8.10 illustrates this fact for an experiment where both streptomycin and copper were added. This figure shows the six replicates for the control, streptomycin- and copper-treated experiments for *Daphnia magna,* and total algal biovolume. The striking similarity of each of the six control runs to each other and of each of the six treatment runs to each other suggests that these data would serve as a good test of a simulation model. If the model fits the data well for the control runs and the treatment runs it can be looked at as predictive.

There are, of course, subjective aspects to model corroboration here. For example, model calibration can lead to several different combinations of parameters fitting the data equally well, the choice of parameters depending on who does the calibration. The interpretation of the goodness of fit of the model to the data depends on who is doing the comparison. Also some model parameters were calibrated to the same data used to evaluate the model. These subjective aspects of model evaluation and model building will, as far as we can see, be with us for a long time (perhaps forever). By developing a set of process alternative hypotheses and choosing the best to build a model, this study could be seen as a test case for examining whether a predictive model can be built for a system at the level of complexity of an ecosystem. The modelers (Swartzman and Rose, 1984; Rose, 1985) built a model (MICMOD) and tested its predictive ability using these microcosm experiments.

The process equations chosen for the model were taken from the aquatic ecosystem modeling literature. The method used to calibrate the model and to choose between alternative process equations was to calibrate a series of experiments of increasing complexity. This reduced the criticism that the model was calibrated to the same data that it was validated against. These experiments included single species phytoplankton growth experiments and paired species growth experiments all conducted with high initial nutrient levels (the same as in the "full" microcosms) and low initial levels ($\frac{1}{10}$ the high nutrient levels).

A compartment diagram of the model is shown in Fig. 8.11. The 10 phytoplankton species have been categorized into functional groups including small green algae, colonial greens, large greens, filamentous greens, blue-green algae, and diatoms. These represent the major algal groups in lake ecosystems. *Daphnia* were categorized into three-size classes, with only the largest size class being reproductive. Initial parameter values were taken from a literature review of process experiments made with the microcosm algae, where these experiments existed. Sometimes parameters were measured by the investigators. The protocol followed in calibrating the model and in choosing among alternative process equations was to eliminate process alternatives when they could not be calibrated

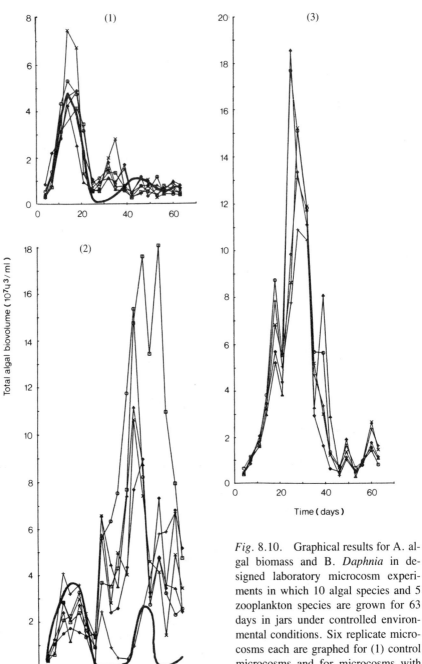

Fig. 8.10. Graphical results for A. algal biomass and B. *Daphnia* in designed laboratory microcosm experiments in which 10 algal species and 5 zooplankton species are grown for 63 days in jars under controlled environmental conditions. Six replicate microcosms each are graphed for (1) control microcosms and for microcosms with additions of (2) streptomycin and (3) copper. Model output from the MIC-MOD model is shown for the control and streptomycin-treated microcosms

(A)

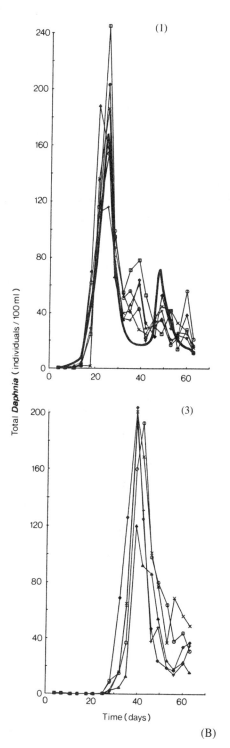

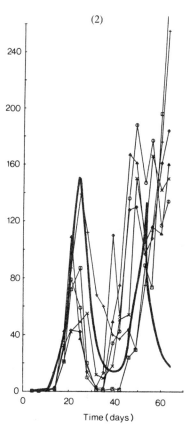

Fig. 8.10. (continued) as bolder time traces. The model time traces for the control experiment had only zooplankton parameters changed during calibration and the streptomycin-treated microcosm run had only streptomycin effect parameters changed during calibration.

(B)

239

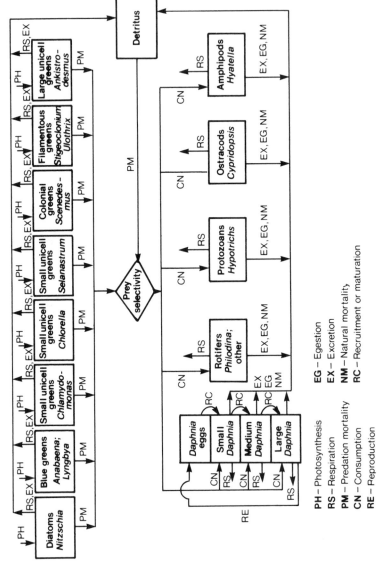

PH – Photosynthesis   EG – Egestion
RS – Respiration   EX – Excretion
PM – Predation mortality   NM – Natural mortality
CN – Consumption   RC – Recruitment or maturation
RE – Reproduction

Fig. 8.11. Compartment diagram showing biomass flows through phytoplankton and zooplankton in the simulation model MICMOD representing laboratory microcosm experiments. From Swartzman, G. L., and K. A. Rose (1984). Simulating the biological effect. In *Ecological Modeling*, Elsevier. Used by permission of Elsevier Science Publishers BV.

to fit the data or where a simpler alternative seemed to perform equally well over all the single species and paired species experiments. An example is the effect of nutrients on growth.

You may recall that the San Joaquin model (see Chap. 5) represented phytoplankton photosynthesis as a function of the external nutrient concentration. A number of alternatives have photosynthesis depend on internal nutrient concentrations. For the microcosm experiments, it was not possible to replicate plankton dynamics with the external nutrient control of photosynthesis, largely because the plankton continued to grow long after the external nutrient levels were drastically reduced. The plankton were apparently growing using internal nutrient stores accumulated in excess of present needs during times of high nutrient concentrations. Thus, the equation having external nutrient control over photosynthesis was eliminated as a possibility.

After calibration of the model to the simple experiments the phytoplankton model was evaluated by comparing predictions with results of an experiment having all phytoplankton species grown together without grazing. The zooplankton parameters were then calibrated to the full control microcosm experiment without changing any of the phytoplankton parameter values. Finally, the toxicant-treated microcosm experiments were used for model corroboration.

The results of this experiment in model prediction are equivocal. The model fits the control experiment well. Figure 8.10A(1) shows total algae and *Daphnia* densities for both the model and the six control experiments. It is apparent from this figure that the model tracked the general pattern of change of these biota over time as well as producing results that are quantitatively close to the data. Although they are not shown here, the model also provided similar fits to all the phytoplankton species except blue-green algae, to the nutrients and to all the zooplankton species except the protozoa, where an early peak and drop in the model lagged several days behind that shown by the data.

Figure 8.10A(2) shows total algae and *Daphnia* densities for the streptomycin-treated microcosm. Here results were less promising, although good fits were obtained for *Daphnia*, the dominant zooplankton, and for most of the phytoplankton species. The main problem areas were the phytoplankton *Scenedesmus*, which had a large bloom later in the experiment that was not replicated by the model, and with ammonia, which was produced in large quantities between days 10 and 30 in the experiment but not in the model.

To model the direct effect of streptomycin an assumption had to be made about how toxicant concentration affected phytoplankton growth. The simplest possible assumption was tried, namely that growth is reduced as a linear function of streptomycin concentration. This ignores all consideration of physiological mechanisms, absorbency of streptomycin by algae, and the fate of the streptomycin. Also, no lagged effects relating to exposure time were included. The only exception was for blue-green algae, which are known to be killed outright by streptomycin. Blue-green algae mortality rate (rather than growth rate) was proportional to streptomycin concentration. The parameters for the sensitivity of each of the phytoplankton species to the toxicant were obtained from single species bioassay experiments. It is known, however, that the effect of the toxicant depends on the plankton density. Therefore, the sensitivity of the plankton will be different in the full microcosm than in the single species bioassays. This was included by varying the sensitivity parameters through calibration while keeping the same ranking and relative sensitivity of the phytoplankton species to streptomycin concentration.

Considering the simplicity of the toxicity assumptions, the fit of the model to the experimental data was good. The areas where the model did not fit the data do not appear to result from having the wrong parameter values but rather from some process operating in the system that is not included in the model. There is no existing process in this model that could produce such a large amount of ammonium. Also the *Scenedesmus* bloom corresponds to a time of high *Daphnia* densities. What is preventing the *Daphnia* from feeding on these *Scenedesmus*? Are there some explanations for these phenomena?

Our strengths are our weaknesses. Because of the reliable data set, if the model does not fit the data in a particular place, then the problem must be with the model. In the final analysis, the fit of the model to the data ranged from excellent to poor, with most biota having excellent fits. The discrepancies raised some questions that can be addressed through further modeling as well as through experiments designed to examine those phenomena leading to the discrepancies. In this sense, the areas where the model does not predict the experimental results are more interesting than those that fit because they point to phenomena that are not well understood. Although many modeling efforts appear to do this they are not supported by such a reliable data set.

8.8   Designed model experiments

Often the power of a simulation model is in its ability to examine various process hypotheses in a whole system framework, rather than its ability to predict the behavior of ecosystems. Alternative process evaluation can be done with simulation experiments that compare model behavior under alternative process equations. An excellent example is provided by Dale and Swartzman (1984), who use a simulation model to examine the changes expected in an aquatic ecosystem when it is subject to thermal loading (increase in water temperature) from power plant cooling systems. Large power plants use a large volume of water to cool the generators, and can raise the water temperature by up to 10°C for areas of around 1 km² adjacent to the plant. Dale and Swartzman used a simulator to examine what plankton responses would be favored under elevated temperatures by producing higher average biomass throughout the year. The responses to temperature considered were the effects of temperature on phytoplankton growth, zooplankton grazing, and zooplankton respiration rates.

Before giving the details of the temperature response experiments, let us give a brief introduction to the model. The model included phytoplankton, herbivorous and carnivorous zooplankton, and detritus and considered nitrate, ammonia, organic nitrogen, phosphate, and organic phosphorous as nutrients. Phytoplankton growth was equal to the product of light, temperature, and nutrient effects minus grazing and respiration. In this sense the model is like the San Joaquin model. It is deterministic, has a daily time step, and a run time of a year. Surface light and temperature were read in as driving variables from data obtained from a monitoring station on Lake Ontario. As such, the model represents this lake and specifically the region around the Nine Mile Point power plant near Oswego, New York. Temperature was increased by 8.3°C to represent thermal loadings from typical northern temperate power plants. Despite the simplification of lumping all phytoplankton into a single group the effect of having the phytoplankton community dominated by different functional groups with different characteristics (model parameters or process equations) could be examined with this model by changing phytoplankton parameters to represent different plankton groups.

For phytoplankton growth, three alternative temperature hypotheses were considered. These are compared in Fig. 8.12. The linear curve was used by Di Toro et al. (1971) in the San Joaquin model. The skewed normal distribution, so called because the equation is similar to a normal distribution but has different parameters on either side of the temperature that is optimum for growth, was taken from Lehman et al. (1975) and represents the known phenomenon that at high temperatures photosynthesis declines. The third equation, due to Eppley (1972), is an exponential or Q_{10} equation. Eppley admitted that individual species have a temperature photosynthesis response of the skewed normal type. As temperatures change over a season, however, the phytoplankton community is dominated by species with different temperature optima. Thus, if the combined species temperature response is considered as an overlay of the individual species response, something more like the Q_{10} than the skewed normal will arise. Because dominance by a smaller subset of phytoplankton than usual (when tem-

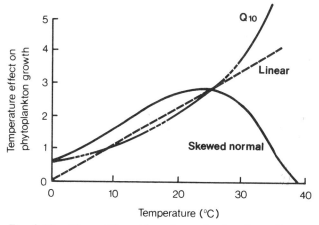

Fig. 8.12.   Three different hypotheses for the effect of temperature on phytoplankton growth. All curves have been parameterized to have the same value at 25°C. From Dale and Swartzman (1984). Reprinted by permission of Academic Press.

peratures are raised) is being considered, however, the skewed normal may be more applicable than the Q_{10}.

Although zooplankton grazing rate has been linked to temperature, not much is known about the exact nature of the relationship. To encompass the range of relationships in the modeling literature a linear equation with different slopes was used to represent low, medium, and high responsiveness of grazing rate to temperature. For the temperature effect on zooplankton respiration three alternatives were considered. These are shown in Fig. 8.13. They are a low linear effect from the San Joaquin model of Di Toro et al. (1971), a medium effect represented by a gamma-like function taken from MacCormick et al. (1972), and a Q_{10} effect representing high temperature control taken from Kremer and Nixon (1978).

A simulation comparison of these alternative equations was made by picking all equation combinations from the simulation "library." Runs were made under both average and thermally loaded temperature conditions. For each run a table was constructed giving the total production, average standing crop, maximum and minimum values, and the time of year of maxima and minima for each biotic functional group and each nutrient species. Table 8.10 is an example of such a table, with medium temperature effects on zooplankton grazing and respiration rates and a high effect of temperature on phytoplankton growth.

As expected, there were significant differences between the runs with different equation combinations. Some combinations led to results that were so far removed from the data on Lake Ontario for each of the biota (the model was compared with the average of 5 years of data plus and minus one standard

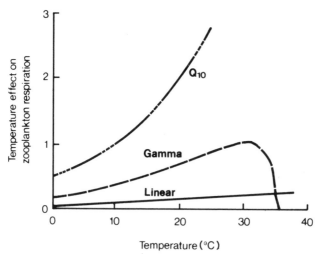

Fig. 8.13. Three different hypotheses for the effect of temperature on zooplankton respiration. The linear relationship was used by Di Toro et al. (1971), the Q_{10} relationship by Kremer and Nixon (1978), and the dashed line is a modified gamma function used by MacCormick et al. (1972). The parameters in these curves are those used in the original models. From Dale and Swartzman (1984). Reprinted by permission of Academic Press.

Table 8.10   Example table from a simulation run of the model of Dale and Swartzman (1984) under thermally loaded conditions.

	Total production	Average standing crop	Maximum value	Day of maximum	Maximum value	Day of minimum
Phytoplankton	102.972	0.565	0.682	136	0.074	79
Herbivorous zooplankton	12.962	0.071	0.253	203	0.002	265
Carnivorous zooplankton	14.066	0.077	0.199	230	0.004	154
Organic nitrogen		0.291	0.247	190	0.090	1
Ammonia		0.045	0.044	365	0.011	187
Nitrate		0.219	0.249	1	0.001	189
Organic phosphorus		0.011	0.009	163	0.004	232
Phosphate		0.015	0.016	89	0.000	141

deviation where each year was considered a replicate) that it was considered impossible for biota having such a combination of characteristics to exist on Lake Ontario. These runs were removed from future consideration as realistic alternatives.

Cluster analysis, a statistical technique for organizing objects into groups according to the similarity of their attributes (see Gordon, 1981 for a description of the method), was applied to the attributes in Table 8.10 for those combinations of temperature effects that were not eliminated by the criterion of whether the behavior was realistic under normal temperature conditions. The results of the cluster analysis is shown as a dendrogram in Fig. 8.14 where combinations that are close together in the "tree" have similar attributes. This diagram compares runs under both normal and thermally loaded conditions. The runs made differ somewhat from the options shown in Fig. 8.12 for the effect of temperature on phytoplankton growth. The linear function was not used. Instead Q_{10} curves with high, medium, and low Q_{10} values were used along with skewed normal curves with parameter values producing a medium and low option. These latter two cases are subscripted "sm" and "sl," respectively, in Fig. 8.14. Notice that of the 90 runs made originally (45 combinations apiece for both normal and thermally loaded conditions) only 18 remain in the subset of feasible combinations.

The best way to summarize results like those in Fig. 8.14 is to return to the runs, now categorized into classes of similar behavior, and to give a biological interpretation to the differences. This was done by Dale and Swartzman, and we paraphrase their discussion here. Simulation runs without thermal loading are characterized by a rapid phytoplankton increase (bloom) during April followed by a rapid decline, with a second smaller phytoplankton peak following in mid-September. The slowdown of growth is caused by phosphorus limitation, but the later decline is caused by grazing. Predation by carnivorous zooplankton

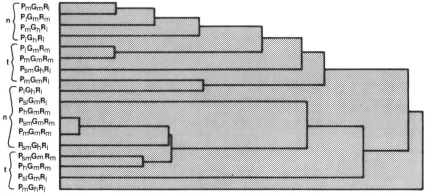

Fig. 8.14. Cluster analysis of simulation runs of the Dale–Swartzman model with runs made with different combinations of hypotheses about the effect of temperature on various processes and with both normal and raised ($+8.3°C$) temperature conditions. Subscripts l, m, sm, sl and h are used to distinguish alternative forms of processes. P denotes phytoplankton growth hypotheses (Fig. 8.12), R denotes zooplankton respiration hypotheses (Fig. 8.13), and G denotes grazing hypotheses. t and n are used to denote runs done with elevated and normal temperatures. From Dale and Swartzman (1984). Reprinted by permission of Academic Press.

in mid-summer reduces herbivorous zooplankton, allowing a second phytoplankton bloom in August, which is enhanced by late summer nutrient turnover in the lake. Early winter reduction in plankton levels is primarily caused by falling temperatures in the lake.

The marine biologist Eppley (1972) has pointed out that temperature is rarely a limiting factor for growth but rather sets the upper limit to maximum growth. This is borne out by the thermally loaded model runs. For those process alternatives judged realistic (by the data comparison test under normal temperature conditions), the thermally loaded runs produced an increase in average phytoplankton production. Because temperatures are elevated, the spring bloom occurs earlier in the year. Higher grazing rates result in zooplankton "tracking" the phytoplankton bloom more closely than in the cooler temperature cases. Sometimes this tracking is so rapid that small oscillations appear in the plankton. This implies that both grazing and growth are rapid, with the growth rates of the phytoplankton at the elevated temperatures being enough to assure stable coexistence of them and their predators despite the oscillations.

Rather than being phosphorous limited, as in the nonthermally loaded case, the first bloom is directly limited by the grazers. The second peak, which now occurs earlier in the year, is limited directly by the nitrogen availability, whereas the later bloom in the unloaded case has enough nutrients because of lake turnover. The decrease in the thermally loaded second bloom is caused by decreased nutrients, reduced light levels because of shortened days and self-shading by the plankton itself. This is not true in the unloaded case because the bloom is not as large and self-shading is, therefore, not as important.

From the above observations, a shift in species composition would occur toward phytoplankton that can use the conditions prevalent in the thermally loaded case. This suggests dominance by phytoplankton species that are less limited by nitrogen, but more limited by phosphorous, and also a shift toward those species able to grow well under lower light conditions. These characteristics are typical of blue-green algae and these are indeed the types of algae that dominate in thermally loaded waters (Patrick, 1969).

To examine the hypothesis that algae better adapted to low light conditions would outproduce those dominating the lake under normal conditions, the parameters controlling the effect of light on photosynthesis were changed toward having a lower light optimum but higher growth at lower light levels. Because the model used Steele's equation for the effect of light on photosynthesis this was done by reducing $kphtlt_{op}$ and increasing the exponent $kextlt$ as shown in Fig. 8.15. With these changes the magnitude of the second phytoplankton bloom was increased, which supports the hypothesis.

Some of the differences between normal and thermally loaded model runs can also be linked to the thermal effects alternatives described above. For example, phytoplankton production is lower in the thermally loaded than the unloaded case when a Q_{10} relationship is used for grazing instead of a linear relationship because the Q_{10} relationship has higher grazing rates at high temperatures and lower grazing rates at average temperatures than the linear grazing equation. Thus, grazing rate is increased relative to the average temperature case. Similarly, in all cases having skewed normal phytoplankton growth, the phytoplankton bloom occurs earlier than in the other cases in a thermally loaded run because the skewed normal distribution gives higher growth rates at intermediate temperatures.

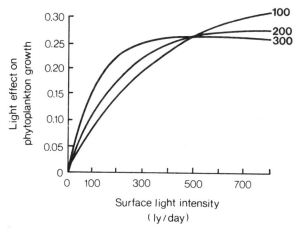

Fig. 8.15.   Comparison of Steele's equation for the effect of light on photosynthesis with optimal light intensities $kphtlt_{op}$ of 100, 200, and 350 ly/day. Maximum photosynthesis rate $kpht_{mx}$ values for these curves were 0.266, 0.83, and 1.0. These were chosen for each curve such that they all had the same value at a surface light intensity of 500 ly/day. From Dale and Swartzman (1984). Reprinted by permission of Academic Press.

The above example shows an important aspect of model evaluation using the model to explore behavior under perturbed conditions. By examining model performance with alternative process equations under both normal and thermally loaded conditions the authors were better able to distinguish between the alternatives as well as shed some light on the response of an ecosystem to perturbation. A better test of model performance was also possible because a wider range of environmental conditions than usual were explored.

8.9 Summary

This chapter has considered the question we must continue to ask about simulation models once we have built them: How good are they? A number of methods were considered including graphical, tabular, statistical, experimental, and simulation approaches.

As mentioned, the evaluation of simulation models is subjective. Model evaluation procedures increase or reduce our confidence that the model has value. This topic borders on the philosophical and many of the papers written about model evaluation are philosophical in tone. We avoid most of the philosophical rhetoric here, although we consider it interesting and give references to it in the annotated bibliography.

Much of the chapter discusses sensitivity analysis methods. These methods represent a hybrid between statistics and simulation, between model corroboration and model experimentation. They can tell us how sensitive various measures of model output are to changes in parameter values both singly and in combinations. How to interpret the output of a sensitivity analysis is a complicated issue. Is a model good if model output is insensitive to parameter change or should we welcome finding out that model output is sensitive to a particular set of parameters as information that further research is needed to estimate these parameters? As the ugly duckling found out "It depends on how you look at things."

The tools used for model evaluation are many. The approach to evaluation often depends on the problem at hand and on the type of information available. In most applications, field data are often highly variable and leave questions when discrepancies arise between model and data about which one is in error. We have taken a case of a model for a microcosm system, where the quality of the data is high, and have graphically compared model and data. There we found that replicating the details of species composition changes over time in a multispecies system is an unrealistic objective, although the overall timing and magnitude of changes in total biota can fit quite well.

Individual processes, such as the feeding process discussed in this chapter, can be examined in detail to produce a model that accurately represents that process and can be used for that process in an ecosystem model.

Finally, we discuss conducting perturbation experiments with a model as a means of model evaluation. Placing a model into a wider than normal range of environmental conditions offers a strong test of how well the postulated mechanisms for relevant processes hold up outside of the range of normal conditions.

In our example of a thermally loaded ecosystem, we postulated the parameter set for phytoplankton that would dominate in a thermally loaded ecosystem and compared that with known shifts in species composition in thermally loaded systems as a measure of model corroboration.

8.10   Annotated Bibliography

Box, G. E. P., W. G. Hunter, and J. S. Hunter: *Statistics for Experimenters: An Introduction to Design, Data Analysis and Model Building,* Wiley, New York, 1978.
  A good review of fractional factorial designs, their application and theoretical basis.

Caswell, H.: The validation problem, in *Systems Analysis and Simulation in Ecology, vol. 4,* ed. B. C. Patten, Academic Press, New York, 1976.
  A good discussion of the philosophy of ecological model corroboration (validation).

Cleveland, W. S., and R. McGill: Graphical perception and graphical methods for analyzing scientific data. *Science,* 229:828–833, 1985.
  Gives an excellent discussion of how to and how not to present graphical information for digestion by others. You will be amazed.

Cochran, W. G., and G. M. Cox: *Experimental Designs, 2nd ed.,* Wiley, New York, 1957.
  A classic text on design of experiments. Contains tables for many experiment designs including fractional factorials.

Costanza, R.: Articulation, accuracy and effectiveness of mathematical models: A review of freshwater wetland applications. *Ecol. Model.,* 27:45–68, 1985.
  A review of model corroboration (validation) that is very up to date, with good examples.

Dale, V. H., and G. L. Swartzman: Simulating the effects of increased temperature in a plankton ecosystem: A case study, in *Algae as Ecological Indicators,* ed. L. Elliot Shubert, Academic Press, London, 1984, pp. 395–427.
  A designed simulation experiment similar to a factorial design to explore how increased temperature might affect the species composition of algae in an aquatic ecosystem.

Gardner, R. H., D. D. Huff, R. V. O'Neill, J. B. Mankin, J. B. Carney, and J. Jones: Application of error analysis to a marsh hydrology model. *Water Resour. Res.,* 16:659–664, 1980a.
  A Monte Carlo sensitivity analysis method different from that presented in this text.

Gardner, R. H., R. V. O'Neill, J. B. Mankin, and D. Kumar: Comparative error analysis of six predator–prey models. *Ecology,* 61:323–332, 1980b.
  A Monte Carlo sensitivity comparison of six relatively small models, but having functions that are general.

Gardner, R. H., R. V. O'Neill, J. B. Mankin, and J. H. Carney: A comparison of sensitivity analysis and error analysis based on a stream ecosystem model. *Ecol. Model.,* 12:173–190, 1981.
  Compares results from the classical Tomovic sensitivity analysis with the Monte Carlo latin hypercube sampling method reported in this chapter.

Garratt, M.: *Statistical Techniques for Validating Computer Simulation Models,* U.S. IBP Grassland Biome Technical Report 286, Colorado State University, Fort Collins, Colo., 1975, p. 68.
  Reviews problems of statistically comparing models with data traces and suggests methods to deal with some of them.

Hedgpeth, J. W.: Models and muddles some philosophical observations. *Helgol. Wiss. Meeresuntersh,* 30:92–104, 1977.

An astute account of some of the fuzzy thinking and reporting in simulation modeling today.

Hornberger, G. M., and R. C. Spear: An approach to the preliminary analysis of environment systems. *J. Environ. Manage.,* 12:7–18, 1981.

An approach to sensitivity analysis in which the co-variance structure of the parameters is intuited from those parameter value combinations that lead to realistic model output.

Iman, R. L., and W. J. Conover: Small sample sensitivity analysis techniques for computer models, with an application to risk assessment. *Commun. Statist. Theor. Meth.,* A9:1749–1842, 1980.

Gives a discussion of the use of latin hypercube sampling for sensitivity analysis.

Kitchell, J. F., D. J. Stewart, and D. Weininger: Applications of a bioenergetics model to yellow perch (*Perca flavescens*) and walleye (*Stizostedion vitrem vitrem*). *J. Fish. Res. Board Can.,* 34:1922–1935, 1977.

This is a process-oriented fish energetics model that uses sensitivity analysis as a guide to future research.

O'Neill, R. V., R. H. Gardner, and J. B. Mankin: Analysis of parameter error in a nonlinear model. *Ecol. Model.,* 8:297–311, 1980.

Another Monte Carlo sensitivity analysis example.

Park, R. A., and C. D. Collins: Realism and ecosystem models. *Perspect. Comput.,* 2:18–27, 1982.

A report on the application of a single model to many sites.

Reckhow, K. H., and S. C. Chapra: Confirmation of water quality models. *Ecol. Model.,* 20:113–133, 1983a.

A discussion of model corroboration (validation) with good examples. Required further reading for anyone interested in the nitty gritty problems in model evaluation.

Steinhorst, R. K., H. W. Hunt, G. S. Innis, and K. P. Haydock: Sensitivity analyses of the ELM model, in *Grassland Simulation Model,* ed. G. S. Innis, Ecological Studies, 26, Springer-Verlag, New York, 1978, pp. 231–255.

Develops sensitivity analysis using a fractional factorial design. The parameters are grouped together into megaparameters to reduce the number of runs.

Swartzman, G. L.: A comparison of plankton simulation models emphasizing their applicability to impact assessment. *J. Environ. Manage.,* 9:145–163, 1979b.

Compares 14 plankton simulation models in a single simulation framework. Discusses the pros and cons of the various models in their realism and ease of parameter fitting.

Tomovic, R.: *Sensitivity Analysis of Dynamic Systems,* McGraw-Hill, New York, 1963.

A treatment of classical sensitivity analysis that will not work for most ecological simulation models because they are too complex and have too many variables and interactions. It is useful for smaller models and minor perturbations from the parameter estimates.

8.11   Exercises

Sensitivity analysis

8.1.   Derive the main effect of parameter B and the AB interaction for a 2^3 factorial design in a fashion analogous to that done for parameter A and the BC interaction in Eq. 8.1 and 8.2.

8.2. Conduct a fractional factorial design sensitivity analysis of Steele's marine ecosystem model discussed in the Chap. 6 exercises. Vary parameters by both a fixed 20 percent and by one standard deviation. Choose the direction to vary a parameter by a coin toss but keep the same direction for both the 20 percent and the standard deviation runs. The following parameters are to be included.

Steele model sensitivity parameters

Parameter	Nominal value	Estimated standard deviation
$kpht_{mx}$	0.2	0.05
kfd_{mx}	1.6	0.6
$kfdwt$	-0.3	0.1
$kfdpy_{hs}$	175.0	75.0
$kphtnit_{hs}$	96.0	40.0

For output variables use the average annual phytoplankton and zooplankton biomass and the total annual phytoplankton and zooplankton production. Which are the most sensitive parameters and interactions and how relatively sensitive are they? Are there any major differences in sensitivity between the 20 percent change and the one standard deviation change methods?

8.3. Conduct a single parameter at a time sensitivity study of the marine ecosystem model and compare results to each of the other studies. Is any additional information obtained from the other methods? is the relative sensitivity of the parameters changed?

8.4. (continuation) Repeat the sensitivity analysis of the model in Ex. 8.2 but with the direction of change of the parameters reversed. Are your results affected?

Chapter 9

Management Applications

9.1 Introduction

In this chapter we discuss management applications of ecological simulation models starting with simple models such as Leslie matrix models and moving to more complex models. We compromise between exposing you to a large number of applications in a sketchy fashion and having only a few examples treated in great detail. We give a few examples, mostly using models already introduced in earlier chapters, to give a flavor of the methods used and of the problems in applying models to managed ecosystems. This is followed by a more extensive (but less detailed) review to give a broader picture of the management problems amenable to simulation modeling, the types of models used, and how effective these models have been. We begin with Leslie matrix models used to consider the effects of harvest on the population dynamics of the harvested species.

9.2 Models of harvested populations

Blue whales

Many natural animal populations have come under harvest pressure from humans and others have received insults from introduced toxicants that have affected their survival and, sometimes, their fecundity. Let us first consider the case of the blue whale. Blue whales (*Balaenoptera musculus*) have been the target of intensive whaling pressure, which has reduced their population to an endangered level. Usher (1976) gave a survival–fecundity matrix for the blue whale in its current, nonharvested state. The time step for this Leslie matrix model is biennial (2 years). Thus, the first age class contains both 0 and 1 year olds. The last age class consists of whales 12 years and older. This accounts for the 0.8 survivorship in the lower right hand corner of the matrix, because whales are long-lived and, therefore, will accumulate in the oldest age class. This is contrary to

Leslie's original assumption of no survival for the oldest age class. The survival–fecundity matrix is:

$$\mathbf{A} = \begin{bmatrix} 0 & 0 & 0.19 & 0.44 & 0.5 & 0.5 & 0.45 \\ 0.87 & 0 & 0 & 0 & 0 & 0 & 0 \\ 0 & 0.87 & 0 & 0 & 0 & 0 & 0 \\ 0 & 0 & 0.87 & 0 & 0 & 0 & 0 \\ 0 & 0 & 0 & 0.87 & 0 & 0 & 0 \\ 0 & 0 & 0 & 0 & 0.87 & 0 & 0 \\ 0 & 0 & 0 & 0 & 0 & 0.87 & 0.80 \end{bmatrix}$$

This matrix has a dominant eigenvalue $\lambda_1 = 1.099$. The dominant eigenvector, when normalized, gives the stable age distribution for this matrix. Normalizing to 1000 animals in the first age class the stable age distribution is:

$$\mathbf{u}_1^T = [1000 \quad 792 \quad 627 \quad 497 \quad 393 \quad 311 \quad 908]$$

In its unharvested condition the blue whale population would eventually settle into a biennial population growth ratio of 1.099. Because blue whale population is so slow growing we would suspect it to be susceptible to harvest or other sources of mortality that would lower survival below the 0.87 estimated for the unharvested population. Figure 9.1 shows how changes in survival affect the dominant eigenvalue and the stable age distribution for this population. From this figure you can see that a 13 percent reduction in adult survival (from 0.87 to 0.75) would reduce the dominant eigenvalue to 1.0, which is the point at which the population remains stationary. Thus, a biennial harvest of 13 percent from each age class would be enough to prevent recovery of the depleted blue whale population. Notice that harvest has the feature of reducing the contribution of older age classes to the population. Because the population age distribution is normalized to 1000 age 0 and 1 animals, increasing harvest actually increases the proportion of younger individuals in the population. This shift in age distribution toward younger individuals is termed *juvenation* in fisheries, and has been observed in many heavily harvested populations. Juvenation is a sign of stress on the population.

One approach to harvest management is to impose a lower age limit or size limit for harvesting animals. We can examine the effect of such a harvest strategy with blue whales by considering having harvest only on whales 12 years and older. This is done in the model by reducing survivorship for 12-year and older adults from its present value of 0.80. Figure 9.2 shows how the dominant eigenvalue and eigenvector would be affected by reducing older adult survival. Total elimination of individuals over 12 would not cause the population to go into a decline. Figure 9.2 shows that harvesting older individuals has the effect of increasing the relative proportion of younger individuals in the population. This effect is more pronounced on the oldest unharvested individuals. These results are typical of the effect of harvesting on any animal population, differing only in degree and not in direction from results for more productive populations such as fish. Note that the whale population is buffered when only older whales

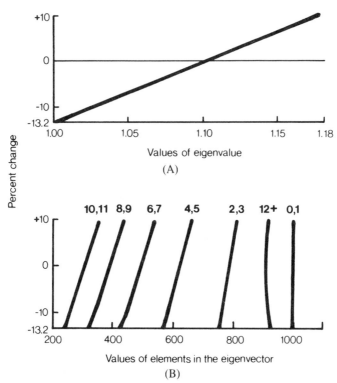

Fig. 9.1.   Diagram showing how small changes in the probability of adult survival (up to about 10%) affects both A. the dominant eigenvalue and B. the stable age distribution (dominant eigenvector) in a Leslie matrix model for the blue whale. The age distributions are normalized so that the youngest age group (0–1) has 1000 individuals. From Usher (1976). Used by permission of Academic Press, London.

are harvested by having a large proportion of the fecundity in the younger age classes. We are not suggesting in this example that older blue whales should be harvested. Blue whales are an endangered species and should be protected. We believe a moratorium on blue whale harvest is necessary to rebuilding the population.

Striped bass

A dangerous risk for any population can result if the harvesting is conducted largely on immature animals. Next we treat an example of this nature where a striped bass (*Morone saxatilis*) population is subjected to increased mortality as juveniles because of power plant cooling systems operating in their spawning region. These power plants entrain larval fish in their cooling system and impose mortality on them. Many simulation models have been developed to predict the impact of entrainment and impingement (juvenile fish being pressed on a screen

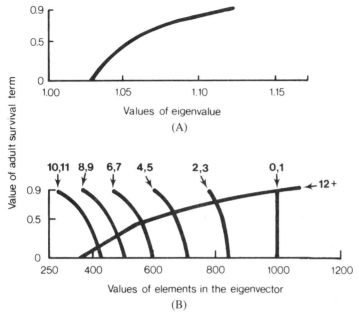

Fig. 9.2. Diagram showing how changes in survival of adult blue whales 12 years and older affect both A. the dominant eigenvalue and B. the stable age distribution (dominant eigenvector) in a Leslie matrix model. From Usher (1976). Used by permission of Academic Press, London.

placed at the inflow to the power plant cooling waters and not being able to escape) on the striped bass population. The population effect of entrainment and impingement was examined using data for striped bass on the Hudson River to estimate survivorship and fecundity without power plant operation. The fecundities (eggs × 10^5) and the survivorship are:

	Striped bass survival and fecundity						
age i	0	1	2	3	4	5	6
f_i	0.0	0.0	0.0	0.0	0.0	0.0	0.0
s_i	2.2e-5	0.4	0.6	0.8	0.8	0.8	0.8
age i	8	9	10	11	12	13	14
f_i	5.9	5.9	7.2	8.0	8.8	8.6	8.4
s_i	0.8	0.8	0.8	0.8	0.8	0.8	0

The 0 for s_{14} implies no survival after the final age class. These parameter values were chosen such that the dominant eigenvalue of the survival–fecundity matrix is 1.0, which means that the population is stationary under average harvest rates before power plant mortality is included.

Entrainment and impingement introduces mortality on young-of-the-year fish which, in a linear model like the Leslie matrix, would reduce survivorship from hatching to age 1. Such induced mortality on the first age class has a stronger

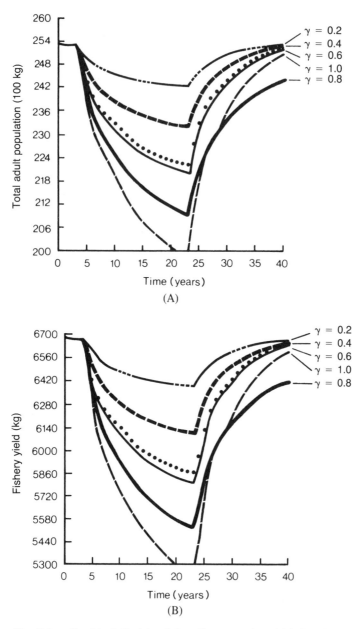

Fig. 9.3.   Graphical display of the effect over time of induced mortality because of entrainment and impingement in power plant cooling systems on A. Total adult population biomass and B. fishery yield. Power plant induced annual mortality rates γ are given in a range from 0.2 (light mortality) to 1.0 (heavy mortality). Dotted lines denote the best estimate of induced mortality rate and, therefore, represent a most likely effect. The underlying scenario is 20 years of power plant operation followed by a 20-year recovery period.

effect on the population than an equivalent level of mortality in any other single age class. There is a complication, however, with the striped bass. It has been proposed (Lawler, 1972) that mortality on striped bass is not constant but instead depends on the bass density. This density-dependent mortality results in more individuals surviving than would be predicted by constant mortality rates.

Most of the models built to study the effects of power plant operation on striped bass in the Hudson River modeled the movement of fish as a hydrodynamic process and had mortality a nonlinear function of fish density. The results from these models were used as inputs to the survival term from ages 0 to 1 in the Leslie matrix from which long-term projections were made. This approach was extremely wasteful of computer time, because it required running the hydrodynamic model each year with different numbers of eggs to get survival to age class 1. Because only numbers of eggs changed from year to year in these simulations a collection of eggs and resulting surviving young-of-the year pairs were generated by running the model at several different initial egg numbers. A two parameter Beverton–Holt (1957) curve having equation

$$xrecrt = \frac{xstock}{\alpha + \beta \times xstock}$$

$xstock$ = the stock level (number of eggs)
$xrecrt$ = the number of recruits (1-year-olds)
α, β = parameters of the Beverton-Holt curve

was fit to these numbers. Under this stock-recruit or stock-progeny relationship the number of recruits replaces survivorship from age 0 to 1 in the Leslie model. The model is no longer linear because it includes the nonlinear effect of density-dependent mortality in the first year of life. Figures 9.3A and 9.3B show the effect of entrainment mortality on the average adult stock level and the average annual harvest by the fishery when different entrainment mortality rates are imposed. This 40-year scenario has the power plant operating for 20 years, the average lifetime of a nuclear power plant, and then shut off for the remaining 20 years. Different stock-recruit curves were used, one for when the power plant was off and several, depending on the rate of mortality, γ, when the power plant was operational. These curves reflected the effect of entrainment on young-of-the-year survival. The parameters for the curve with the power plant off were $\alpha = 1.88 \times 10^3$ and $\beta = 2.0 \times 10^{-8}$, whereas with the best estimate for the entrainment rate γ (γ close to $= 0.6 \ yr^{-1}$) they were $\alpha = 2.08 \times 10^3$ and $\beta = 2.2 \times 10^{-6}$. As Fig. 9.3 shows, reductions in both annual fishery yield and adult stock can result from power plant operation, even when compensation (density-dependent mortality) is present. The level of density-dependence chosen corresponds to best estimates from the hydrodynamic models. Recovery following power plant retirement is rapid. The reductions observed in this model are less than they would be from the Leslie matrix model with the same entrainment rates because no compensation occurs with Leslie models.

9.3   Eutrophication models

A most common application of simulation models is for exploring the effects of large nutrient additions to aquatic ecosystems, *eutrophication,* on primary production and plankton dynamics. Simulation models have been used here since the mid-1960s and there is now a vast modeling literature on the effects of eutrophication on aquatic biota and the effectiveness of various lake restoration projects. A prototype for eutrophication models is the San Joaquin model considered in Chap. 5. That same model, or minor variants thereof, was applied to a number of other sites including the Potomac Estuary, Lake Erie, and Lake Ontario. Other applications of similar models were made for Shagawa Lake, Minnesota and Saginaw Bay, Lake Huron.

Let us first consider the San Joaquin estuary system in some detail. The major management concern in this system was the effect of diversion of water from the San Joaquin river for irrigation further south in California. This proposed diversion plan would result in reduced river flow as well as the possibility of increased light transmission owing to a reduction in suspended sediment concentration. In addition, it was expected that municipal, industrial, and agricultural growth in the region would increase the nutrient input. The nutrient input for 1980 was estimated to be 159,000 pounds of nitrogen per day (compared with 12,500 pounds per day in 1966). A simulation run of the model with the increased 1980 nutrient conditions for Antioch (a region further down river from Mossdale; the segment used for the driving variable data in Chap. 6) showed an appreciably increased phytoplankton bloom, with a maximum chlorophyll *a* concentration of 90 µg/L versus 40 µg/L in 1966. This simulation run (Fig. 9.4) used flow conditions from 1966. At Suisun Bay, yet further downriver, the

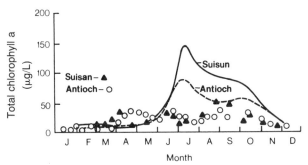

Fig. 9.4.   Simulation experiment with the San Joaquin model exploring the effects of nutrient enrichment (eutrophication). Results are shown for Antioch and Suisun Bay. River flow conditions for 1966 are used in the simulation along with nutrient loading conditions for 1980, a time when nutrient loadings on the estuary had increased significantly. Data values for 1966 are given for Antioch and Suisun Bay. From O'Connor, D. J., D. M. DiToro, and R. V. Thomann, "Phytoplankton Models and Eutrophication Problems." *In* Russell, C. S., ed., *Ecological Modeling in a Resource Management Framework,* RFF Working Paper QE-1 (Washington, D.C., Resources for the Future, Inc., 1975) pp. 174, 176, 179. Reprinted by permission of Resources for the Future, Inc., copyright holder.

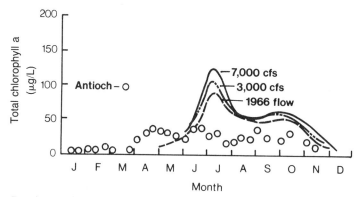

Fig. 9.5.   Simulation experiment with the San Joaquin model exploring the combined effect of nutrient enrichment and lowered river flow, because of water diversion for agricultural purposes. Results are shown for the Antioch river segment. Data values for 1966 for Antioch are given. From O'Connor et al., 1975. Reprinted by permission of Resources for the Future, Inc., copyright holder.

effect of nutrient enrichment on plankton dynamics was projected by the model to be even more drastic.

The possible effect of decreased flow was explored by reducing the flow by both 3000 and 7000 cfs (cubic feet per second) except during the slack summer flow period when flow was not allowed below 2000 cfs, a level that would presumably be maintained under any water diversion plan. These runs, all with 1980 projected nutrients (Fig. 9.5), resulted in significant increases in total chlorophyll over the case with increased nutrient but no water diversion (Fig. 9.4). Increased water clarity (the other likely effect of water diversion) was simulated by reducing the light extinction coefficient due to water by 30 percent below the 1966 level. This resulted in an earlier phytoplankton bloom that persisted much longer than the 1966 case.

Various control measures for the removal of nitrogen from wastewater discharges were considered. An 80 percent reduction was considered to be a technologically and economically feasible level. Figure 9.6 shows the results of this

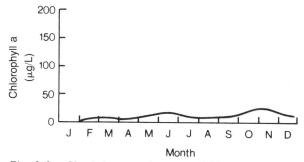

Fig. 9.6.   Simulation experiment exploring the effect of a nutrient waste treatment program to remove 80 percent of the effluent. Results use 1966 flow conditions. From O'Connor et al., 1975. Reprinted by permission of Resources for the Future, Inc., copyright holder.

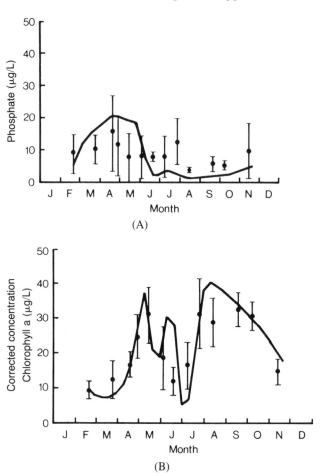

Fig. 9.7.   Graphical comparison of A. phosphate and B. phytoplankton chlorophyll *a* for a dynamic plankton model of Saginaw Bay, Lake Huron. Model time traces are compared with data shown with plus and minus half a standard deviation error bars. Data values have been corrected to represent the same depth considered by the model. Data means are shown by closed circles. From Bierman et al. (1974).

control with 1966 flows. Since the level of phytoplankton is less than that observed in 1966 the modelers reasoned that some diversion would be possible along with control measures. It is important to note, and this is emphasized in all the management modeling literature, that the results of these scenario runs, although they are quantitative, are not accurate and must be looked at as only advisory.

Another major eutrophication problem is the outbreak of blooms of noxious smelling, blue-green algae in the U.S. Great Lakes. Many models have been developed to investigate the sources of these outbreaks. We focus on a model by Bierman (1976) for Saginaw Bay, Lake Huron. The model is similar in

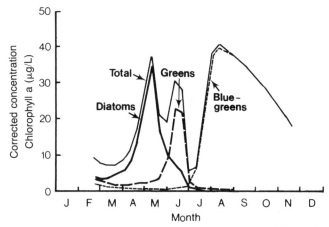

Fig. 9.8. Graphical output from the Saginaw Bay model showing seasonal succession of phytoplankton groups. Total algae is shown as the envelope of time traces for each of the functional groups. From Bierman (1976).

structure to the San Joaquin model except that it separates phytoplankton into diatoms, greens, and blue-green algae rather than having only a single phytoplankton group. The model processes are also similar to the San Joaquin model, except that the effect of nutrient limitation on photosynthesis depends on intracellular rather than environmental nutrient concentrations. This assumption is probably more realistic.

Literature reviews of the effects of nutrients on photosynthesis and how they have been represented in simulation models are found in Swartzman and Bentley (1979) and Rose (1985). Because diatoms require silicon as an essential nutrient, silicon is also represented as a nutrient in the Bierman model. Whereas in the San Joaquin system nitrogen was the nutrient that, if controlled, resulted in a reduction in phytoplankton growth, in Saginaw Bay the dynamics of phosphorous play a dominant role in the blue-green bloom. This change is partly because some of the blue-green algae are nitrogen fixing and are, therefore, not nitrogen limited and partly because, in the nitrogen rich Great Lake waters, phosphorous tends to be the limiting nutrient. Figure 9.7 compares phosphate and phytoplankton chlorophyll *a* concentration from the Bierman model with field data for 1974. The good fits result from calibration of the model to these data. Figure 9.8 shows the change in phytoplankton dominance over the year. The blue-green bloom occurs in late summer, after the diatom and green blooms.

The Bierman model was used to investigate what processes primarily influence the blue-green bloom. The three processes most influential in controlling the bloom are phosphorous recycling owing to phytoplankton death and zooplankton excretion, the rate of sinking of blue-green algae relative to greens (the blue-greens are assumed to sink more slowly), and the phosphorous affinity of the blue-green algae (the phosphorous uptake half-saturations for blue-greens is higher than for other algae, i.e., their phosphorous uptake rate is more limited

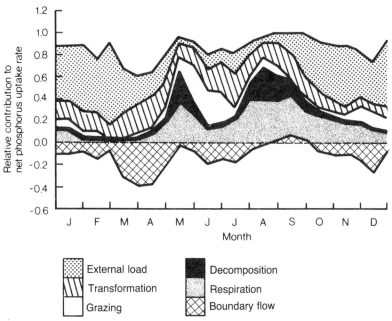

Fig. 9.9. Graphical analysis from the Saginaw Bay model showing the relative importance throughout the year of grazing, transformation, decomposition, external loading,
and boundary flow from the main body of Lake Huron, to phosphorous available for
phytoplankton uptake. *Boundary flow* represents a net phosphorous transfer between the
lake and Saginaw Bay and may be either positive or negative. From Bierman et al.
(1980).

at low phosphorous levels). Shutting off phosphorous recycling processes, increasing the blue-green algae sinking rates (from 0.15 to 0.40 m/day), and
increasing the phosphorous half saturations (from 30 to 60 μgP/L) all result in
reduction of the blue-green algal bloom to insignificant levels.

The model was next used to evaluate the relative contribution of phosphorous
from sources internal and external to the ecosystem, with an eye to examining
how much of the phosphorous loading comes from external sources. The results
gave some indication of the magnitude and timing of phosphorous reduction
needed. The relative contribution of respiration, grazing, transformation, boundary flow (phosphorous inflow and outflow across the boundary with the rest of
Lake Huron), and external loading to phosphorous uptake by phytoplankton, an
index of phytoplankton growth, is shown in Fig. 9.9. By controlling the external
load some reduction in blue-green algal production can be achieved. This reduction is shown in Fig. 9.10, where gross biomass production as a function of
percentage phosphorous reduction is shown, along with the phosphorous load
goal set by the International Joint Commission for the Great Lakes. This goal
was based both on the simulated reduction achieved in blue-green algal blooms
by reducing external phosphorous loading in the Bierman model and on corre-

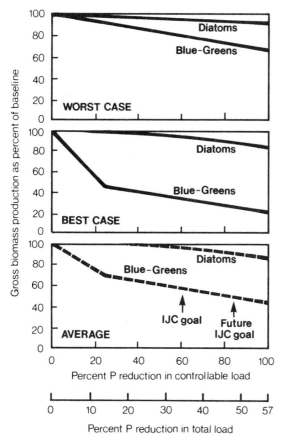

Fig. 9.10.   Results of a simulation experiment with the Saginaw Bay model in which the external loading of phosphorous was reduced by various percentages. Gross annual biomass production as a percentage of the 1974 baseline model run are graphed for blue-green algae and diatoms as a function of the percentage reduction in external phosphorous loading. Best and worst case refer to boundary flow assumptions between Saginaw Bay and the rest of Lake Huron. The IJC is the International Joint Commission for the Great Lakes, that sets water quality standards. From Bierman et al. (1980).

lations between blue-green biomass and taste and odor parameters found from an extensive field data set on Saginaw Bay. Two striking features apparent from Figs. 9.9 and 9.10 are:

1. Phosphorus reduction by control of external loading is more possible at some seasons (e.g., autumn and winter) than others (spring) and
2. Blue-green algae are much more strongly affected by phosphorous reduction than diatoms are.

The second feature is primarily caused by the blue-green bloom being later in the year than the other blooms and, therefore, at a time of phosphorous limitation. Fortunately, reducing external loads of phosphorous is more feasible in the fall, when blue-green algae blooms occur, than in the spring and summer, when green and diatom blooms occur.

In summary, eutrophication models have been used to examine the effects of nitrogen and phosphorous loading on plankton production in aquatic ecosystems. In some cases phytoplankton have been divided into functional groups and the seasonal dynamics of these groups has been examined. The role of various processes in controlling the blooms is often examined and some proposed clean-up or nutrient input control programs are examined using the model in a scenario mode. The models are particularly effective at suggesting standards for nutrient loading and linking them to plankton blooms and at indicating the possible effectiveness of various control measures and when these might be most expeditiously applied. These models have played a major role in the setting of nutrient loading guidelines and limits.

9.4   Impact assessment

Simulation models have been frequently used to examine the effects of various perturbations on biota. Some of these perturbations are an incidental result of man's activity in another sector, such as the entanglement of fur seals (*Callorhinus ursinus*) in fishermen's discarded net fragments. Others are the results of our management of an ecosystem, such as the effect of grazing and of irrigation projects on primary production in grasslands. Still others are the results of the interaction of two natural biota. Sometimes humans are trying to preserve one of them, such as the effect of insect defoliation on coniferous forest trees. In other cases humans have introduced, either accidentally or purposely, a species and is interested in the effect on resident species, such as the case of fish introduction of rainbow trout (*Salmo gairdnerii*) to Lake Washington in Seattle, Washington. We give examples of each of these four simulation model applications.

9.4.1   Fur seal entanglement

The fur seal population on the Pribilof Islands has been dropping steadily since the mid-1970s. This reduction has been hypothetically linked to a number of causes, including the harvest of subadult males on the island (which has been going on for over 200 years, making this an unlikely possibility), the reduction in fish food resources by the expanded activity of commercial fisheries in the Bering Sea since the late 1960s, and the entanglement of seals in net fragment discarded or lost by fishermen. The seals often cannot extricate themselves from such debris, which then results in mortality because of strangulation, severe infection, or starvation as the seals continue to grow through the restraining netting.

Fowler (1982) developed a simple differential equation model to investigate the possible effect of entanglement on the fur seal population. He divided the total population, irrespective of age, into two categories, entangled seals *xent* and total unentangled seals xn_{tt}. The model assumes a constant entanglement rate *kent* and a mortality rate (entanglement caused plus natural) *kmrtent* for entangled seals. Because some seals can free themselves from entangling debris, the model further assumes that the entanglement rate applies only to seals that do not become unentangled.

The differential equation for the rate of change of population numbers of entangled seals is:

$$\dot{xent} = kent \times xn_{tt} - kmrtent \times xent$$

xent = number of entangled seals in the population
kent = entanglement rate
xn_{tt} = total unentangled seal population
kmrtent = mortality rate of entangled seals

If the number of entangled seals is in equilibrium (e.g., $\dot{xent} = 0$) a relationship between entanglement rate and mortality rate can be derived. This is:

$$kent = kmrtent \times \frac{xent}{xn_{tt}} \tag{9.1}$$

An estimate of the fraction of entangled animals in the harvest, which was assumed to represent the frequency of entangled animals in the entire population, was available from data taken during the annual harvest of sudadult male (age 2–6) seals on their summer rookery grounds at St. Paul Island, Alaska. These data suggest that about 0.4 percent of the males are currently entangled in net debris (i.e., $\frac{xent}{xn_{tt} + xent} = 0.004$). This, along with Eq. 9.1, implies that the entanglement rate is 0.4 percent of the mortality rate for entangled seals.

Because it is not well known how long animals live after being entangled, the mortality rate of entangled seals is not known. Fowler, therefore, gave estimates of entanglement rate for different mortality rates. He calculated these estimates for different assumed time periods over which 75 percent of the entangled animals die. These ranged from an estimate of 2 months to 1 year. Additional entanglement mortality is presumed to result from entanglement in large net fragments—animals that are not observed on the Pribilof Islands because they do not live long enough to return there after being entangled. Based on data from net fragments washed ashore on Amchitka Island (in the Aleutian Islands, Alaska) and on St. George Island (the other major Pribilof Island besides St. Paul), Fowler reckoned that large netting material is about 4 times as abundant as small netting material. He, therefore, estimated the entanglement rate to be four times that estimated from the above model, which includes only small net fragment entanglement. Table 9.1 shows estimates from Fowler's model of the annual percent mortality of seals because of entanglement in net fragments. Estimates range from 3 to 18 percent annual mortality, depending on how long the seals survive after being entangled.

Table 9.1 Estimates of mortality rates for the Pribilof Island fur seal population caused by entanglement in debris. From Fowler (1982).

Months for 75% mortality	Mortality rate	Monthly survival in small net fragments	Annual entanglement rate (%)		
			in small fragments	in large fragments	Total percent entanglement
2	8.32	0.50	3.33	13.31	17.75
4	4.16	0.71	1.67	6.65	8.87
6	2.77	0.79	1.11	4.44	5.92
8	2.08	0.84	0.83	3.33	4.44
10	1.66	0.87	0.67	2.66	3.55
12	1.39	0.89	0.56	2.21	2.96

The Fowler model was revised by Swartzman (1984) to account for the age structure in the population and to include the hypothesis that young subadult males may be more susceptible to net entanglement because of their lack of experience in avoiding danger and their frequently being observed playing around flotsam.

The Swartzman extension to the Fowler model was an age-structured model with density-dependent pup mortality on land. Equations for entangled and unentangled animals are as follows:

$$xn_{tt_i} = -(kent_i + kmrt_i + kmrtlg_i) \times xn_{tt_i} \qquad (9.2)$$

$$xent_i = kent_i \times xn_{tt_i} - (kmrtent_i + kmrtlg_i + kmrt_i) \times xent_i \qquad (9.3)$$

xn_{tt_i} = total number of unentangled age i seals
$xent_i$ = number of entangled age i seals
$kent_i$ = entanglement rate in small net debris for age i seals
$kmrt_i$ = mortality rate for unentangled age i seals
$kmrtlg_i$ = age i seal mortality from enganglement in large net debris
$kmrtent_i$ = age i seal mortality due to entanglement in small net debris

The extensions to the Fowler model include separating natural mortality from entanglement caused mortality for the entangled animals, explicitly distinguishing entanglement in large from that in small net fragments, with entanglement in large fragments resulting in a different mortality rate that accrues on both previously entangled (in small net fragments) and unentangled animals, and making all rates age-class specific.

By assuming the population to be in equilibrium, both Eq. 9.2 and Eq. 9.3 can be set equal to zero and the entangled population expressed as a function of the unentangled population. The result is $xent = \dfrac{kent_i \times xn_{tt_i}}{\beta_i}$, where $\beta_i = kmrtent_i + kmrtlg_i + kmrt_i$ is the sum of all mortality rates for age i entangled seals. The total fecund female stock may be expressed as a function of the number of pups leaving land at the end of the suckling period, xn_{tt_0}, and the parameters introduced earlier, with the addition of age-specific fecundity. The fecund stock is thus given by:

$$xstock = xn_{tt_0} \sum_{i=r}^{n} kbirth_i \times kent_{rt_i} \times ksurv_i \times ksex_{rt}$$

$xstock$ = total fecund female seal stock
xn_{tt_0} = number of pups leaving land after suckling period
$ksurv_i$ = fraction of pups leaving land that survive to age i
$ksex_{rt}$ = sex ratio of pups at birth
$kbirth_i$ = age specific fecundity (pups per female)
r = the age seals begin to mature
n = the total number of age classes
$kent_{rt_i}$ = age specific correction factors to include entangled seals in fecundity

The $kent_{rt_i}$ is the ratio of total equilibrium number of seals to the number of unentangled seals in age class i (i.e., $kent_{rt_i} = 1 + \dfrac{kent_i}{\beta_i}$ obtained from Eq. 9.3 at equilibrium). The $ksurv_i$ are given by $ksurv_i = \prod\limits_{j=0}^{i} e^{-\alpha_j}$ where α_j is the sum of all the loss terms in Eq. 9.3.

Data collected on St. Paul Island suggests that survival of pups on land is density dependent (Fowler, 1982). This can be expressed as a relationship between pups leaving land and fecund stock. This relationship was expressed by:

$$xn_{tt_0} = xstock \times (1 - e^{-k_1} \times e^{k_2 xstock}) \qquad (9.4)$$

k_1, k_2 = parameters controlling the shape of the pup survivorship curve

This equation can be written as $xn_{tt_0} = xstock \times h(xstock)$ where h is the function in Eq. 9.4. The equilibrium solution for the number of female pups leaving land can be computed analytically according to a method presented by Reed (1980) and Getz (1980) and is given by:

$$xn_{tt_0}^{eq} = \frac{1}{\sum\limits_{i=r}^{n} kbirth_i \times ksurv_i \times kent_{rt_i} \times ksex_{rt}} \times$$

$$h^{-1}\left(\frac{1}{\sum\limits_{i=r}^{n} kbirth_i \times ksurv_i \times kent_{rt_i} \times ksex_{rt}} \right)$$

where h^{-1} is the inverse of the function h given in Eq. 9.4.

Results for the equilibrium number of pups leaving land and the total fecund stock analogous to results in Table 9.1, is given in Table 9.2 for cases where entanglement mortality accrues on only 1- to 3-year olds and on the entire

Table 9.2   Equilibrium number of pups leaving land, and number of pups born as a function of the level of entanglement for cases where all ages and only age 1–3 seals are assumed to be entangled.

Cases	Time for 75% mortality (mo)	Pups born × 10^5	Pups leaving land × 10^5
No entanglement		5.30	3.63
All ages entangled	12	3.57	3.16
All ages entangled	2–6	0.00	0.00
age 1–3 entangled	12	4.95	3.67
age 1–3 entangled	6	4.50	3.60
age 1–3 entangled	4	3.85	3.32
age 1–3 entangled	2	0.00	0.00

From Swartzman (1984).

population. Apparently, the importance of entanglement depends a great deal on which age classes are affected and on the mortality rate of entangled seals, both of which are not well known at this time.

9.4.2   Grassland and forest perturbations

We turn next to examples where models were used to investigate the effects of both natural and human-induced perturbations on an ecosystem. Both chronic and acute perturbations are considered. Both examples involve models developed as part of the International Biological Program (IBP). The first examines the effect of grazing and irrigation on plant production in a short grass prairie, while the second investigates the effect of defoliation on water and carbon dynamics in an eastern U.S. deciduous forest dominated by yellow poplar *Liriodendron tulipifera*.

Grazing and irrigation effects

The ELM model (Innis, 1978) was developed by a team of scientists with the expressed purpose of helping to better understand the dynamics of grassland ecosystems. Figure 9.11 gives a Forrester-type flow diagram of the major state variables and flows in this model. The model is composed of submodels including water, temperature, nitrogen, and phosphorous flow modules for the abiotic elements, and producer, plant phenology, mammal, insect, and decomposer modules for the biotic elements. We do not describe but only discuss the process equations. The model is complex, both in compartmental and mechanistic detail. Its time step is daily. The model equations and behavior are described in detail in Innis (1978).

   Producer dynamics is controlled by primary production, which depends on plant biomass, temperature, and soil moisture, and by consumption, largely by cattle. Plant species are divided into functional groups including warm-season

Table 9.3   The simulated effect of irrigation and grazing in the ELM grassland model compared with field data from selected perturbation experiments.

Treatment	Water total (cm)	Net primary production (g dry-wt/m^2)	Gross primary production (g dry-wt/m^2)	Live peak (g dry-wt/m^2) Model	Live peak (g dry-wt/m^2) Data
Baseline run	27.1	707.0	1086.0	180.0	—
Water added to keep tension > -0.8 bars	88.0	3494.0	5151.0	620.0	388.0
Grazing at 0.082 cows/ha for 6 months	27.1	715.0	1087.0	139.0	182.0
Grazing at 0.2 cows/ha for 6 months	27.1	823.0	1294.0	169.0	170.0
Grazing at 0.3 cows/ha for 140 days	27.1	552.0	901.0	119.0	—

grasses, cool-season grasses, forbs, shrubs, and cacti. Primary production is computed in a fashion similar to that introduced in Chap. 4, and consumption is similar to that discussed for fish in Chap. 4 but with greater detail for consumption by cattle, because so much work has been done on cattle energetics and diets.

This model was used to investigate the effect of an irrigation project designed to keep the soil moisture tension greater than -0.8 *bars*[1] throughout the year, and of grazing at three levels; light grazing (0.08 cows/ha for 6 months), moderate grazing (0.2 cows/ha for 6 months) and intensive, heavy grazing (0.3 cows/ha for 140 days). These grazing alternatives represent cattle grazing practices on the Pawnee National Grassland in northeastern Colorado. In all simulations, weather data for 1972 were used as driving variables, and the results for a number of model output indices were compared with the control (no cattle grazing or irrigation) case. The results are shown in Table 9.3.

Irrigation had a striking effect on the system, as well it should because the shortgrass prairie, with an average rainfall of less than 30 cm, is severely water limited. Net primary production was increased by a factor of 5, whereas moisture, coming from both natural sources and introduced, increased by a factor of 3. Simulations show that the species composition changes markedly after irrigation, with a shift toward grasses, although the data did not show as large a change as the simulation. Despite the marked change in primary production, the model shows that increases in secondary production after irrigation are insignificant, except for the grasshoppers. Apparently, this is a result of the inability of most terrestrial consumers to respond rapidly to changes in primary production, unlike aquatic ecosystems.

[1]See note on p. 284.

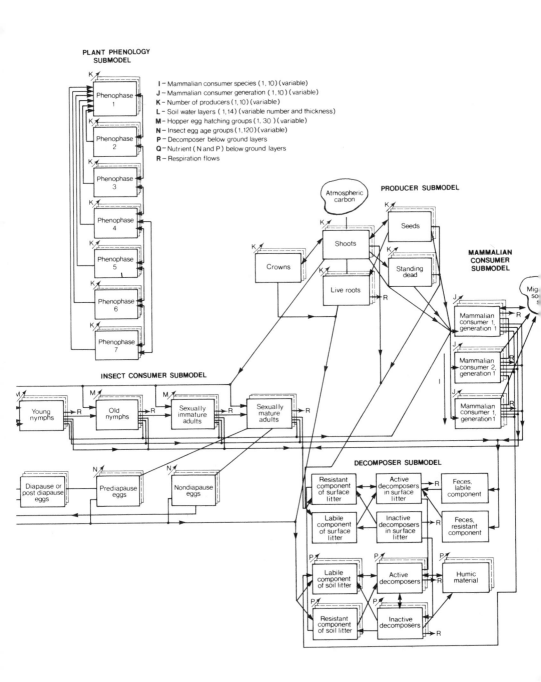

Fig. 9.11.   Compartment flow diagram for the whole ecosystem grassland model ELM. The model is subdivided into temperature, nitrogen, phosphorous, water, producer, plant phenology, insect consumer, mammal consumer, and decomposer submodels. From Innis (1978). Used by permission of Springer-Verlag, New York.

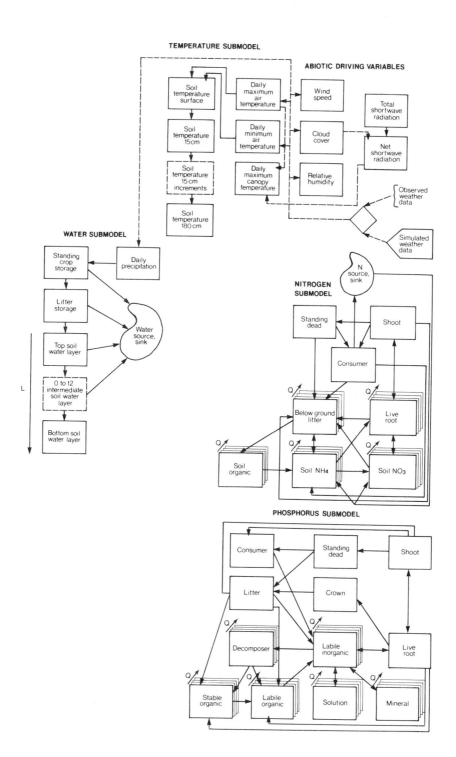

TEMPERATURE SUBMODEL

ABIOTIC DRIVING VARIABLES

WATER SUBMODEL

NITROGEN SUBMODEL

PHOSPHORUS SUBMODEL

271

Grazing, especially moderate grazing, appears to stimulate net primary production. This model prediction is also borne out by the field data on plots where grazing experiments were conducted. Heavy cattle grazing appears to shift species composition away from shrubs and toward warm season grasses.

This study was unusual in that experiments were performed on the ecosystem to which the predictions of some of the perturbation experiments could be compared. These comparisons were only made with data on the peak biomass of each of the plant functional groups in the model. Model predictions for the effect of grazing on peak plant biomass were closer to data than predictions of the effect of irrigation. The model greatly overpredicted the growth of warm season grasses in response to irrigation. Otherwise, results were respectable, except for the model underpredicting the growth of cool season grasses in the cattle grazed system. This might be linked to a low estimate of cool season grass growth in the model control run, which was used for calibration. Unfortunately, no comparisons were made to consumer data, specifically to test the model prediction that grasshopper production would respond markedly to irrigation.

Defoliation in coniferous forests

A model called CONIFER (Swartzman, 1979a) was used to investigate the effects of defoliation, both chronic and acute, on forest growth and water balance in an old growth coniferous forest stand. Figure 9.12 shows the major compartments and flows for water, carbon, and nutrients in the stand. The model is a melding of a photosynthesis submodel, with photosynthesis controlled by light (a function of stand biomass density), nutrients and soil moisture status, a carbohydrate allocation submodel among the various tree parts (foliage, branches, roots, etc.), and water flow and evapotranspiration submodels that interface with the photosynthesis submodel through a transpiration index of moisture stress. A daily time step is used for the water and energy modules, whereas the carbon flow processes are on a weekly basis.

A series of experiments on the effects of acute defoliation, caused for example, by the application of a chemical defoliant, were compared with the control case. Table 9.4 shows several indices of ecosystem behavior from model simulations for each of 2 years. Acute defoliation levels from 10 to 90 percent were simulated at different times by direct removal of foliage at these times. All acute defoliations occurred only in the first year. The simulations began on day 130, because this was a day when good data were available to use as an initial condition.

Acute defoliation has a striking effect on evaporation and transpiration, both of which are reduced, especially after drastic defoliation (i.e., 90%). This drop extended into the second year after defoliation as well. The evaporation reduction is caused by reduced needle interception of water, which now directly enters the soil instead of evaporating. Reduced transpiration after defoliation is caused by the reduced needle surface area, needed to drive the transpiration process (transpiration in the model is proportional to needle surface area). Snow cover

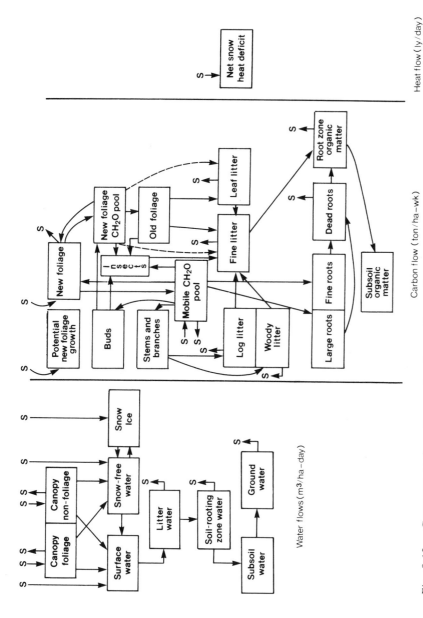

Fig. 9.12. Compartment flow diagram for the coniferous forest ecosystem model CONIFER. The diagram shows the separate flows for water, carbon, and heat. From Swartzman, G. L., Modeling material and energy, *J. Ecol. Modeling*, 1979. Used by permission of Elsevier Science Publishers, B. V.

273

Table 9.4 The simulated effect of various levels of acute defoliation on various ecosystem indices in the CONIFER model.

Index	Year	No. defoliation	10% new foliage day 182	25% new foliage, 10% old foliage day 182	25% new foliage, 25% old foliage day 182	25% new foliage, 25% old foliage day 135	25% new foliage, 25% old foliage day 240	50% new foliage, 50% old foliage day 182	90% new foliage, 90% old foliage day 182	50% old foliage day 350
					Defoliation/Perturbation					
Evaporation (m³/ha/yr)	1	3200	3190	3130	3070	3060	3080	2900	2330	3030
	2	2570	2560	2490	2490	2490	2490	2370	1850	2370
Transpiration (m³/ha/yr)	1	3340	3320	3290	3210	3220	3210	2980	1660	3030
	2	2385	2365	2365	2340	2340	2340	2265	1760	2265
Percent time snow is on ground	1	25.1	25.1	24.5	24.0	24.0	24.0	23.7	13.2	23.4
	2	48.9	48.9	48.7	48.4	48.4	48.4	47.1	38.8	47.0
Net carbon assimilation by new foliage (ton/ha/yr)	1	1.53	1.55	1.70	1.97	2.10	1.67	2.22	1.70	1.53
	2	1.19	1.18	1.31	1.63	1.63	1.64	2.40	6.30	2.40
Net carbon assimilation by old foliage (ton/ha/yr)	1	32.2	32.4	32.8	32.7	32.7	31.7	31.1	16.0	27.9
	2	26.5	26.3	26.1	25.6	25.7	25.7	24.1	13.9	24.2
Change in old foliage carbon (ton/ha)	1	0	-0.12	-0.63	-1.28	-1.14	-1.14	-2.28	-4.10	-2.28
	2	0.03	-0.06	-0.11	0.17	0.06	0.09	0.02	0.25	0.02
Change in CH₂O pool carbon (ton/ha)	1	1.06	1.08	1.16	0.69	0.49	0.23	-1.36	-13.11	-2.82
	2	-0.11	-0.42	-1.43	-1.76	-1.69	-1.48	-3.21	-3.10	-1.97
Maximum new foliage carbon mass (ton/ha)	1	1.40	1.40	1.40	1.40	1.11	1.44	1.40	1.44	1.44
	2	1.50	1.33	1.11	1.11	1.11	1.11	0.74	0.32	0.74

274

also decreases, because increased light levels at the ground surface promote more rapid snow melt. In all, except the most severe defoliation cases, foliage appears to be able to recover well in the second year. This is accomplished, however, through a reduction of the trees' carbohydrate stores as can be seen by examining the carbohydrate pool carbon totals for the second year. The worst time for defoliation appears to be day 182, the height of the growing season. Greater reductions in carbohydrate pool result from defoliation on this day than for the same level of defoliation on day 135 and day 240.

The carbohydrate pool stores are a measure of the trees' ability to resist future defoliation. Notice that defoliation actually stimulates growth of new foliage in the first year because of the improved light status resulting from the thinning of the stand, but this improved condition does not carry into the second year, when carbohydrate pools must pay for the foliage loss of the previous growing season.

A second set of simulations mimic the effects of chronic defoliation caused by factors such as an ongoing insect infestation. Three different biennial defoliation levels, 25, 50, and 75 percent, are simulated. The equivalent weekly defoliation rates are 0.005, 0.01, and 0.02, respectively. Table 9.5 shows the

Table 9.5 The simulated effect of chronic defoliation at different levels on water and carbon dynamics from the CONIFER model.*

Ecosystem index	Year	Percent reduction in foliage at end of year			
		0%	25%	50%	75%
Evaporation	1	3140	3100	2970	2735
(m³/ha/yr)	2	2500	2450	2290	1890
Transpiration	1	3290	3230	3070	2630
(m³/ha/yr)	2	2320	2230	2030	1875
Percent time snow	1	24.5	24.2	24.0	21.5
on ground	2	48.4	47.6	45.6	37.9
Net carbon assimilation	1	1.64	1.74	1.91	1.90
by new foliage	2	1.50	1.72	2.43	4.19
(ton/ha/yr)					
Net carbon assimilation	1	32.5	32.2	31.0	27.2
by old foliage	2	25.0	23.6	19.6	15.6
(ton/ha/yr)					
Old foliage carbon	1	3.86	3.43	2.47	1.33
(ton/ha)	2	3.41	2.76	1.51	0.49
Carbohydrate pool carbon	1	16.5	16.0	14.4	10.2
(ton/ha)	2	14.4	12.6	7.7	2.3
Maximum new foliage	1	1.33	1.24	0.94	0.60
carbon mass	2	1.25	1.10	0.73	0.36
(ton/ha)					

*The simulations are over a 2-year time span.

effect of chronic defoliation on the same indexes of ecosystem performance as those used for acute defoliation. The results are similar to, but more striking than the acute defoliation cases. Snow cover, transpiration, and evaporation are reduced. Unlike the acute defoliation case, new foliage cannot recover from defoliation to achieve higher production than the control case. The mobile carbohydrate pool is seriously reduced by the 75 percent defoliation case. It appears that this stand could not sustain defoliation at this level for another growing season. This fits with the cyclical character of many insect pests that maintain peak populations for at most two growing seasons. This appears to be essential to the future survival of the host trees.

9.5   Modeling species introductions

More and more lake sports fisheries are being managed through species introductions. The success of a fish species introduction can depend on a variety of factors, including the growth rate of the introduced species, its spawning habitat needs, the existence of predators, and the presence of a great enough size range and abundance of prey to allow for coexistence of these prey with the introduced predator fish. Because of the complexity of interactions between resident species and of the life history of the introduced fish, simulation modeling is an important tool for investigating the success of a fish species introduction. Sometimes more than one species are introduced into the same system or a newly introduced species can potentially prey on an existing sports fishery. This was the case with the introduction of rainbow trout (*Salmo gairdnerii*) into Lake Washington near Seattle, Washington and their potential for preying on parr (yearlings) of sockeye salmon (*Oncorhynchus nerka*), who are resident in the lake for a 15-month period before moving out to the open ocean. A model was developed to investigate the probable influence of rainbow trout on sockeye salmon populations. The rainbow trout additions to Lake Washington have been substantial, more than 200,000 fish per year over a 5-year period. The fish have been observed to grow rapidly on a diet consisting mostly of *Daphnia* and chironomid larvae, but have not been observed to breed in Lake Washington. A seasonal sports fishery on the trout has removed large quantities of these fish. The concern of the state agency regulating sockeye salmon was that the trout may grow fast enough and remain in sufficient numbers, despite the fishery, to feed heavily on sockeye salmon parr. The state agency responsible for the trout introductions (obviously a different agency) countered that there has been no observed predation by trout on sockeye in Lake Washington, although the literature show that rainbow trout do predate on sockeye parr in other, similar, lake ecosystems.

The model was built to generate scenarios of different levels of introductions, different sizes of sockeye runs into the lake, and different levels of fishing intensity to show how much predation trout could inflict on sockeye under these different conditions. It is assumed that predation by rainbow trout is size-selective and that the size at which this predation begins is similar to the size range of rainbow trout predating on sockeye salmon in other lakes. Figure 9.13 is a

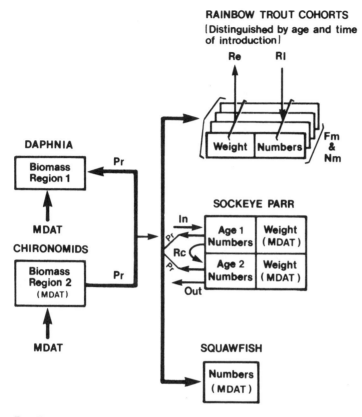

Fig. 9.13. Compartment flow diagram for a model investigating the effect of the introduction of rainbow trout into Lake Washington on sockeye salmon parr. This diagram shows the increased compartment complexity as the biota get closer to the most important biota in the system; in this case the rainbow trout.

flow diagram for the model. The focus is primarily on the energetics, population dynamics, and feeding of the rainbow trout and on the population dynamics of the sockeye salmon, with salmon weight change being a driving variable independent of food availability. This approach was used because the primary focus was on the growth rate of the trout and on determining at what size they would start feeding on sockeye salmon and how much they would eat. The major predator on sockeye before the trout introductions, the northern squawfish (*Ptychocheilus oregonensis*), was represented in the model by numbers and by a feeding rate. Other food sources for the trout were *Daphnia* and chironomids, whose densities were read in as driving variables on a monthly basis (the model time step). These densities were assumed not to change in response to consumption (i.e., quantities are great enough that predation effects can be averaged into the data on density of these species). Because the introduced lake trout have not been observed to breed in Lake Washington, trout fecundity is ignored in the model.

Trout feeding selectivity was related to the size of the prey relative to trout size according to the lognormal function of Andersen and Ursin (1977).

$$gprefwt = e^{\dfrac{-\left(log\left(\frac{xwt}{xpreywt}\right) - kprefpy\right)^2}{2 \times kpref_{vr}}}$$

gprefwt = size selective feeding preference for a specific size (weight) prey
xwt = fish weight
xpreywt = prey weight
kprefpy = most preferred predator to prey size logarithmic ratio
kpref$_{vr}$ = variance in fish prey size preference

In this equation selectivity depends on the weight ratio of the predator to its prey, with selectivity being a maximum (relative weight of 1.0) when the log-

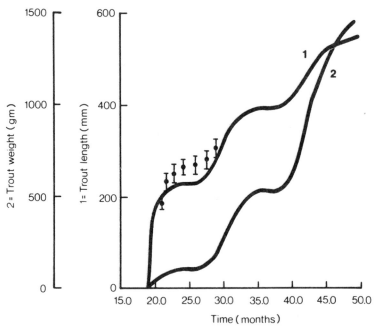

Fig. 9.14.   Graphical comparison of model output for the length and weight of a cohort of rainbow trout released into Lake Washington on July 20, 1981 with data on the length of recaptured marked fish from this cohort. The closed circles represent the mean lengths and the bars plus and minus one standard error around the mean.

arithm of this ratio equals *kprefpy*. Selectivity drops when the ratio is both above and below this parameter, with the sharpness of the decline depending on the "variance" parameter $kpref_{vr}$. Fish ration is also affected by the temperature and total prey abundance as well as by the trout weight (see Chap. 4). Evaluation of the fish growth submodel was done by comparing the simulated growth over a 2-year-period with growth of fish marked by the Washington State Game Department and recaptured in the fishery. Agreement between the model and data for two cohorts of releases was good. Figure 9.14 shows the results for one of these cohorts. The history of releases from 1981 to 1983 was simulated by releasing cohorts of trout during the month of release at their average released weight. The size of the sockeye parr run for each year was also obtained from field data as were the density of *Daphnia* and chironomids. Fishing mortality rate was estimated from creel censuses (sample surveys of sports fish catch per fisherman).

Table 9.6 shows the results of a simulation experiment to examine the effect of having trout releases of different sizes ranging from the present 200,000 fish per year at an average weight of 5 g to having up to 5 million fish per year at an average weight of 15 g. Each of these alternatives were considered with three different fishing mortality rates; the present best estimate, reduced (one-half the best estimate) and increased (double the best estimate). Table 9.6 shows, for each 2-year simulation, the numbers of sockeye at the end of each year, with their percentage survival, the trout numbers and their percentage survival, the sockeye biomass taken by all predators over the 2 years and the *Daphnia* biomass taken by rainbow trout.

The latter category was included because there was some concern that if releases of trout were high enough they could severely reduce the *Daphnia* biomass in the lake. Although the model was not built to investigate this phenomenon and treats *Daphnia* as a trout food source unaffected by predation, we could give an order of magnitude estimate of consumption compared with average standing stock of *Daphnia*. Such an estimate shows that the trout can plausibly reduce *Daphnia* numbers significantly. Table 9.6 shows that survival of sockeye can be severely reduced by the trout, especially in the simulations where trout fishing pressure is reduced. Sockeye survival is less sensitive to the number of trout introduced, although when numbers of trout are increased by an order of magnitude the reduction in sockeye survival is large.

9.6   Ecosystem risk analysis

Producing deterministic results from environmental impact models has had the undesirable effect of implying that impact effects are known (deterministic). A more appropriate option is to cast impact predictions as probabilities and to refer to the *risk* of various possible untoward outcomes. Such a risk model was developed by O'Neill et al. (1982) to assess the potential impact of toxicants on ecosystems. The method features the use of a generic lake ecosystem model called SWACOM (Standard WAter COlumn Model) having 10 general phyto-

Table 9.6 Simulated effect of changing trout fishing pressure and of different numbers of introduced fish on trout growth and sockeye and Daphnia populations in Lake Washington.*

Fishing	Annual trout introduced	Weight of initial trout	Year	Sockeye ($\times 10^5$)	Trout ($\times 10^4$)	Sockeye biomass taken (2 yr) ($g \times 10^4$)	Daphnia biomass taken (2 yr) ($g \times 10^6$)
Regular	200,000	5	1	8.6 (3.0%)	61.0 (3.1%)	1.40	0.913
			2	14.0 (10.0%)	0.02 (0.10%)		
Regular	500,000	5	1	8.6 (3.0%)	1.54 (3.1%)	3.35	2.28
			2	12.6 (8.0%)	0.06 (0.13%)		
Regular	1,000,000	5	1	8.6 (3.0%)	3.08 (3.1%)	6.28	4.56
			2	10.4 (7.0%)	0.12 (0.12%)		
Regular	2,000,000	5	1	8.6 (3.0%)	6.2 (3.1%)	11.0	9.13
			2	6.3 (4.0%)	0.23 (0.12%)		
Regular	500,000	15	1	8.5 (3.0%)	1.03 (5.0%)	2.71	2.01
			2	13.1 (9.0%)	0.39 (0.20%)		
Regular	5,000,000	15	1	7.8 (2.8%)	10.3 (2.0%)	17.5	20.1
			2	1.9 (1.6%)	0.39 (0.08%)		
High (2 ×)	2,000,000	5	1	8.6 (3.0%)	0.46 (0.23%)	0.30	5.07
			2	15.1 (11.0%)	0.0015 (0.0%)		
High (2 ×)	500,000	5	1	8.5 (3.0%)	0.11 (0.25%)	.075	1.27
			2	15.3 (11.0%)	0.0004 (0.0%)		
Low (0.5 ×)	2,000,000	5	1	8.6 (3.0%)	22.6 (11.3%)	34.0	15.5
			2	1.9 (1.3%)	2.8 (1.4%)		
Low (0.5 ×)	500,000	5	1	8.6 (3.0%)	5.65 (12.0%)	17.2	3.9
			2	3.7 (2.1%)	0.71 (1.4%)		

*Numbers in parentheses are percentages of sockeye salmon runs or trout introduced surviving.

plankton groups, 5 zooplankton, 3 forage fish, and a single carnivorous fish species. Nutrients, light, and temperature are model driving variables. Rather than simulating any specific biota the model has generalized species groups that mimic the expected behavior in a northern dimictic (mixing completely from top to bottom twice a year) lake.

Phytoplankton growth is modeled as a multiplicative function of light, nutrients, and temperature effects. Self-shading is accounted for, as in the San Joaquin model, by integrating photosynthesis over the euphotic zone with the extinction coefficient a function of the phytoplankton biomass. Parameters for the phytoplankton include such factors as optimal temperature, nutrient half saturation for growth (assuming a Michaelis function), and the optimum temperature for photosynthesis. Because nutrients are driving variables in the model, their status cannot be changed by factors such as phytoplankton uptake. The grazers and higher predator feeding rates depend on the amount of food available and on the temperature. These functions are described in detail in DeAngelis et al. (1975). Metabolism is a function of temperature for the animals and natural and predation mortality are limiting factors on their population dynamics.

The main feature we are interested in here is how the model deals with toxicant effects. The general hypothesis is that any or all of the processes that control biomass and population dynamics of the biota may be affected by the toxicant. As such, a toxicant *effects matrix* is set up in parallel to the model parameters. In total there are 18 populations with six parameters for each of the phytoplankton and five for each consumer. A toxicant effect is associated with each of these parameters for each population.

To simulate the effect of toxicity each parameter is multiplied by the corresponding element of the effects matrix for each population, which has a value of 1.0 for no toxicity and toxicant effect values less than or greater than 1.0 depending on whether the toxicant is expected to increase or decrease the rate associated with that parameter for that population. For example, for a consumer population, respiration might be expected to increase, whereas feeding rates might decline under toxicity. The result is an overall direct effects syndrome associated with the values in the effects matrix.

Because the effect of toxicants on many processes is uncertain the values for the effects matrix are generated by Monte Carlo methods by multiplying the average effect for each parameter and population by a normally distributed random variable. For example, if it is expected that a particular toxicant would increase the mortality rate on a population by an average of 40 percent, then the element in the effects matrix corresponding to the mortality rate parameter would be 1.4. If it is estimated that the uncertainty in this increase is about 20 percent (i.e., the increase can be anywhere between 20 and 60%), then the 1.4 value is altered in each Monte Carlo run by a normally distributed random number with a mean of 0.0 and a standard deviation of 0.1 (so that 95% of the values will be between 1.2 and 1.6).

Because it is known that this toxicant will not decrease the mortality rate, and the normal distribution can generate negative numbers large enough such that a modified (Monte Carlo generated) mortality rate effect less than 1 can result, a folded normal approximation is used. This approximation mirrors numbers that are on the wrong side of the 1.0 cutoff level. Thus, a final value of 0.98 would be mirrored or folded to a value of 1.02. Because this usually happens only to a small fraction of the total numbers generated the error in approximating the normal will not be serious.

Because a Monte Carlo method is used, output from a large number of runs generates a frequency distribution over time for each model variable. To reduce this large amount of output to manageable size various indices of ecosystem performance are defined. Examples are the average phytoplankton biomass, the maximum phytoplankton biomass, and the average productivity to respiration ratio. This same approach was adopted by Innis (1978) in the ELM model and by Dale and Swartzman (1984) in studying thermal effects on aquatic ecosystems. Toxicant effects are defined in terms of the risk (probability) of various of these indexes reaching some undesirable ecosystem endpoint.

An example is given of the effects of phenol and quiniline. As with the microcosm toxicity study single species bioassays on various test biota were used. Here they provided estimates for the effects matrix and for the uncertainty levels for these parameters. The risk of various undesirable ecological endpoints in the phenol and quiniline examples is shown in Table 9.7, and the frequency distribution of the effect of these toxicants on phytoplankton and zooplankton biomass and on the production to respiration ratio is given in Fig. 9.15. The results in Table 9.7 come from distributions like those in Fig. 9.15. Because of the uncertainties in the underlying model and its generic nature, O'Neill et al. suggest that the method is better as a measure of relative risk of deleterious toxicant effects than of absolute risk.

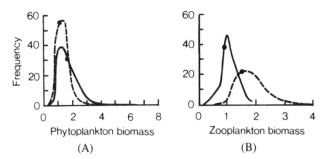

Fig. 9.15. Frequency distribution of A. average annual phytoplankton biomass and B. average annual zooplankton biomass resulting from 500 Monte Carlo runs of the SWACOM model with input of phenol (solid line) and quinoline (dashed line). The closed circles on the lines indicate the deterministic solution to the model with average parameter values. Biomasses given are relative and are thus unitless. Reprinted with permission from *Environmental Toxicology and Chemistry 1,* O'Neill, R. V., et al., copyright 1982, Pergamon Press, Ltd.

Table 9.7   Simulated probability (risk) of arriving at an undesirable ecosystem state, defined by various ecological endpoints, resulting from addition of both phenol and quinoline into the environment.*

Ecological endpoint	Phenol	Quinoline
5% increase in maximum phytoplankton biomass	0.16	0.05
20% decrease in phytoplankton biomass	0.16	0.41
20% decrease in zooplankton biomass	0.07	0.002
20% decrease in forage fish biomass	0.21	0.84
50% decrease in carnivorous fish biomass	0.31	0.88
10% increase in total respiration	0.32	0.78
10% decrease in productivity:respiration ratio	0.37	0.82

*Results are generated by 500 Monte Carlo runs of the SWACOM model. Percentages in the endpoints are relative to the standard (no toxicant) run.

9.7   Summary

This chapter has reviewed a broad range of management applications for models. Although the applications are varied, the structure of the models used was similar. The most common use of these models was in scenario mode. Here various management options were tried and the results were tabulated to be made available to managers as an advisory tool. They are the logical consequences of applying the model to a new situation. They do not purport to be quantitatively accurate but only give the relative effect of various options. The applications are split between those examining the possible effects of some preexisting impact phenomenon and those helping to decide at what level a particular intervention can best be applied. With eutrophication these interventions are sewage diversion and water flow management and with fish introductions the intervention is the best level of introduction to maintain a healthy fishery yet protect other important stocks in the system.

Until recently, all the simulation models in the literature built for management applications were advisory. They were not being used to directly manage a system. However, now simulation models are being used for the direct management of agricultural lands (Lemmon, 1986) and have served to improve yields substantially. Some other ecological simulation applications have influenced decisions about land development and some models, for example the Hudson River striped bass models, have been brought into the court as evidence for or against a particular development project. The eutrophication models have played a role in setting the goals for nutrient loading levels in the Great Lakes. In these last two cases the decision process resulted from the distillation of many models that converged toward each other in their predictions as better information became available and as the modelers benefited from interaction with each other. Whether this is the most efficient way to build and improve models is not at all clear, because it can lead to much redundancy and wasted effort. On the other hand, it does highlight the major uncertainties about mechanisms and clearly shows where more information is needed.

Finally, we discussed the use of models to generate predictions about the risk of adverse outcomes rather than projecting an actual outcome. Models of this sort are being used more frequently as the danger of taking deterministic model output as predictions of ecosystem outcomes is becoming apparent.

Note

1. Bars are a unit of pressure, with 1 bar being a force of 1 million dynes/cm^2. One bar is about the pressure of the atmosphere at sea level.

9.8   Annotated Bibliography

Beanlands, G. E., and P. N. Duinker: *An Ecological Framework for Environmental Impact Assessment in Canada,* Federal Environmental Assessment Review Office, Hull, Quebec, 1983.
    A Canadian government perspective on using ecological models for impact assessment.
Beddington, J. R., and R. M. May: Harvesting natural populations in a randomly fluctuating environment. *Science,* 197:463–465, 1977.
    A simple model used as an example to show how harvesting can have counterintuitive indirect effects on unharvested populations in the ecosystem.
Beyer, J. E.: *Aquatic Ecosystems: An Operational Research Approach,* University of Washington Press, Seattle, 1981.
    Gives a good discussion of problems in modeling marine ecosystems as well as a discussion of quantitative problems in fishery management.
Broecker, W. S., T. Takahashi, H. J. Simpson, and T. H. Peng: Fate of fossil fuel carbon dioxide and the global carbon budget. *Science,* 206:409–418, 1979.
    Global carbon cycling has been an important application of models to study the greenhouse effect and possibly nuclear winter. This is a good review.
Chen, C. W., J. D. Dean, S. A. Gherini, and R. A. Goldstein: Acid rain model—Hydrologic module. *J. Environ. Engin.,* 108:455–472, 1982.
    This model, entitled ILWAS, was developed to examine the movement of acidic rainwater through a forested watershed. It exemplifies the application of ecological models to acid rain problems.
Frankiel, F. N., and D. W. Goodall, eds.: *Simulation Modeling of Environmental Problems,* Wiley, Chichester, 1978.
    An international review on the use of simulation models in management.
Hilborn, R., C. J. Walters, R. M. Peterman, and M. J. Stanley: Models and fisheries: A case study in implementation. *N. Am. J. Fish. Manage.,* 4:9–13, 1984.

An interesting review of the involvement of managers in modeling and what they can learn.

Holling, C. S.: *Adaptive Environmental Assessment and Management,* Wiley, New York, 1978.

Contains a detailed account of an approach developed by Holling and his associates at the University of British Columbia for using models as an aid in resource management decision making. The method relies on workshops for model development and uses this process for manager education.

Innis, G. S. ed.: *Grassland Simulation Model,* Ecological Studies, 26, Springer-Verlag, New York, 1978.

Reports on the results of simulated perturbation experiments, some of which were actually done as field experiments.

Kessell, S. R.: *Gradient Modeling, Resource and Fire Management,* Springer-Verlag, New York, 1979.

How to use modeling in forest fire management. This is not really an ecological simulation model but a physical model that uses detailed wood inventory information along with physical driving variables to predict the spread of fires.

Lassiter, R. R.: *Modeling Dynamics of Biological and Chemical Components of Aquatic Ecosystems,* EPA Ecological Research Series, EPA-66013-75-012, Corvallis, Oregon, 1975.

Shows what the EPA is doing to model the fate of toxicants in the environment. The EPA has made the code for this model available to the public.

Lawler, J. P.: *The Effect of Entrainment at Indian Point on the Population of the Hudson River Striped Bass,* Testimony Before the Atomic Safety and Licensing Board in the Matter of Indian Point Unit No. 2, USAEC Docket No. 50-247, 1972.

Modeling on the Hudson River striped bass population and how it might be affected by entrainment of fish larvae in power plant cooling systems. An example of how models have been used for advocacy.

Lemmon, H.: Comax: An expert system for cotton crop management. *Science,* 233:29–33, 1986.

A case where a simulation model has been used to manage a large cotton plantation with significant increase in profit. This is the first example case of this sort in the literature. The simulation model is part of an expert system to make it accessible to the user.

Levin, S. A., and K. D. Kimball: *New Perspectives in Ecotoxicology,* Ecosystem Research Center Report NO. 14a, Cornell University, Ithaca, New York, 1983, p. 158.

An edited report on the use of models for studying toxicant fates and effects in natural ecosystems.

O'Neill, R. V., R. H. Gardner, L. W. Barnthouse, and G. W. Suter: Ecosystem risk analysis: A new methodology. *Environ. Toxicol. Chem.,* 1:167–177, 1982.

Uses a generic simulation model to predict toxicant effects in natural ecosystem. Also produces risk using Monte Carlo simulation.

Reckhow, K. H., and S. C. Chapra: *Engineering Approaches for Lake Management, Vol. 1 Data Analysis and Empirical Modeling,* Ann Arbor Science, Ann Arbor, Mich., 1983b.

An excellent review of how to approach simulation models and data for lake ecosystem management.

Russell, C. S. ed.: *Ecological Modeling in a Resource Management Framework,* Resources for the Future, Inc, Washington, D. C., 1975.

A collection of models put together by Resources for the Future to highlight the broad range of model applications possible.

Vaughan, B.: Problems in evaluating radiation dose via terrestrial and aquatic pathways. *Environ. Health Perspect.*, 42:149–161, 1981.

A class of mostly linear dose response models are ably reviewed in this article.

West, D. C., S. B. McLaughlin, and H. H. Shugart: Simulated forest response to chronic air pollution stress. *J. Environ. Quality*, 9:43–49, 1980.

Uses forest succession models to explore long-term effects of chronic air pollution on forest succession.

9.9   Exercises

Eutrophication interaction with other ecosystem perturbations

9.1.   In the San Joaquin model the inflow and outflow of water in the river segment played a major role in determining the plankton dynamics. In this problem the dynamics of that model are compared with an ecosystem where inflow and outflow are limited such as a small pond. Shut off both the inflow and the outflow in the San Joaquin model and compare model behavior with that in the original model.

9.2.   (continuation) Next, simulate an altered flow regime as might result from a dry or wet year by halving or doubling the model flow rate driving variable. How does the system respond to a wet year following a dry year or vice-versa?

9.3.   Examine the effects of thermal loading in the San Joaquin model that would result from the cooling system of a power plant. Uniformly raise the temperature daily by 5°C over the entire year. Does this more strongly effect the timing or the magnitude of plankton blooms?

Defoliation

9.4.   The defoliation model in this chapter examined the effects of defoliation on short-term growth and biomass. In this problem we examine the effect of defoliation on long-term succession dynamics.

   a. Simulate defoliation in the Hubbard Brook model. Defoliate by reducing leaf area *xarealf* ($x(3,i)$) for each species annually by 10 percent in subroutine *update*.

   b. Make the defoliation species-specific by assuming that the defoliator hits the broadleaf plants (maple, birch, and cherry) exclusively and imposes a 10 percent annual reduction in leaf area on them. How does this change the succession sequence, if at all?

Nuclear winter

9.5.   Recent literature has stressed the potential danger to ecosystems of nuclear war that would, through intense fire storms and the resultant particulate matter in the atmosphere, lead to nuclear winter and the extinction of many biota. In this problem the Hubbard Brook model is used to simulate the effects of nuclear winter on forest succession. Assume that particulate matter in the atmosphere reduces solar radiation by 50 percent and lowers the temperature by an average of 8°C for a period of 3 years. Begin the Hubbard Brook model in the conditions of run 2 in Chap. 6 (Table 6.3) and simulate 3 years with half the original solar radiation and with the number of degree-days higher than 4.5°C reduced to 650. After 3 years return to normal temperature and solar radiation conditions. How does nuclear winter affect long-term succession?

Chapter 10

Spatial and
Optimization Models

10.1 Introduction

The topics we have covered so far give a solid underpinning to ecological simulation tools and to the application of simulation techniques to help understand and manage ecosystems.

Two kinds of models have been omitted. Their total absence in a book on ecological simulation neglects a major branch of the literature. These topics are spatial and optimization models. In this chapter we give an introduction to each of these topics to give a flavor of the applications they consider. We only scratch the surface here. A treatment allowing you to make models of this type would require a text in itself.

Most models treated in this book have viewed the ecosystem as a single spatially homogeneous compartment, perhaps having flows into and out of a spatial segment. Spatial compartments have, however, been considered explicitly in some model applications in the literature and implicitly in others. In discussing random walk models in Chap. 7 we introduced a method for representing predator search on a spatial grid. There a single predator searched, according to several decision rules, for spatially distributed prey. The simplicity of the model lay in having only a single predator. At the ecosystem level we are dealing with the movement and interaction of whole populations. Different methods are called for. Two ecosystem level applications involving spatial processes are presented in this chapter. The first is a model of the invasion of Lake Gatun, Panama, by peamouth bass (*Cichla ocellaris*), the second is a spatially segmented model of the Bering Sea. In both models we focus solely on spatial processes and not on the entire models.

In the area of optimization models we discuss both models representing optimization in the use of resources by animals and optimization models in which humans are wanting to achieve some objectives relative to a managed ecosystem.

288

The question about whether biota in some sense behave optimally has long intrigued ecologists. It has given rise to optimal foraging theory (Charnov, 1976), which claims that animals choose their prey to maximize their energy intake (or perhaps to minimize their foraging time). This topic has reached simulation models in defining the process of prey selection. An application to fish foraging is considered here.

The final topic in this chapter is the use of optimization techniques in combination with simulation to help in the management of ecosystems over time by achieving some objectives over a finite planning horizon; a period of time into the future short enough that we can have enough information that forecasting is believable but long enough that some consideration can be given to the effect present decisions have on that future. Several examples are discussed, including the use of dynamic programming with a model of old field succession, a grazing land management model in Australia, and a fishery management model that uses optimal control theory to manage a fishery over some planning horizon.

10.2   Spatial models

Most ecological simulation models in the literature deal with spatial variability in the most simple fashion imaginable; either by ignoring it and assuming spatial homogeneity or by dividing the system into segments with flows between segments driven by such quantities as river flow or diffusion. In this latter type of model the biota are presumed to be passive in their movement between segments. What is done, however, in cases where the biota are moving or migrating actively? A small number of models in the literature represent movement as an active process. We have encountered one approach in Chap. 7 for a predator–prey model. At the ecosystem level such an approach would prove cumbersome. One approach to spatial migration by an introduced fish species was developed by Swartzman and Zaret (1983).

10.2.1   Fish migration in Lake Gatun

The problem addressed was to model the advance of *Cichla ocellaris* through Lake Gatun from 1970 to 1974 after its accidental introduction in 1966 into the Chagres River, a tributary of the lake. The advance of this predator through Lake Gatun was recorded (Fig. 10.1) and its effect on the native fish species was striking, with all but one group of native species eliminated in the lake. Migration of *Cichla* was characterized by the formation of large schools of predatory juveniles that advanced in fronts up the lake. Larvae, smaller juveniles, and adult *Cichla* did not participate in this advance. The goal of the model was to replicate not only the advance of *Cichla* but also the elimination of their prey and the proper growth pattern over time of the *Cichla* (correct weight at age). To do this a model of *Cichla* cohorts was developed, with each prey group represented by a single or two size classes. The flow diagram and major

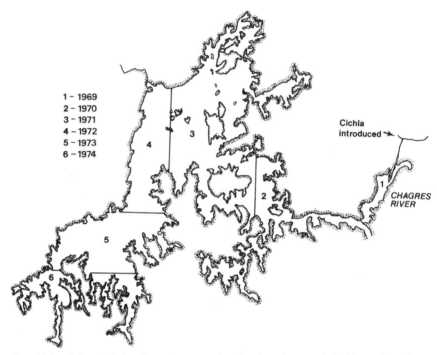

1 - 1969
2 - 1970
3 - 1971
4 - 1972
5 - 1973
6 - 1974

Cichla
introduced

CHAGRES
RIVER

Fig. 10.1.   Map of Lake Gatun Panama, showing the advance of *Cichla ocellaris* from its introduction into the lake in 1970 through 1974.

processes are given in Fig. 10.2. We focus here on the migration process that is represented in an implicit fashion.

Instead of dividing the lake into segments and explicitly imposing the migration rates on *Cichla* from one segment to the next, migration was implicitly represented from the standpoint of the predator as an increase in prey density at a rate such as to mimic the known migration rate of the juvenile *Cichla* on the lake. As *Cichla* entered new areas, the average prey density they encountered was the original lake density. Thus, prey densities during the migration period increased at a rate equal to the migration rate times the original densities of each of the relevant prey species. The initiation and cessation of migration were keyed to the weight of the juvenile *Cichla* cohorts, with migration beginning as fish reached 200 g and ending at a weight of 600 g. By representing migration in this fashion the spatial dynamics in the model were greatly simplified. Furthermore, by gearing the timing of migration to fish weight and not time of year, the model could be corroborated by observing whether the model migration period agreed with the observed migration period, which it did.

There is a price to pay for simplifying the migration process. By having spatial migration implicit in the model the authors diverged from the common assumption in spatially homogeneous models that the state variables are the average

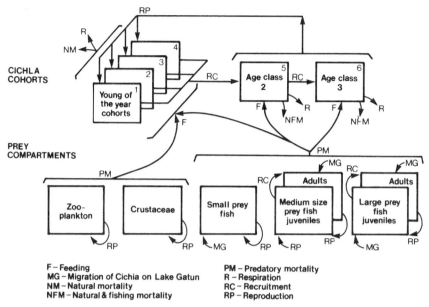

Fig. 10.2.   Compartment flow diagram for the CICHLA model of energetics, feeding, and population dynamics of *Cichla ocellaris* introduced into Lake Gatun. *Cichla* young-of-the-year are subdivided into four cohorts based on the time of spawning, whereas older fish cohorts are combined into age classes.

over the area under consideration or are at a typical or average point within the ecosystem. This model did not look at any typical point. The point moved with migration. This meant that the average point was different for each predator group, which is undesirable unless sampling proceeds in the same way that the fish proceed. Because of this assumption, the prey densities represented a composite of prey increases owing to migration and predation induced prey reduction over the entire area covered by the migration. This made the interpretation of model output less than obvious. For example, Fig. 10.3 shows the density of small- and medium-sized prey fish in the model. Notice that prey densities actually appear to increase for a period of time in the first and second years of the simulation. This is caused by the increase in density resulting from migration being greater than the predation rate on these fish—the fish are moving into an area faster than they are depleting their prey. The difference in timing of this increase between these two size groups is due to the size selectivity of *Cichla* predation, which depends on the relative size of predator and prey using the lognormal size selective feeding function of Andersen and Ursin (1977; see Chap. 9). As the *Cichla* grow their most preferable prey size also increases. Thus, they will prey most heavily on small fish earlier in the migration period (when the *Cichla* weigh less) and more heavily on medium-sized fish later in the season.

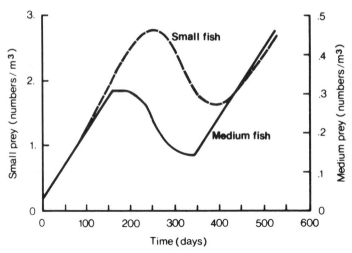

Fig. 10.3.   Graphical results from the CICHLA model showing the density of small prey fish (*Atherinids*) and medium size prey fish (*characins*) over time in a 2-year run of the model. Changes in prey density in CICHLA reflect both the predation rate by *Cichla* and the migration rate of *Cichla* into previously unexploited habitats.

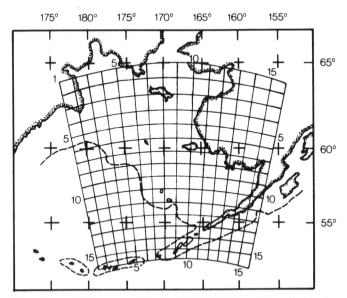

Fig. 10.4.   Grid map used for the DYNUMES model applied to the eastern Bering Sea. The map is drawn on a Mercator projection. Latitude and longitude lines demarcate the x and y axes, respectively. The continental shelf break in the Bering Sea is shown with a dashed line.

10.2.2   Bering Sea fish migration

A model like the *Cichla* model can only be applied when the migration pattern is known and can be looked at as in some sense linear. For larger areas such as the eastern Bering Sea a more explicit approach to migration was adopted by Laevastu and Larkins (1981) in a model called DYNUMES (Dynamic Numerical Ecosystem model).

This model represents the dynamics of a large number of fish species in the eastern Bering Sea as it is influenced by their prey (plankton and forage fish) and their predators (marine mammals, birds, and humans via fishing). The model is a trophic "top down" model in that initial fish biomass estimates were obtained from population estimates of marine mammals and birds and of their annual consumption of fish.

The eastern Bering Sea is divided in the model into a grid of equal area segments of 92.5 km on a side. Figure 10.4 shows the spatial resolution of this model. Migration is simulated by combining a directed migration process with a biomass density related diffusion process. Migration is a two-dimensional process with each grid segment able to communicate with each adjacent grid segment at each time step (weekly). Migration speeds in the "x" and "y" direction are provided for each species. Migration of each species at each time step is in the direction of greater biomass density of their prey species. The underlying hypothesis is that fish tend to migrate to areas where food is abundant.

To compute migration between segments a finite difference approximation is used. The biomass difference between all adjacent segments is computed. This difference is multiplied by the swim speed for that species and the time step (7 days) and is then divided by the length of the grid segment. This is then multiplied by a model specific smoothing coefficient (less than 1.0) to prevent the biomass in any segment from going negative. In addition to this directed migration there is also diffusion from each compartment to adjacent compartments depending only on the segment length, the time step, and the biomass in that segment. The diffusion process works much like the outflow process in the San Joaquin model except that flows occur equally to each adjacent segment instead of downstream. Of course, on top of the migration and diffusion functions there is feeding, growth, fishing, and other mortality, which control the spatial biomass distribution and growth of the fish species over time.

Figure 10.5 shows the model generated distribution in August of yellowfin sole, a commercially important species of flatfish in the eastern Bering Sea and of herring, a major forage fish. The similarity between these two spatial distributions is striking and reflects the migration model's moving fish toward areas of plankton abundance, which in turn attracts predators on these smaller fish. This is demonstrated in Fig. 10.6, which shows consumption of herring and of zooplankton. Areas of high herring density tend to be areas of high zooplankton consumption. A similar analogy exists between yellowfin sole density and herring consumption.

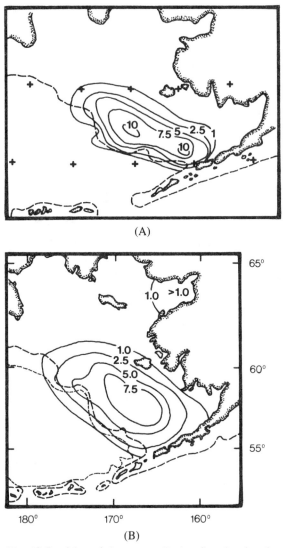

Fig. 10.5. Map of the eastern Bering Sea showing density contours produced by the DYNUMES model for August of A. yellowfin sole and B. herring. Densities are in units of tons/km². From *Marine Fisheries Ecosystem*—Laevastu and Larkins published by Fishing News Books Ltd. England.

10.3 Optimization models

We turn to optimization models as they are used in an ecological simulation model framework. First we treat an example from classical ecology; the selectivity of predators for prey types. In Chap. 9 we presented the size selective feeding hypothesis in connection with the feeding of rainbow trout on sockeye

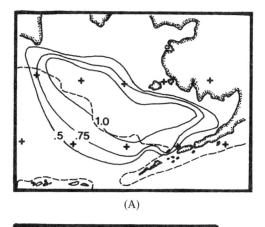

(A)

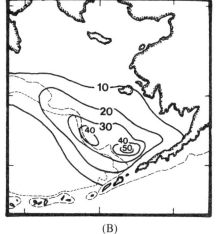

(B)

Fig. 10.6.   Map of the eastern Bering Sea showing contours of consumption of A. herring and B. zooplankton for August. Herring consumption is given in units of tons/km^2 whereas zooplankton consumption is represented as a percentage of mean zooplankton standing stock. From *Marine Fisheries Ecosystem*—Laevastu and Larkins published by Fishing News Books Ltd. England.

salmon. Here we consider feeding selection arising from the predator's internal objectives; if such a thing can be said to exist. In this case the objective is to maximize energy intake. We then turn to management applications where some conceived objective is placed on the system and human "controls" are adopted based on this objective relative to model performance.

10.3.1   Optimal foraging and prey selection

It has often been observed that fish appear to select prey according to their relative size. This phenomenon has led to a number of hypotheses to explain it. One is that the fish actually select their prey such as to optimize their energy intake per unit handling time. Another is that, because fish can see larger prey at greater distances, the encounter rates, and, therefore, the pursuit rates, on larger prey are larger than those on smaller prey. Of course both of these hypotheses only consider prey small enough that they can be feasibly taken by the predator. We describe the *optimal foraging hypothesis* here. It is an example where a general theory has been developed about a process and illustrates the difficulties in corroborating such a theory.

Optimal foraging theory perceives a predator as dividing its time between searching for, pursuing, and eating (if captured) prey. Prey are classified according to species and, here, into discrete size (length) classes. The theory asserts that the predator will choose to pursue prey such as to maximize its energy intake rate. Each prey type (species and size class) is encountered at a different rate.

The energy intake rate if all prey encountered would be pursued is equal to:

energy intake rate = [search time × encounter rate × capture success × prey density × energy intake per capture]/[search time + prey density × encounter rate × handling time × search time]

This is equal to the search time times the prey energy intake per unit search time divided by the total search time plus prey handling time. Because search time is in both the numerator and denominator of this expression it can be canceled out. The mathematical expression of this equation is:

$$gintak = \frac{\sum_{i=1}^{n} xprey_i \sum_{j=1}^{m_i} generg_{ij} \times gcapeff_{ij} \times genctr_{ij} \times glenprp_{ij}}{1 + \sum_{i=1}^{n} xprey_i \sum_{j=1}^{m_i} genctr_{ij} \times ghantim_{ij} \times glenprp_{ij}} \qquad (10.1)$$

n = total number of prey types
m_i = number of size classes for prey type i
$gintak$ = energy intake per unit time
$xprey_i$ = density of prey species i
$glenprp_{ij}$ = proportion of species i in length class j
$generg_{ij}$ = energy intake per captured individual of class ij
$gcapeff_{ij}$ = fraction of pursued prey of type ij captured
$genctr_{ij}$ = encounter rate of prey type ij
$ghantim_{ij}$ = time taken to pursue, capture, and eat an item of prey type ij

The handling time $ghantim_{ij}$ is the sum of the pursuit and eating times:

$$ghantim_{ij} = gpurtim_{ij} + gcapeff_{ij} \times geattim_{ij}$$

The optimal foraging theory gives criteria for how the predator should choose which prey groups to pursue such as to maximize equation (10.1). It can be shown that the energy intake rate will increase as long as the ratio of energy content of a prey taken over the handling time of that prey type $\frac{generg_{ij}}{ghantim_{ij}}$ is greater than some critical threshold, which depends on the parameters and prey densities in equation (10.1). Thus, the optimal solution says to always pursue prey having a ratio greater than this critical value and eschew prey with a smaller ratio. The derivation of this criterion and how to compute it is given in Charnov (1976). The above equations are a simplification of equations derived by Eggers (1977) in an application of optimal foraging to fish feeding, in that they assume that the encounter rates, capture efficiency, energy content, and handling time of each prey type are size-class specific whereas Eggers assumes they are a continuous function of prey size. To generalize the equations the sums over j would be replaced by integrals.

As a test of the optimal foraging hypothesis Werner and Hall (1974) examined the relationship between size distributions of *Daphnia* in the diets of blue-gill sunfish and the *Daphnia* size distribution in the water column. They found that at low *Daphnia* densities the sunfish consumed all *Daphnia* encountered (the size distributions in sunfish stomachs were the same as those in the water column). At high-prey densities, however, the sunfish consumed the larger size classes with greater frequency and the smaller size classes with lesser frequency. The implications of this experiment are that, although encountered at the same rate, the smaller *Daphnia* were not pursued because they were below the energy per unit handling time threshold for optimizing energy consumption rate.

Although the Werner and Hall experiment was seen as supporting the optimal foraging hypothesis, Eggers pointed out that another explanation is equally plausible—namely that encounter rates and relative size of prey alone determine prey selection. This idea was developed in work by O'Brien et al. (1976). At low prey densities the sunfish would most commonly encounter prey singly whereas at higher densities multiple prey might be within the visual field of the fish at a given time. The larger prey, being visually more stimulating, would thus tend to be pursued more commonly than the smaller prey. O'Brien modeled this hypothesis and produced results that agreed well with Werner and Hall's data without invoking the optimal foraging hypothesis.

We are uncertain about the role of optimality in ecological processes. The explanatory potential of the theory is attractive, but alternative hypotheses often cannot be discounted, except perhaps by careful experimentation. Furthermore, even if animals do attempt to optimize, their objectives may be more sophisticated than simple energy optimization and constraints may exist on the processes. Also, some of the parameters, such as capture success, may be difficult to measure in the field and may be different between the field and the laboratory.

10.3.2   Management applications overview

Optimization has been extremely important in the management of ecosystems. Ecosystem managers such as forest, range, and fishery managers and National Park personnel are beset with problems of conflicting demands on the resources they manage. Natural resource industries such as timber companies and fishermen have to deal with the effects that their industry has on its resource base, so that the resource can be truly kept in perpetuity as a renewable resource. To deal with these problems resource managers have, over the past few years, begun to rely on optimization techniques, which had already proven their worth for industrial planning and military applications. These are techniques for optimizing some *objective function* relative to the resource subject to *constraints*—financial, manpower, and resource based. The objective may be to maximize harvest or profit or to minimize cost or time to accomplish a task.

Optimization methods have been applied most successfully to natural resources in the area of timber harvesting; to determine harvest rotation schedules (frequency of cutting) or to determine the best order to harvest and replant a large area consisting of many forest stands. The success of these applications derives from the relative straightforwardness of the objectives, the relative amount of control the manager has over the resource, the long-term data set available to model the resource's growth, and the predictability of the resource's growth and establishment.

In other application areas one or more of these four features are lacking and this either makes the applications less effective or the conclusions less certain. For example, in fisheries many stocks have highly variable recruitment, which makes prediction of the future of the resource uncertain. Even in cases where fish stocks are influenced through hatchery programs, the establishment success is highly variable from year to year and depends on factors (such as river flow and the abundance of predators) beyond the control of the manager. Another example is arid grazing lands where uncertainty about the occurrence of drought makes animal stocking subject to a risk of losing the herd. Other cases have multiple objectives that are often conflicting. For example, forest managers have to balance the demands of timber harvest, hiking trail systems, wildlife management, fire control, and other recreational uses of the land. In newly developed areas mistakes relative to native biota can easily be made because there are limited or no data about the biota.

In cases of multiple resources where the manager is interested in achieving some objectives over a period of time (a planning horizon) it is often useful to have some dynamic model of the managed system to which the optimization technique can be applied. For example, Swartzman and Van Dyne (1972) combined a simulation model of herbivore–plant dynamics in an arid grazing land in Australia with an optimization technique called *linear programming* to attempt to achieve maximum profit from a property over a 12-year planning horizon, while meeting constraints of minimum viable herds of cattle and sheep, of keeping the herbage above a level required to resist drought conditions, and of maintaining control over native herbivores (rabbits and kangaroos) that compete with the cattle and sheep for forage. Because the linear programming algorithm was *static* (no explicit time considerations) rather than *dynamic,* the solution achieved was not guaranteed to be the true optimum over the entire planning horizon. The solution, however, was evaluated by comparing it with other management practices (such as constant stocking levels) now maintained on these properties and was found to be superior, especially under drought conditions.

Another example of optimizing over a planning horizon, with fisheries, dealt with the management of the South African anchovy (*Engraulis capensis*). This system was examined by Getz (1985) who used an optimization technique called *Pontryagin's maximization principal* combined with a modified Leslie matrix population dynamics model (with a nonlinear stock–recruit relationship) to maximize yield over a 10-year planning horizon. Average annual yield for the optimal solution was compared with constant fishing effort and constant escapement (harvest at a level to assure that a constant number of fish escape harvest and spawn each year) policies and was found to be superior, both at maintaining stock levels above desired minima and at maximizing long-term average harvest rates.

We treat our next example in greater detail because it is simple enough to be described more completely than the last two examples. It deals with grazing land management in India.

10.3.3   Indian grazing example

An unusual aspect of this optimization application to grazing land management is that, although the land is grazed by cattle, the cattle are not harvested, because the land is in India, where cows are sacred and cattle harvest is sacrilegious. The objective for land managers is to maintain as much high quality herbage as possible over the long-term. The stocking level of cattle in each year is under the manager's control. The options are ungrazed, light (two 15-day grazing periods a year with 10 cows/ha), moderate (2-month long grazing periods with a stocking level of 15 cows/ha), or heavy (a 2-month 15 cows/ha period followed by 10 cows/ha stocking and scraping of the land until the arrival of the next monsoon) grazing.

The grassland is located near Benares on the upper Gangetic plain, and is subject to a 4-month rainy season (June–September) with the other months being extremely dry. Annual rainfall is about 110 cm of which 90 percent comes during the rainy season. The perennial grasses consist of seven species, with the dominant being *Dichanthium annulatum*. A linear donor-controlled difference equation model was constructed by Swartzman and Singh (1974) to represent the annual transfer of percent cover of the study area between the species for ungrazed, lightly grazed, moderately grazed, and heavily grazed conditions. The transfer fractions for the transition matrices for each grazing regime were estimated from data collected over a 3-year period at enclosures on the study site. Land not covered by any of the seven species was considered to be bare space. The existence of bare space can be important to management of this grassland because during the rainy season annual grass species, which form an important part of the grazing forage, grow on this bare space. Too much bare space is undesirable, however, because it leads to heavy erosion during the rainy season. Figure 10.7 shows the simulated succession sequence over 30 years of nongrazing. The system becomes rapidly dominated by *Dichanthium annulatum*, although later *Desmodium triflorum*, which requires shade for growth, appears as a co-dominant. The assumption in the model is that the transfer matrix does not change over time. The steady state condition derived from the ungrazed

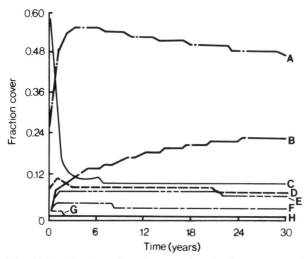

Fig. 10.7. Fraction of total cover of each of seven perennial grass species plus bare space in an ungrazed succession sequence beginning with a heavily grazed initial condition. These time traces represent model output from a linear, donor controlled succession model for a Benares, India grassland. Curve A = *Dichanthium annulatum,* B = *Desmodium triflorum,* C = bare space, D = *Bothriochloa pertusa,* E = *Indigofera linifolia,* F = *Alysicarpus monilifer,* G = *Convolvulus pluricaulus,* and H = *Evolvulus alsinoides.* From Swartzman and Singh (1974). Used by permission of Blackwell Science Publications Ltd.

model gave results that were close to the species mix in long-term ungrazed plots.

Management objectives and model

Maintenance of long-term quality of the herbage and use of the grazing land was represented by five criteria. These were the total biomass use (by grazing and scraping), the percent cover by *Dichanthium annulatum* (the ungrazed system dominant and a measure of range land condition), the percentage of bare space (linked to annual production and erosion), the diversity index (another measure of grazing land health), and the percent cover by legumes (a measure of the nutritional status of the grazing land). Three of the species, *Desmodium triflorum, Alysicarpus monilifer,* and *Indigofera linifolia,* are legumes. The diversity index is calculated by:

$$diversity = \sum_i p_i \times log(p_i)$$

where p_i is the percentage of biomass in species i. Annual production was considered in computing plant utilization by using a relationship between bare space, annual plant production, and fraction of annual plants used under each grazing regime.

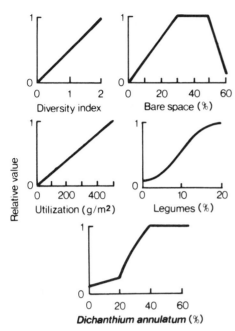

Fig. 10.8. Value standardization curves for five measures of grazing land quality. Each measure has been normalized to be between 0 and 1 and reflect management criteria for the Benares, India grassland. Used by permission of Blackwell Science Publications Ltd.

To place each of the criteria in the objective function into the same framework a series of value curves were developed, relating the levels of each of these criteria to their value to the manager. These are shown in Fig. 10.8. The goal then became to find the sequence of grazing strategies over a 12-year period that maximized the total value of the indices associated with the objectives. The optimal solution was found using *dynamic programming*. A *branch and bound algorithm* was employed.

Dynamic programming was developed by Richard Bellman to solve dynamic optimization problems where decisions around controllable variables are taken in stages (i.e., at discrete times). There are many dynamic programming algorithms and each application seems to require a different solution algorithm, although there are some common rules. To describe the branch and bound algorithm used in simple terms we consider two aspects of the problem; the state of the system and the value of the objective function. Here the state is the total biomass on the site (more exactly, it is the biomass of each species but the state was simplified to a single state; the total biomass) and the objective function is the sum of the values of the five criteria.

The algorithm operates under the principal that at each stage any decision that leads to a state that is both inferior and has a lower value of the objective function than some other decision can be eliminated as a possible optimal decision at that stage. The algorithm proceeds by brute force, starting at the first decision stage, eliminating all inferior decisions at that stage and then taking the second stage decisions, starting from each of the remaining states (those left after all inferior cases have been eliminated at the previous stage). The number of possible decision sequences proliferates rapidly under this algorithm, unless most of the decisions are eliminated at each stage. For a four-decision problem, such as described here (three grazing states and no grazing), with 12-decision stages, the total possible number of decision sequences is 4^{12}, an enormous number. This is why the method works best with a small number of decision alternatives at each stage.

Optimal solution

Table 10.1 shows the first two stages of the algorithm. The heavy grazing alternative was eliminated at the first stage because it gave less value V and less total plant biomass P than the moderate grazing case. No other case could be eliminated at this stage because each one resulted in a stage-value combination that was higher in one of these attributes than any other decision sequence could produce. At the second stage all decisions following nongrazing in the first year except nongrazing in the second year were eliminated because both value and biomass were inferior to some other option (here it was the moderate–no grazing option).

The final optimal solution is shown in Fig. 10.9. It is moderate grazing in every year with a single year of light grazing thrown in. This year was necessary to prevent legumes, which are dropping over time under persistent moderate grazing, from becoming seriously low. You can see from Fig. 10.8 that when the legume percentage falls below 15 percent the value associated with it drops

Table 10.1   Branch and bound algorithm applied to grazing problem (first 2 years).

			Strategy		
			1	$V = 5.996, P = 734$	
	1	$V = 3.032, P = 482$	2	$V = 7.017, P = 428.83$	eliminate
		No grazing	3	$V = 7.761, P = 305.1$	eliminate
			4	$V = 7.549, P = 6.4$	eliminate
			1	$V = 6.99, P = 527.1$	
Start	2	$V = 3.954, P = 275.1$	2	$V = 7.93, P = 302.6$	
		Light grazing	3	$V = 8.64, P = 214$	eliminate
			4	$V = 8.35, P = 4.36$	eliminate
			1	$V = 7.84, P = 446.2$	
	3	$V = 4.655, P = 194.2$	2	$V = 8.76, P = 253.3$	
		Moderate grazing	3	$V = 9.29, P = 178.4$	
			4	$V = 8.89, P = 3.55$	eliminate

4    $V = 4.384, P = 3.91$ eliminate
Heavy grazing

Year 1   Year 2

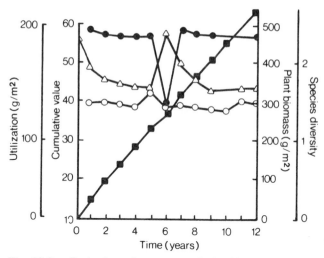

Fig. 10.9.   Optimal grazing strategy obtained by dynamic programming for a 12-year planning horizon for the Benares grassland model. Output for each of the criteria upon which the objectives are based is shown. The optimal solution calls for moderate grazing for all years except year 6 when the system was lightly grazed. Legend is ■ - cumulative value; ○ - species diversity; Δ - plant biomass; • - utilization. Used by permission of Blackwell Science Publications Ltd.

precipitously. Moderate grazing tends to force legumes down to about 12 percent of the total forage, whereas light grazing pushes it up over 20 percent.

This example shows the complications inherent in a multiobjective optimization and suggests why complicated dynamic problems are not commonly managed by optimization, even when the underlying model is simple.

10.4   The end

We have finally come to the end of this text. We have covered a vast amount of material, some of it in detail and some only superficially. Those of you who are interested in continuing further will find the keys to the literature at the end of each chapter useful in aiding your further growth. For many of you we hope this is just the beginning. Simulation modeling today needs to broaden its horizons. We must develop innovative ways of communicating model results. We must continue to develop the data bases so necessary to building and testing these models. We must continue to educate others to the strengths and weaknesses of this tool. Creative applications in any field require a good overview. The modeling literature has burgeoned so rapidly that it is difficult to get this overview in any application area solely by reading the literature. The need is great for review articles in specific fields of ecological simulation. We think that this book is a step in the right direction by providing keys to the literature. We hope that we have provided enough background that you can read the literature confidently and critically and that you carry application and theory beyond its present confines, or at least know enough to appreciate when a true advance has been made.

10.5 Annotated Bibliography

Abramsky, Z., and G. M. Van Dyne: Field studies and a simulation model of small mammal inhabiting a patchy environment. *Oikos,* 35:80–92, 1980.
A good example of model–field study interaction with spatial considerations.

Ackoff, R. L., and M. W. Sasieni: *Fundamentals of Operations Research,* Wiley, New York, 1968.
Includes a review of linear programming that is comprehensible without linear algebra. Also reviews many other operations research methods including queueing theory.

Charnov, E. L.: Optimal foraging, the marginal value theorem. *Theor. Popul. Biol.,* 9:129–136, 1976.
The original optimal foraging paper

Clark, C. W.: *Mathematical Bioeconomics: The Optimal Management of Renewable Resources,* Wiley, New York, 1976.
A readable introduction to optimal control theory with special emphasis on management of fishery and forestry resources over time.

Eggers, D. M.: The nature of prey selection by planktivorous fish. *Ecology,* 58:46–59, 1977.
Excellent in its attempt to combine model predictions, experimental evidence, and ecosystem theory in a study of fish feeding strategies.

Getz, W. M.: Optimal and feedback strategies for managing multicohort populations. *J. Optim. Theory Appl.,* 46:505–514, 1985.
Treats a fishery problem as a nonlinear control problem, using Pontryangin's maximization principal and the method of gradients for solution.

Goh, B. S.: *Management and Analysis of Biological Populations,* Elsevier, New York, 1980.
Applications of optimization models to biological populations

Hadley, G.: *Linear Programming,* Addison-Wesley, Reading, Mass, 1962.
A review of linear programming requiring linear algebra.

Kremer, J. N., and S. W. Nixon: *A Coastal Marine Ecosystem,* Ecological Studies, 24, Springer-Verlag, New York, 1978.
This model of Narragansett Bay has an approach to including spatial heterogeneity that worked well in this case.

Laevastu, T., and H. A. Larkins: *Marine Fisheries Ecosystem: Its Quantitative Evaluation and Management,* Fish News Books Ltd., Farnham, England, 1981.
An overview of the massive modeling mission of Laevastu on the Bering Sea and points beyond.

Simons, T. J.: Analysis and simulation of spatial variations of physical and biochemical processes in Lake Ontario. *J. Great Lakes Res.,* 2:215–233, 1976.
This paper makes the point very well that it is easier to fit a model to spatially aggregated data than to try to model various spatial segments and their interaction.

Swartzman, G. L., and G. M. VanDyne: An ecologically based simulation–optimization approach to natural resource planning. *Annu. Rev. Ecol. Syst.,* 3:347–395, 1972.
Combines a simulation with a look ahead optimization to try to achieve a good long-term management strategy for an arid land grazing system in Australia.

Appendix 1

Matrices

Matrices

Matrices were developed by Hamilton and Cayley around 1865 and were first used in applied mathematics by Heisenberg in 1925. For an initial introduction we can think of a matrix as an array of numbers organized into rows and columns. The position of each number in the array is identified as being in a row i and a column j. The *dimensions* or order of a matrix are its numbers of rows and columns. An $m \times n$ matrix has m rows and n columns. All matrices in the text are denoted by a bold upper case letter. An $n \times m$ matrix $\mathbf{A}$ is written as:

$$\mathbf{A} = \begin{bmatrix} a_{11} & a_{12} & \cdots & a_{1m} \\ a_{21} & a_{22} & \cdots & a_{2m} \\ \cdot & \cdot & & \cdot \\ \cdot & \cdot & & \cdot \\ \cdot & \cdot & & \cdot \\ a_{n1} & a_{n2} & \cdots & a_{nm} \end{bmatrix}$$

where the a_{ij}'s are the *elements* of the matrix $\mathbf{A}$.

The *transpose* of an $n \times m$ matrix $\mathbf{A}$, denoted $\mathbf{A}^T$ is the $m \times n$ matrix whose first row is the first column of $\mathbf{A}$, second row is the second column of $\mathbf{A}$, etc.

A *vector* is a matrix with only one column (a column vector) or only one row (a row vector). All vectors in the text are column vectors and are denoted by a bold lower case letter. The transpose is used to denote a row vector, i.e.,

$$\text{if } \mathbf{x} = \begin{bmatrix} x_1 \\ x_2 \\ \cdot \\ \cdot \\ \cdot \\ x_n \end{bmatrix} \text{ then } \mathbf{x}^T = (x_1 \ x_2 \ \cdots \ x_n)$$

305

Operations on matrices, such as addition, subtraction, multiplication, and division, may be performed just as they are for *scalars* (single numbers), although the rules for these operations are different.

Matrix addition and subtraction

In matrix addition or subtraction matrices of the same dimension are combined by adding or subtracting their corresponding elements. Matrix addition and subtraction is not defined for matrices of different dimensions. Thus if **A, B,** and **C** are all $n \times m$ matrices then **C** = **A** + **B** is defined by:

$$c_{ij} = a_{ij} + b_{ij} \quad \text{for each } i = 1, \ldots, n \text{ and } j = 1, \ldots, m$$

For example:

$$\begin{bmatrix} 2 & 1 \\ 3 & 2 \end{bmatrix} + \begin{bmatrix} 4 & 5 \\ 1 & 0 \end{bmatrix} = \begin{bmatrix} 2+4 & 5+1 \\ 3+1 & 2+0 \end{bmatrix} = \begin{bmatrix} 6 & 6 \\ 4 & 2 \end{bmatrix}$$

Scalar multiplication

A matrix is multiplied by a scalar by multiplying each element by the scalar. If **B** = $k \times$ **A**, then $b_{ij} = k \times a_{ij}$ for each i and j. For example:

$$3 \begin{bmatrix} 2 & 1 \\ 3 & 2 \end{bmatrix} = \begin{bmatrix} 6 & 3 \\ 9 & 6 \end{bmatrix}$$

Multiplication

Two matrices are multiplied by summing the elementwise product of each of the rows of the first matrix (the *premultiplier*) with each of the columns of the second matrix (the *postmultiplier*). Two matrices may be multiplied only if the number of columns in the premultiplier matrix equals the number of rows in the postmultiplier matrix. The resulting product matrix has the same number of rows as the premultiplier and the same number of columns as the postmultiplier. Thus

$$\mathbf{C}_{n \times m} = \mathbf{A}_{n \times r} \mathbf{B}_{r \times m}$$

with

$$c_{ij} = \sum_{k=1}^{r} a_{ik} b_{kj} \quad \text{for each } i = 1, \ldots, n \; j = 1, \ldots, m$$

For example:

$$\begin{bmatrix} 3 & 2 & 1 \\ 2 & 1 & 2 \end{bmatrix} \begin{bmatrix} 1 \\ 3 \\ 2 \end{bmatrix} = \begin{bmatrix} 3 \times 1 + 2 \times 3 + 1 \times 2 \\ 2 \times 1 + 1 \times 3 + 2 \times 2 \end{bmatrix} = \begin{bmatrix} 11 \\ 9 \end{bmatrix}$$

Determinants

Before we can describe matrix division we need to describe the *determinant* of a matrix. The formal definition of a determinant is rather complicated and beyond the scope of this book. The determinant of a matrix is a scalar involving sums of products of the elements of a matrix, where each product has a value from each row and each column. Determinants only exist for square matrices. The determinant of the matrix **A** is usually denoted $|\mathbf{A}|$ or $\det(\mathbf{A})$.

For 2×2 and 3×3 matrices simple formula exist for the determinant, it is the sum of the products of elements in the upper left to lower right diagonals minus the sums of products of lower left to upper right diagonals. That is:

$$\det\begin{bmatrix} a_{11} & a_{12} \\ a_{21} & a_{22} \end{bmatrix} = a_{11}a_{22} - a_{21}a_{12}$$

As an example the determinant of the 3×3 matrix is:

$$\det(\mathbf{A}) = \begin{vmatrix} 3 & 0 & 2 \\ -1 & 1 & 2 \\ 1 & -2 & -2 \end{vmatrix}$$

$$= (3 \times 1 \times (-2) + (-1) \times (-2) \times 2 + 1 \times 0 \times 2)$$

$$- (1 \times 1 \times 2 + -1 \times 0 \times (-2) + 3 \times (-2) \times 2)$$

$$= 8$$

For larger matrices the determinant can be calculated recursively by the formula:

$$|\mathbf{A}| = \sum_{j=1}^{n} (-1)^{i+j} a_{ij} A_{ij} \qquad \text{for any row } i \text{ of } \mathbf{A}_{n \times n}$$

where A_{ij} is the *cofactor* of the a_{ij} element, which is the determinant of the matrix left when row i and column j is crossed out of the original matrix **A**.

Matrix division and inverses

Instead of dividing a matrix **A** by another matrix **B** we instead multiply **A** by the inverse of **B** (denoted by $\mathbf{B}^{-1}$). This is analogous to doing division with scalars by multiplying by one over the divisor. The inverse of matrix **A** is that matrix that yields the *identity matrix* **I** when multiplied by **A**. The identity matrix is a square matrix with 1's on the diagonal and 0's off the diagonal. For example

$$\mathbf{I}_{3 \times 3} = \begin{bmatrix} 1 & 0 & 0 \\ 0 & 1 & 0 \\ 0 & 0 & 1 \end{bmatrix}$$

Thus the inverse of **A**, $\mathbf{A}^{-1}$, gives $\mathbf{A}\mathbf{A}^{-1} = \mathbf{I}$. Also $\mathbf{A}^{-1}\mathbf{A} = \mathbf{I}$; it does not matter which is the premultiplier and which is the postmultiplier.

Inverses only exist for square (n × n) matrices that must be of full *rank*. The rank of a matrix is equal to the number of linearly independent rows (which is the same as the number of linearly independent columns). A full rank matrix is one in which all rows (or columns) are linearly independent. This means that no combination of multiplying rows (or columns) of the matrix by scalars and adding or subtracting them from other rows (or columns) can result in a row (or column) consisting entirely of zeros. The determinant of a matrix can be used to check if the matrix has an inverse. A square matrix of less than full rank will have a determinant of zero, whereas a matrix of full rank will have a nonzero determinant.

The inverse of a matrix is best calculated using a computer especially if the matrix is large. For a small matrix the inverse can be calculated using the method of cofactors. If the matrix **B** is the inverse of matrix **A**, then elements b_{ij} of **B** are given by:

$$b_{ij} = -1^{i+j} \frac{A_{ji}}{D} \qquad \text{for } i = 1, \ldots, n \ \ j = 1, \ldots, n$$

where D is the determinant of matrix **A**, and A_{ji} is the cofactor of element a_{ji} of matrix **A**.

Let us compute the inverse of the matrix we found the determinant of above, using the cofactor formula.

$$b_{11} = \frac{(-1)^2 A_{11}}{8}, \qquad A_{11} = (1 \times (-2)) - ((-2) \times 2) = 2,$$

$$b_{11} = 2/8 = 1/4$$

$$b_{12} = \frac{(-1)^{1+2} A_{21}}{8}, \qquad A_{21} = (0 \times (-2)) - ((-2) \times 2) = 4,$$

$$b_{12} = -1 \times 4/8 = -1/2$$

$$b_{13} = \frac{(-1)^{1+3} A_{31}}{8}, \qquad A_{31} = (0 \times 2) - (1 \times 2) = -2,$$

$$b_{13} = -2/8 = -1/4$$

$$b_{21} = \frac{(-1)^{2+1} A_{12}}{8}, \qquad A_{12} = (-1 \times -2) - (1 \times 2) = 0,$$

$$b_{21} = 0$$

$$b_{22} = \frac{(-1)^{2+2} A_{22}}{8}, \qquad A_{22} = (3 \times (-2)) - (1 \times 2) = -8,$$

$$b_{22} = -8/8 = -1$$

By similar calculations we obtain for the other elements:

$$b_{23} = -1, b_{31} = 1/8, b_{32} = 3/4, \text{ and } b_{33} = 3/8$$

The inverse matrix of **A** is, therefore:

$$\mathbf{B} = \begin{bmatrix} 1/4 & -1/2 & -1/4 \\ 0 & -1 & -1 \\ 1/8 & 3/4 & 3/8 \end{bmatrix}$$

We can check the correctness of **B** as the inverse of **A** by matrix multiplication. When done this shows that $\mathbf{AB} = \mathbf{BA} = \mathbf{I}$; hence $\mathbf{B} = \mathbf{A}^{-1}$ and $\mathbf{A} = \mathbf{B}^{-1}$. The inverse of the inverse of a matrix is the original matrix.

Appendix 2

Eigenanalysis

Definition of eigenvalues and eigenvectors

Eigenanalysis is the technique of finding *eigenvalues* and *eigenvectors* of square matrices. To illustrate the concept let us consider the 2×2 matrix

$$A = \begin{bmatrix} 1 & 1 \\ -2 & 4 \end{bmatrix}$$

We want to see what happens to a two-dimensional vector, $\mathbf{v}$, when we pre-multiply it by $\mathbf{A}$. That is

$$\mathbf{Av} = \begin{bmatrix} 1 & 1 \\ -2 & 4 \end{bmatrix} \begin{bmatrix} v_1 \\ v_2 \end{bmatrix} = \begin{bmatrix} v_1 + v_2 \\ -2v_1 + 4v_2 \end{bmatrix} = \mathbf{u}$$

In this multiplication the vector $\begin{bmatrix} v_1 \\ v_2 \end{bmatrix}$ has been changed to the vector $\mathbf{u} = \begin{bmatrix} v_1 + v_2 \\ -2v_1 + 4v_2 \end{bmatrix}$. For example the vector $\mathbf{v}_1 = \begin{bmatrix} 1 \\ 0 \end{bmatrix}$ becomes $\mathbf{u}_1 = \begin{bmatrix} 1 \\ -2 \end{bmatrix}$ and $\mathbf{v}_2 = \begin{bmatrix} -3 \\ -1 \end{bmatrix}$ becomes $\mathbf{u}_2 = \begin{bmatrix} -4 \\ -2 \end{bmatrix}$ under this transformation. This is illustrated graphically in Fig. A2.1, where you can see that both vectors have been changed in both length and direction by the transformation of matrix premultiplication by matrix $\mathbf{A}$. If the vector $\mathbf{v}_3 = \begin{bmatrix} 1 \\ 2 \end{bmatrix}$ is premultiplied by $\mathbf{A}$, we obtain the vector $\mathbf{u}_3 = \begin{bmatrix} 3 \\ 6 \end{bmatrix}$, which is equal to the original vector multiplied by the scalar 3. This transformation is also shown in Fig. A2.1, which indicates that the resultant vector is changed in length but not in direction. When premultiplied by $\mathbf{A}$ the vector $\begin{bmatrix} 1 \\ 2 \end{bmatrix}$ is one of a family of vectors, that differ from each other only by a

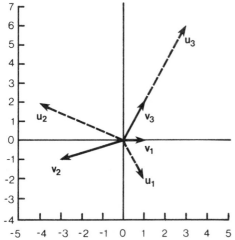

Fig. A2.1.   Graph the change in vectors $\mathbf{v}_i$ to $\mathbf{u}_i$ by the matrix transformation $\mathbf{A}\mathbf{v}_i = \mathbf{u}_i$ for $i = 1, 2, 3$. The vectors $\mathbf{v}_1$ and $\mathbf{v}_2$ change direction and length whereas $\mathbf{v}_3$, an eigenvector of $\mathbf{A}$, changes in length only.

scalar multiple, that give another vector of this class. These are termed the *eigenvectors* of the matrix $\mathbf{A}$. The scale factor 3 associated with the transformation of the eigenvector is called an *eigenvalue* of matrix $\mathbf{A}$. Transformation of eigenvectors by their mother matrix results in a change only in magnitude and not in direction of the vector. Note that there may be more than one family of eigenvectors associated with a matrix. Also, only square matrices have eigenvectors.

Now that we have introduced the concept of eigenvectors. Let us more formally define them. An eigenvalue, eigenvector pair of $\mathbf{A}$ is any real or complex number-vector pair denoted as $(\lambda, \mathbf{v})$, such that $\mathbf{A}\mathbf{v} = \lambda\mathbf{v}$.

Because $\mathbf{A}\mathbf{v} = \lambda\mathbf{v}$ is equivalent to $\mathbf{A}(k\mathbf{v}) = \lambda(k\mathbf{v})$ for all scalars k, we see that there are an infinite number of eigenvectors associated with a particular eigenvalue.

Finding eigenvalues

We now show how to find the eigenvectors and eigenvalues for a matrix. The method arises from considering the definition of eigenvectors and eigenvalues. They must satisfy $\mathbf{A}\mathbf{v} = \lambda\mathbf{v}$ or $(\mathbf{A}\mathbf{v} - \lambda\mathbf{I}\mathbf{v}) = (\mathbf{A} - \lambda\mathbf{I})\mathbf{v} = 0$. If the matrix $\mathbf{A} - \lambda\mathbf{I}$ has an inverse, then premultiplying both sides of the above equation by this inverse would give $\mathbf{v} = 0$. This is not what we want to find, because we are looking for nonnull eigenvectors $\mathbf{v}$. If the matrix does not have an inverse then we can hope that nonnull vectors $\mathbf{v}$ exist that satisfy the condition. This can only happen if the determinant of $\mathbf{A} - \lambda\mathbf{I}$ equals zero. We can use this condition, that the determinant of $\mathbf{A} - \lambda\mathbf{I} = 0$, to solve for the eigenvalues of matrix $\mathbf{A}$. This condition $\det(\mathbf{A} - \lambda\mathbf{I}) = 0 = \Phi_{\mathbf{A}}(\lambda)$ is called the *characteristic equation* for $\mathbf{A}$ and is an nth degree polynomial in λ for an $n \times n$ matrix. Therefore, there are at most n distinct values of λ that satisfy $\Phi_{\mathbf{A}}(\lambda) = 0$.

As an example let us find the eigenvalues and eigenvectors of the matrix **A** considered above. The characteristic equation of this matrix is

$$\det(\mathbf{A} - \lambda\mathbf{I}) = \det\left(\begin{bmatrix} 1 & 1 \\ -2 & 4 \end{bmatrix} - \begin{bmatrix} \lambda & 0 \\ 0 & \lambda \end{bmatrix}\right)$$

$$= \det\begin{bmatrix} 1 - \lambda & 1 \\ -2 & 4 - \lambda \end{bmatrix}$$

$$= (1 - \lambda)(4 - \lambda) + 2 = \lambda^2 - 5\lambda + 6 = 0$$

There are two distinct solutions to this quadratic equation, which are $\lambda = 3$ and $\lambda = 2$. These two solutions are identified as $\lambda_1 = 3$ and $\lambda_2 = 2$. Each of these eigenvalues is associated with an eigenvector that can be solved for by solving the set of equations $\mathbf{Av} = \lambda\mathbf{v}$ with λ_1 or λ_2 substituted for λ in the equation. Because we already know an eigenvector associated with λ_1 (we used it as an example above) let us solve for an eigenvector of λ_2. From the defining equation we have:

$$\begin{bmatrix} 1 & 1 \\ -2 & 4 \end{bmatrix}\begin{bmatrix} v_{21} \\ v_{22} \end{bmatrix} = 2\begin{bmatrix} v_{21} \\ v_{22} \end{bmatrix}$$

This gives two equations in two unknowns; however, the two equations are redundant (they are not linearly independent). The first equation tells us that $v_{21} = v_{22}$ and so does the second equation. This emphasizes the point made earlier that each eigenvector is a member of a family. Any vector with $v_{21} = v_{22}$ works. We can arbitrarily choose the vector $\begin{bmatrix} 1 \\ 1 \end{bmatrix}$ as our representative eigenvector.

An $n \times n$ matrix has at most n eigenvalues and families of eigenvectors. In some cases there are fewer than n. The following is such a case.

Let us find the eigenvalues for the matrix;

$$\begin{bmatrix} 2 & 2 & 1 \\ 1 & 3 & 1 \\ 1 & 2 & 2 \end{bmatrix}$$

solving the characteristic equation gives us:

$$\Phi_\mathbf{A}(\lambda) = \det\begin{bmatrix} 2 - \lambda & 2 & 1 \\ 1 & 3 - \lambda & 1 \\ 1 & 2 & 2 - \lambda \end{bmatrix} = \lambda^3 - 7\lambda^2 + 11\lambda - 5 = 0$$

There are two distinct roots for this polynomial equation: $\lambda_1 = 5$ and $\lambda_2 = 1$ and the second eigenvalue appears twice in the solution to the polynomial and is called a twofold repeated root. When an attempt is made to solve for the eigenvector corresponding to λ_2 it is found that of the three equations in three unknowns two of them are redundant. In this case all three equations tell us that $v_{21} + 2v_{22} + v_{23} = 0$. We can find two linearly independent vectors that satisfy this equation. Linear independence of two vectors means that no one of them can be expressed as a linear combination of the other. This implies that one is not a scalar multiple of the other, as all eigenvectors associated with a unique eigenvalue (e.g., $\lambda_1 = 5$) would be. For example, the two eigenvectors $\mathbf{v}_2^{(1)} = \begin{bmatrix} 2 \\ -1 \\ 0 \end{bmatrix}$ and $\mathbf{v}_2^{(2)} = \begin{bmatrix} 1 \\ 0 \\ -1 \end{bmatrix}$ are linearly independent and are associated with

$\lambda_2 = 1$. They both satisfy $v_{21} + 2v_{22} + v_{23} = 0$. Thus, a twice repeated eigenvalue has two families of linearly independent eigenvectors associated with it.

Appendix 3

Review of Probability and Statistics

Probability

We define a *random variable* as a real variable whose value is determined by the outcome of a random experiment. There are two types of random variables, a *discrete random variable* can only take a finite or countably infinite number of discrete values. A *continuous random variable* can take on a continuum of values.

A random variable's probability distribution assigns a probability to all possible outcomes of that random variable. For continuous random variables, the probability distribution is described by the *probability density function* (often abbreviated p.d.f.), which is denoted by $f(x)$, and for discrete random variables by the *probability function,* which is denoted by $p(x)$. The probability distribution for both continuous and discrete random variables is also described by the *cumulative distribution function* (often abbreviated c.d.f.). The c.d.f. for a random variable X is a function such that

$$F(x) = Pr(X \leq x)$$

i.e., the probability that random variable X is less than or equal to x. The probability function for a discrete random variable X is:

$$p(x) = Pr(X = x)$$

and for a continuous random variable X the p.d.f. is defined by

$$f(x) = \frac{dF(x)}{dx} \quad \text{or} \quad F(x) = \int_{-\infty}^{x} f(x) \, dx$$

The expected value or *mean* of a random variable is the weighted average of the possible values of the random variable (weighted by the probability of occurrence). Mathematically, for the discrete random variable X with probability function $p(x)$, the expected value of X, denoted $E(X)$ is:

$$E(X) = \sum_{all\ x} xp(x)$$

and for the continuous random variable Y with p.d.f., $f(y)$, the expected value is:

$$E(Y) = \int_{-\infty}^{+\infty} yf(y)\ dy$$

The mean is a measure of the "center" of a distribution. The *variance* is a measure of the dispersion of a distribution. It is the expected value of the squared deviations from the mean

$$Var(X) = E(X - E(X))^2$$

$$= \sum_{all\ x} (x - E(X))^2 p(x) \quad \text{for a discrete random variable}$$

or

$$= \int_{-\infty}^{+\infty} (x - E(X))^2 f(x)\ dx \quad \text{for a continuous random variable}$$

The positive square root of the variance is called the *standard deviation* and is often used instead of the variance to measure variability. The symbols μ and σ^2 are frequently used to denote the mean and the variance of a distribution, respectively.

Refer to Hoel, Port, and Stone (1971) or Hogg and Craig (1978) for a more thorough discussion of the basic concepts of probability distributions.

Sample statistics

The goal of statistics is to make inferences about a distribution by sampling from it. For a sample $x_1, x_2, \ldots, x_n$ the *sample mean* (denoted by $\bar{x}$) is the equally weighted sum of the observed values,

$$\bar{x} = \frac{1}{n} \sum_{i=1}^{n} x_i$$

Similarly, the *sample variance* (denoted s^2) is the equally weighted sum of squared deviations from the mean,

$$s^2 = \frac{1}{n-1} \sum_{i=1}^{n} (x_i - \bar{x})^2$$

For mathematically optimal purposes, the weights $\dfrac{1}{n-1}$ are used rather than

$\dfrac{1}{n}$. As the sample size n increases, both estimates $\bar{x}$ and s^2 approach their true

values μ and σ^2.

A *histogram* is a graphical representation of a sample in which the range of values is divided into intervals and a bar above each interval is drawn proportional to the number of values from the sample that fall into that interval. An example of a histogram is shown in Fig. 7.5. As the number of observations in the sample gets large and the size of the intervals decreases the histogram looks more and more like the random variable's p.d.f. By comparing the shape of a histogram to various p.d.f.'s we can get an idea from which probability distribution our data arise.

Quantile–Quantile plots

Another graphical method for comparing a sample to a known probability distribution is a *theoretical quantile–quantile plot* often called a theoretical Q–Q plot or probability plot. A *quantile* for a continuous random variable with c.d.f. $F(x)$ is a number $Q_t(p)$ (denoting the theoretical quantile) such that

$$F(Q_t(p)) = p \quad or \quad Q_t(p) = F^{-1}(p)$$

where $F^{-1}(\cdot)$ is the inverse function of $F(\cdot)$. $Q_t(p)$ is the value such that a fraction p of the c.d.f. occurs for values less than or equal to $Q_t(p)$. Some distributions (e.g., the normal) have no closed form for $Q_t(p)$ and it must be approximated, usually on a computer.

Quantiles can also be computed for a sample $x_1, x_2, \ldots, x_n$ and we denote these by $Q_e(p)$ for empirical quantiles. Let $x_{(1)}, x_{(2)}, \ldots, x_{(n)}$ be the sample values sorted from smallest to largest. Then $Q_e(p_i) = x_{(i)}$ where $p_i = \dfrac{i - 0.5}{n}$. In words, a fraction p_i of the sample values are less than or equal to $Q_e(p_i)$.

Plotting $Q_e(p_i)$ against $Q_t(p_i)$ for $i = 1$ to n gives the Q–Q plot. If the sample comes from the theoretical probability distribution chosen, the plot will be approximately a straight line. Departures from a straight line indicate how the sample deviates from the theoretical probability distribution. For a very complete discussion of Q–Q plots, including formulas for calculating theoretical quantile functions, see Chambers et al. (1983).

Q–Q plots are a graphical version of what is known as a *goodness-of-fit test*. There are also many numerical goodness-of-fit tests such as the Kolmogorov–Smirnov test and the chi-square goodness-of-fit test. These tests are discussed in statistics textbooks such as Sokal and Rohlf (1981).

Correlation

The *correlation coefficient*, denoted ρ, for a pair of random variables X and Y is a measure of the linear relationship between X and Y. An estimate of ρ, usually denoted r, from a sample of size n of (x, y) pairs is given by:

$$r = \frac{\displaystyle\sum_{i=1}^{n} x_i y_i - \frac{\displaystyle\sum_{i=1}^{n} x_i \sum_{i=1}^{n} y_i}{n}}{\sqrt{\left(\left(\sum_{i=1}^{n} x_i^2 - \frac{\left(\sum_{i=1}^{n} x_i\right)^2}{n}\right)\left(\sum_{i=1}^{n} y_i^2 - \frac{\left(\sum_{i=1}^{n} y_i\right)^2}{n}\right)\right)}}$$

Both ρ and its estimate r lie between -1 and 1 with a value of 1 indicating perfect positive correlation, a -1 value implies perfect negative correlation (i.e., as X increases Y decreases) and a value of 0 indicates no correlation. Note that r only measures linear relationships, a deterministic nonlinear relationship, such as $y = x^2$, will not result in an r value of 1.

If X and Y are normally distributed, the correlation coefficient is a measure of the dependence between the random variables, with ρ of zero implying independence. For nonnormal random variables, a nonzero correlation implies dependence but a correlation of zero does not imply independence. Most basic statistical texts discuss methods for testing if ρ is equal to some hypothesized value, usually zero. These tests all assume the data are from a normal distribution.

The *partial correlation coefficient* between X and Y adjusting for Z, often denoted $\rho_{xy.z}$, is a measure of the linear relationship between the two random variables X and Y after adjusting for any linear relationship that X and Y might share with the random variable Z. The concept is easily extended to allow adjustment for any number of random variables. For estimation of the partial correlation coefficient, see Snedecor and Cochran (1980).

Another measure of linear association between two random variables is the *rank correlation coefficient* (often called the Spearman rank correlation coefficient after its founder). This is the ordinary correlation coefficient, r, computed from the ranks of the data rather than the data itself. The ranks of an ordered sample $x_{(1)} < x_{(2)} < \cdots < x_{(n)}$ are 1, 2, 3, . . . , n. Like the ordinary correlation coefficient, the rank correlation coefficient can range between -1 and 1. The advantage of using ranks instead of the raw data is that the coefficient will now measure monotonic relationships between the two variables instead of only linear relationships. For example, the variables x and x^2 will have perfect rank correlation. Another advantage of using ranks instead of the raw data is that statistical tests for a significant association have been developed for the rank correlation coefficient. These tests do not require the normality assumption needed above. The partial rank correlation coefficient is the partial correlation coefficient based on ranks.

Regression

Linear regression is a statistical technique to find the line that "best" fits a set of (x, y) data. The x values are the *independent variable* and are assumed to be measured without error. The y's are the *dependent* or "response variable." The statistical model that is assumed is:

$$y_i = \alpha + \beta x_i + \epsilon_i \qquad i = 1, \ldots, n$$

where the ϵ_i's are independent normal random variables with mean 0 and common variance σ^2 and the α and β are to be estimated. By "best" fit we mean the line that minimizes the sum of squared deviations of the y_i's from the line, i.e., we minimize:

$$SS(\alpha,\beta) = \sum_{i=1}^{n} (\alpha + \beta x_i - y_i)^2$$

This fitting technique is known as the method of *least squares*. The resulting estimates, denoted $\hat{\alpha}$ and $\hat{\beta}$, are given by:

$$\hat{\beta} = \frac{\sum_{i=1}^{n} x_i y_i - \dfrac{\sum_{i=1}^{n} x_i \sum_{i=1}^{n} y_i}{n}}{\sum_{i=1}^{n} x_i^2 - \dfrac{\left(\sum_{i=1}^{n} x_i\right)^2}{n}}$$

$$\hat{\alpha} = \frac{\sum_{i=1}^{n} y_i}{n} - \hat{\beta}\frac{\sum_{i=1}^{n} x_i}{n}$$

Multiple linear regression is the extension of linear regression to several independent variables. The regression line is replaced by a regression hyperplane with dimension equal to the number of independent variables in the regression. The general model for p independent variables is:

$$y_i = \alpha + \beta_1 x_{i1} + \beta_2 x_{i2} + \cdots + \beta_p x_{ip} + \epsilon_i \qquad i = 1, \ldots, n$$

Simple formulae for $\hat{\alpha}$ and the $\hat{\beta}_j$'s do not exist and a computer is usually needed to solve the problem.

A *residual* is the difference between the observed y and its predicted value from the regression equation. For example, in simple linear regression the ith residual is:

$$e_i = y_i - (\hat{\alpha} + \hat{\beta} x_i)$$

The residuals are estimates of the error term (ϵ_i) in the regression model. Various graphical displays using the residuals have been developed to check if the data follow the model assumptions. For instance, a normal Q–Q plot can be used to check the normality assumption of the ϵ's. The book by Chambers et al. (1983) contains a wealth of graphical methods for checking regression models.

Often the data do not meet the assumptions of linear regression but a modification or *transformation* of the data does. For example, the log function is often used to linearize an exponential model:

$$y = \alpha e^{\beta x} \, \epsilon$$

Taking the logarithms of both sides:

$$log(y) = log(\alpha) + \beta x + log(\epsilon)$$

Note that this requires the errors ϵ on the original scale to be multiplicative and not additive. Using regression on the transformed model assumes that $log(\epsilon)$ is normally distributed with mean 0 and common variance σ^2.

The area of regression is one of the largest in the field of statistics and we obviously cannot cover even a small fraction of it here. Two good reference books for regression are Draper and Smith (1981) and Weisberg (1985).

Analysis of variance

Analysis of variance or *ANOVA* is a partitioning of the variability in a set of observations into groups of specific factors. In one-way ANOVA the variability in the observations y_{ij} is partitioned into an "among" groups or treatments term and a "within" groups or error term. The model used for this is:

$$y_{ij} = \mu + \alpha_i + \epsilon_{ij} \qquad i = 1, \ldots, k, \qquad j = 1, \ldots, n_i$$

where μ is the overall mean effect, α_i is the ith treatment effect, and ϵ_{ij} are independent normally distributed errors with mean 0 and common variance σ^2. If all the assumptions are valid, then formal statistical tests can be used to find significant differences among the k treatments (see Sokal and Rohlf, 1981).

Factorial experiments

A *factorial experiment* is a method for setting up an experiment when the outcome of the experiment is influenced by factors that may be at any of a number of levels. In the factorial experiment, all possible combinations of factors and levels are used. This allows the estimation of not only the effects of each factor individually (called *main effects*) but also *interactions* between factors. A two-way interaction between factors A and B occurs when the effect of factor A depends upon the level of factor B. There can also be a three-way and higher order interactions among factors. An analysis of variance is used to test for significant main effects and interactions.

A *fractional factorial design* is a factorial experiment where only a fraction of the possible factor-level combinations are used. Because the number of possible combinations can get very large, this design has great appeal. However, the reduction in runs is not without cost. Certain effects cannot be distinguished from each other and are said to be *confounded*. By careful selection of the combinations, effects of interest (usually the main effects and lower order interactions) will be confounded only with effects that can be assumed to be negligible (usually the higher order effects). Designs for fractional factorial experiments and their confounding patterns have been published in statistics books such as Box et al. (1978) and Cochran and Cox (1957).

Annotated Bibliography

Chambers, J. M., W. S. Cleveland, B. Kleiner, and P. A. Tukey: *Graphical Methods for Data Analysis,* Wadsworth, Belmont, Calif., 1983.

Contains a wealth of information on visually analyzing and displaying data. A complete discussion of quantile–quantile plots is presented.

Draper, N. R., and H., Smith: *Applied Regression Analysis, 2nd ed.,* Wiley, New York, 1981.

A huge text on all aspects of regression including nonlinear regression.

Snedecor, G. W., and W. G. Cochran: *Statistical Methods, 7th ed.,* Iowa State University Press, Ames, Iowa, 1980.

The classic general statistical methods book. Although somewhat difficult to read, it is very complete.

Sokal, R. R., and F. J. Rohlf: *Biometry, 2nd ed.,* W. H. Freeman and Co., San Francisco, 1981.

Gives a good, ecologically oriented review of basic statistics including such relevant topics as ANOVA, correlation, and simple linear regression.

Weisberg, S.: *Applied Linear Regression, 2nd ed.,* Wiley, New York, 1985.

A good reference book on multiple linear regression. Presents the latest techniques for regression model building and regression diagnostics.

Appendix 4

Model Code

San Joaquin model code

Phytoplankton subroutines

```
      subroutine pphs (x,g,k,z)
c
c  Compute the phytoplankton photosynthesis rate.
c

      real x(1),g(1),k(1),z(1)

      call plite(x,g,k,z)
      call ppht(x,g,k,z)
      call pphn(x,g,k,z)

c     gpht = gphtlttt * gphttp * gphtnit
      g(1) = g(2) * g(3) * g(4)

      return
      end


      subroutine plite(x,g,k,z)
c
c  Compute the light effect on photosynthesis.
c

      real x(1),g(1),k(1),z(1)
      data e /2.718282/

      call plext(x,g,k,z)
```

```
c      gphttp = kphtmx * e / (gextlt * kd) *
c              (exp(-zinsol / kphtltop * exp(-gextlt * kd)) -
c              exp(-zinsol / kphtltop))
       g(2) = k(1) * e / (g(16) * k(19)) *
     +        (exp(-z(1) / k(2) * exp(-g(16) * k(19)) -
     +        exp(-z(1) / k(2)))

       return
       end



       subroutine plext(x,g,k,z)
c
c  Compute the light extinction as a function of phytoplankton
c  biomass.
c

       real x(1),g(1),k(1),z(1)

c      gextlt = kextlt + .0088 * xbiop / kCChlrt +
c              .053 * (xbiop / kCChlrt) ** .6667
       g(16) = k(20) + .0088 * x(1) / k(3) +
     +        .053 * (x(1) / k(3)) ** 0.66667

       return
       end



       subroutine ppht(x,g,k,z)
c
c  Compute the temperature effect on photosynthesis.
c

       real x(1),g(1),k(1),z(1)

c      gphttp = kphttp * ztp
       g(3) = k(4) * z(2)

       return
       end



       subroutine pphn(x,g,k,z)
c
c  Compute the nitrogen effect on photosynthesis.
c

       real x(1),g(1),k(1),z(1)
```

```
c      gphtnit = xnit / (xnit + kphtniths)
       g(4) = x(3) / (x(3) + k(5))

       return
       end


       subroutine prsp(x,g,k,z)
c
c  Compute the phytoplankton respiration rate.
c
       real x(1),g(1),k(1),z(1)

c      grsp = krsptp * ztp
       g(5) = k(6) * z(2)

       return
       end


       subroutine pinfl (x,g,k,z)
c
c  Compute the inflow of phytoplankton.
c
       real x(1),g(1),k(1),z(1)

c      gflowp = zflow / kevol * (kflowp - xbiop)
       g(6) = z(3) / k(21) * (k(7) - x(1))

       return
       end
```

Zooplankton subroutines

```
       subroutine zgraz(x,g,k,z)
c
c  Compute the zooplankton grazing rate.
c
       real x(1),g(1),k(1),z(1)

       call zgrp(x,g,k,z)
       call zgrt(x,g,k,z)

c      gfd = kfdmx * gfdpy * gfdtp
       g(7) = k(9) * g(8) * g(9)

       return
       end
```

```
      subroutine zgrp(x,g,k,z)
c
c  Compute the phytoplankton density effect on grazing.
c

      real x(1),g(1),k(1),z(1)

c     gfdpy = xbiop / (xbiop + kfdpyhs)
      g(8) = x(1) / (x(1) + k(11))

      return
      end


      subroutine zgrt(x,g,k,z)
c
c  Compute the temperature effect on grazing.
c

      real x(1),g(1),k(1),z(1)

c     gfdtp = kfdtp1 + kfdtp2 * ztp
      g(9) = k(12) + k(13) * z(2)

      return
      end


      subroutine zrsp(x,g,k,z)
c
c  Compute the zooplankton respiration rate.
c

      real x(1),g(1),k(1),z(1)

c     grspz = krsptpz * ztp
      g(10) = k(14) * z(2)

      return
      end


      subroutine zmrt(x,g,k,z)
c
c  Compute the zooplankton mortality.
c

      real x(1),g(1),k(1),z(1)
```

```
c      gmrt = kmrt
       g(11) = k(15)

       return
       end


       subroutine zinfl(x,g,k,z)
c
c  Compute the inflow of zooplankton.
c
       real x(1),g(1),k(1),z(1)

c      gflowz = zflow / kevol * (kflowz - xbioz)
       g(12) = z(3) / k(21) * (k(16) - x(2))

       return
       end
```

Nitrogen subroutines

```
       subroutine nuptk(x,g,k,z)
c
c  Compute the nitrogen uptake by phytoplankton.
c
       real x(1),g(1),k(1),z(1)

c      gnup = knitCrt * gpht * xbiop
       g(13) = k(8) * g(1) * x(1)

       return
       end


       subroutine ndisch(x,g,k,z)
c
c  Compute the nitrogen discharge into the system.
c
       real x(1),g(1),k(1),z(1)

c      gndis = kdisnit / kevol
       g(14) = k(17) / k(21)

       return
       end
```

```
      subroutine ninfl(x,g,k,z)
c
c   Compute the inflow of nitrogen.
c

      real x(1),g(1),k(1),z(1)

c     gflownit = zflow / kevol * (kflownit - xnit)
      g(15) = z(3) / k(21) * (k(18) - x(3))

      return
      end
```

Hubbard Brook forest model code

Growth subroutines

```
      subroutine growth(x,g,k,z,nspec)
c
c   Compute the tree growth rate.
c

      real x(8,6),g(20,6),k(25,6),z(5)

c   first compute all the intermediate variables
      call gsolar(x,g,k,z,nspec)
      call gtemp(x,g,k,z,nspec)
      call gmoist(x,g,k,z,nspec)

      do 10 i = 1, nspec
c         ggrow = kgrowmx * ggrowlt * ggrowtp * gspace
          g(1,i) = k(7,i) * g(2,i) * g(4,i) * g(5,i)
   10 continue

      return
      end


      subroutine gsolar(x,g,k,z,nspec)
c
c   Compute the solar radiation effect on tree growth.
c

      real x(8,6),g(20,6),k(25,6),z(5)

      call gupsr(x,g,k,z,nspec)

      do 10 i = 1, nspec
```

```
c           ggrowlt = kpgrs1 * (1.0 - exp(-kpgrs2 * (glt - kpgrs3)))
            g(2,i) = k(8,i) * (1.0 - exp(-k(9,i) * (g(9,i) - k(10,i))))
   10 continue

      return
      end




      subroutine gupsr(x,g,k,z,nspec)
c
c Compute the solar radiation from trees higher than the current one.
c

      real x(8,6),g(20,6),k(25,6),z(5)

      call gupla(x,g,k,z,nspec)

      do 10 i = 1, nspec
c           glt = zinsol * exp(-kextlt * garealf)
            g(9,i) = z(1) * exp(-k(16,i) * g(3,i))
   10 continue

      return
      end




      subroutine gupla(x,g,k,z,nspec)
c
c Compute the leaf area for trees higher than the current one.
c
c    garealf = sum xarealf(j) * xn(j)
c               over trees j with xht(j) .ge. current xht
c
c    g(3,i) = sum x(3,j) * x(4,j) over trees j with x(2,j) .ge. x(2,i)
c

      real x(8,6),g(20,6),k(25,6),z(5)

      do 20 i = 1, nspec
         g(3,i) = 0.0
         do 10 j = 1, nspec
            if (x(2,j) .gt. x(2,i)) g(3,i) = g(3,i) + x(3,j) * x(4,j)
   10    continue
   20 continue

      return
      end
```

```
      subroutine gtemp(x,g,k,z,nspec)
c
c  Compute the temperature effect on growth.
c

      real x(8,6),g(20,6),k,(25,6),z(5)

      do 10 i = 1, nspec
c          ggrowtp = 4.0 * (ztp - kdegdymn) * (kdegdymx - ztp) /
c                          (kdegdymx - kdegdymn) ** 2
           g(4,i) = 4.0 * (z(2) - k(11,i)) * (k(12,i) - z(2)) /
     +                    (k(12,i) - k(11,i)) ** 2
 10   continue

      return
      end


      subroutine gmoist(x,g,k,z,nspec)
c
c  Compute the moisture effect on growth.
c

      real x(8,6),g(20,6),k,(25,6),z(5)

      call gbasar(x,g,k,z,nspec)

      do 10 i = 1, nspec
c          gspace = 1.0 - gareabs / kareabsmx
           g(5,i) = 1.0 - g(6,i) / k(17,i)
 10   continue

      return
      end


      subroutine gbasar(x,g,k,z,nspec)
c
c  Compute the total basal area in the stand.
c
c      gareabs = pi/4 * sum xn * xdbh ** 2 over all species
c      g(6,i) = 0.785398 * sum x(4,i) * x(1,i) ** 2 over all species
c

      real x(8,6),g(20,6),k(25,6),z(5)

      sum = 0.0
      do 10 i = 1, nspec
           sum = sum + x(4,i) * x(1,i) ** 2
```

```
   10 continue

      sum = 0.785398 * sum
      do 20 i = 1, nspec
         g(6,i) = sum
   20 continue

      return
      end
```

Mortality subroutines

```
      subroutine gmorta(x,g,k,z,nspec)
c
c  Compute the age mortality rate.
c

      real x(8,6),g(20,6),k(25,6),z(5)

      do 10 i = 1, nspec
c        gmrtage = 1.0 - kmrtage) ** xage
         g(7,i) = 1.0 - (1.0 - k(13,i)) ** x(5,i)
   10 continue

      return
      end


      subroutine gmorts(x,g,k,z,nspec)
c
c  Compute the slow growth mortality rate.
c

      real x(8,6),g(20,6),k(25,6),z(5)

      do 10 i = 1, nspec
c        gmrt = 1.0 - (1.0 - kmrt) ** xage
         g(8,i) = 1.0 - (1.0 - k(14,i)) ** x(5,i)
   10 continue

      return
      end
```

Appendix 5

San Joaquin Driving Variables

San Joaquin driving variables

Temperature (ztp) is in °C and flow ($zflow$) is in L/day.

day	ztp	zflow	day	ztp	zflow	day	ztp	zflow
1	8.00	1.218e + 10	51	9.41	7.646e + 09	101	15.80	3.810e + 09
2	8.00	1.210e + 10	52	9.54	7.551e + 09	102	16.00	3.741e + 09
3	8.00	1.201e + 10	53	9.67	7.465e + 09	103	16.10	3.663e + 09
4	8.00	1.192e + 10	54	9.80	7.370e + 09	104	16.20	3.594e + 09
5	8.00	1.184e + 10	55	9.93	7.284e + 09	105	16.40	3.516e + 09
6	8.00	1.175e + 10	56	10.10	7.188e + 09	106	16.50	3.447e + 09
7	8.00	1.166e + 10	57	10.20	7.102e + 09	107	16.60	3.370e + 09
8	8.00	1.158e + 10	58	10.30	7.007e + 09	108	16.70	3.300e + 09
9	8.00	1.149e + 10	59	10.40	6.921e + 09	109	16.90	3.223e + 09
10	8.00	1.140e + 10	60	10.60	6.826e + 09	110	17.00	3.154e + 09
11	8.00	1.123e + 10	61	10.70	6.748e + 09	111	17.10	3.076e + 09
12	8.00	1.115e + 10	62	10.80	6.679e + 09	112	17.10	3.007e + 09
13	8.00	1.106e + 10	63	11.00	6.601e + 09	113	17.20	2.929e + 09
14	8.00	1.097e + 10	64	11.10	6.532e + 09	114	17.30	2.860e + 09
15	8.00	1.089e + 10	65	11.20	6.454e + 09	115	17.30	2.782e + 09
16	8.00	1.080e + 10	66	11.30	6.385e + 09	116	17.40	2.713e + 09
17	8.00	1.071e + 10	67	11.50	6.307e + 09	117	17.50	2.635e + 09
18	8.00	1.063e + 10	68	11.60	6.238e + 09	118	17.50	2.566e + 09
19	8.00	1.054e + 10	69	11.70	6.160e + 09	119	17.60	2.488e + 09
20	8.00	1.045e + 10	70	11.90	6.091e + 09	120	17.70	2.419e + 09
21	8.00	1.037e + 10	71	12.00	6.013e + 09	121	17.70	2.393e + 09
22	8.00	1.028e + 10	72	12.10	5.944e + 09	122	17.80	2.359e + 09
23	8.00	1.020e + 10	73	12.20	5.867e + 09	123	17.90	2.333e + 09
24	8.00	1.011e + 10	74	12.40	5.797e + 09	124	17.90	2.307e + 09
25	8.00	1.002e + 10	75	12.50	5.720e + 09	125	18.00	2.272e + 09
26	8.00	9.936e + 09	76	12.60	5.651e + 09	126	18.10	2.246e + 09

day	ztp	zflow	day	ztp	zflow	day	ztp	zflow
27	8.00	9.850e + 09	77	12.80	5.573e + 09	127	18.10	2.220e + 09
28	8.00	9.763e + 09	78	12.90	5.504e + 09	128	18.20	2.186e + 09
29	8.00	9.667e + 09	79	13.00	5.426e + 09	129	18.30	2.160e + 09
30	8.00	9.590e + 09	80	13.10	5.357e + 09	130	18.30	2.134e + 09
31	8.00	9.418e + 09	81	13.30	5.279e + 09	131	18.40	2.100e + 09
32	8.00	9.331e + 09	82	13.40	5.210e + 09	132	18.50	2.074e + 09
33	8.00	9.245e + 09	83	13.50	5.132e + 09	133	18.50	2.048e + 09
34	8.00	9.158e + 09	84	13.70	5.063e + 09	134	18.60	2.013e + 09
35	8.00	9.072e + 09	85	13.80	4.985e + 09	135	18.70	1.987e + 09
36	8.00	8.986e + 09	86	13.90	4.916e + 09	136	18.70	1.961e + 09
37	8.00	8.899e + 09	87	14.00	4.838e + 09	137	18.80	1.927e + 09
38	8.00	8.813e + 09	88	14.20	4.769e + 09	138	18.90	1.901e + 09
39	8.00	8.726e + 09	89	14.30	4.692e + 09	139	18.90	1.875e + 09
40	8.00	8.640e + 09	90	14.40	4.622e + 09	140	19.00	1.840e + 09
41	8.13	8.554e + 09	91	14.60	4.545e + 09	141	19.10	1.814e + 09
42	8.26	8.459e + 09	92	14.70	4.476e + 09	142	19.10	1.788e + 09
43	8.39	8.372e + 09	93	14.80	4.398e + 09	143	19.20	1.754e + 09
44	8.51	8.277e + 09	94	14.90	4.329e + 09	144	19.30	1.728e + 09
45	8.64	9.191e + 09	95	15.10	4.251e + 09	145	19.30	1.702e + 09
46	8.77	8.096e + 09	96	15.20	4.182e + 09	146	19.40	1.668e + 09
47	8.90	8.009e + 09	97	15.30	4.104e + 09	147	19.50	1.642e + 09
48	9.03	7.914e + 09	98	15.50	4.035e + 09	148	19.50	1.616e + 09
49	9.16	7.828e + 09	99	15.60	3.957e + 09	149	19.60	1.581e + 09
50	9.29	7.733e + 09	100	15.70	3.888e + 09	150	19.70	1.555e + 09
151	19.70	1.529e + 09	201	22.50	8.726e + 08	251	23.90	1.840e + 09
152	19.80	1.495e + 09	202	22.50	8.813e + 08	252	23.70	1.901e + 09
153	19.90	1.469e + 09	203	22.60	8.899e + 08	253	23.60	1.961e + 09
154	19.90	1.443e + 09	204	22.60	8.986e + 08	254	23.40	2.013e + 09
155	20.00	1.408e + 09	205	22.60	9.072e + 08	255	23.30	2.074e + 09
156	20.10	1.382e + 09	206	22.60	9.158e + 08	256	23.10	2.134e + 09
157	20.10	1.356e + 09	207	22.70	9.245e + 08	257	23.00	2.186e + 09
158	20.20	1.322e + 09	208	22.70	9.331e + 08	258	22.80	2.246e + 09
159	20.30	1.296e + 09	209	22.70	9.418e + 08	259	22.70	2.307e + 09
160	20.30	1.270e + 09	210	22.80	9.504e + 08	260	22.50	2.359e + 09
161	20.40	1.236e + 09	211	22.80	9.590e + 08	261	22.40	2.419e + 09
162	20.50	1.210e + 09	212	22.80	9.677e + 08	262	22.20	2.480e + 09
163	20.50	1.184e + 09	213	22.90	2.763e + 08	263	22.10	2.532e + 09
164	20.60	1.149e + 09	214	22.90	9.850e + 08	264	21.90	2.592e + 09
165	20.70	1.123e + 09	215	22.90	9.936e + 08	265	21.80	2.652e + 09
166	20.70	1.097e + 09	216	23.00	1.002e + 09	266	21.70	2.704e + 09
167	20.80	1.063e + 09	217	23.00	1.011e + 09	267	21.50	2.765e + 09
168	20.90	1.037e + 09	218	23.00	1.020e + 09	268	21.40	2.825e + 09
169	20.90	1.011e + 09	219	23.00	1.028e + 09	269	21.20	2.877e + 09
170	21.00	9.763e + 08	220	23.10	1.037e + 09	270	21.10	2.938e + 09
171	21.10	9.504e + 08	221	23.10	1.045e + 09	271	20.90	2.998e + 09
172	21.10	9.245e + 08	222	23.10	1.054e + 09	272	20.80	3.050e + 09
173	21.20	8.899e + 08	223	23.20	1.063e + 09	273	20.60	3.110e + 09
174	21.30	8.640e + 08	224	23.20	1.071e + 09	274	20.50	3.171e + 09
175	21.30	8.355e + 08	225	23.20	1.080e + 09	275	20.30	3.223e + 09
176	21.40	8.061e + 08	226	23.30	1.089e + 09	276	20.20	3.283e + 09
177	21.50	7.776e + 08	227	23.30	1.097e + 09	277	20.00	3.344e + 09

day	ztp	zflow	day	ztp	zflow	day	ztp	zflow
178	21.50	7.491e + 08	228	23.30	1.106e + 09	278	19.90	3.396e + 09
179	21.60	7.197e + 08	229	23.40	1.115e + 09	279	19.70	3.456e + 09
180	21.70	6.912e + 08	230	23.40	1.123e + 09	280	19.60	3.516e + 09
181	21.70	6.998e + 08	231	23.40	1.132e + 09	281	19.50	3.568e + 09
182	21.80	7.085e + 08	232	23.40	1.140e + 09	282	19.30	3.629e + 09
183	21.90	7.171e + 08	233	23.50	1.149e + 09	283	19.20	3.689e + 09
184	21.90	7.258e + 08	234	23.50	1.158e + 09	284	19.00	3.741e + 09
185	22.00	7.344e + 08	235	23.50	1.166e + 09	285	18.90	3.802e + 09
186	22.00	7.430e + 08	236	23.60	1.175e + 09	286	18.70	3.862e + 09
187	22.10	7.517e + 08	237	23.60	1.184e + 09	287	18.60	3.914e + 09
188	22.10	7.603e + 08	238	23.60	1.192e + 09	288	18.40	3.974e + 09
189	22.10	7.690e + 08	239	23.70	1.201e + 09	289	18.30	4.035e + 09
190	22.20	7.776e + 08	240	23.70	1.210e + 09	290	18.10	4.087e + 09
191	22.20	7.862e + 08	241	23.70	1.270e + 09	291	18.00	4.147e + 09
192	22.20	7.949e + 08	242	23.80	1.322e + 09	292	17.80	4.208e + 09
193	22.20	8.035e + 08	243	23.80	1.382e + 09	293	17.70	4.260e + 09
194	22.30	8.122e + 08	244	23.80	1.443e + 09	294	17.50	4.320e + 09
195	22.30	8.208e + 08	245	23.80	1.495e + 09	295	17.40	4.380e + 09
196	22.30	8.294e + 08	246	23.90	1.555e + 09	296	17.30	4.432e + 09
197	22.40	8.381e + 08	247	23.90	1.616e + 09	297	17.10	4.493e + 09
198	22.40	8.467e + 08	248	23.90	1.668e + 09	298	17.00	4.553e + 09
199	22.40	8.554e + 08	249	24.00	1.728e + 09	299	16.80	4.605e + 09
200	22.50	8.640e + 08	250	24.00	1.788e + 09	300	16.70	4.666e + 09
301	16.50	4.856e + 09	321	13.60	8.726e + 09	341	9.16	1.253e + 10
302	16.40	5.046e + 09	322	13.40	8.899e + 09	342	8.92	1.270e + 10
303	16.20	5.244e + 09	323	13.30	9.072e + 09	343	8.68	1.287e + 10
304	16.10	5.435e + 09	324	13.10	9.245e + 09	344	8.44	1.313e + 10
305	15.90	5.625e + 09	325	13.00	9.418e + 09	345	8.20	1.331e + 10
306	15.80	5.815e + 09	326	12.80	9.677e + 09	346	7.96	1.348e + 10
307	15.60	6.005e + 09	327	12.50	9.850e + 09	347	7.72	1.365e + 10
308	15.50	6.195e + 09	328	12.30	1.002e + 10	348	7.48	1.382e + 10
309	15.30	6.394e + 09	329	12.00	1.020e + 10	349	7.24	1.408e + 10
310	15.20	6.584e + 09	330	11.80	1.045e + 10	350	7.00	1.426e + 10
311	15.10	6.774e + 09	331	11.60	1.063e + 10	351	7.10	1.443e + 10
312	14.90	6.964e + 09	332	11.30	1.080e + 10	352	7.20	1.460e + 10
313	14.80	7.154e + 09	333	11.10	1.097e + 10	353	7.30	1.486e + 10
314	14.60	7.353e + 09	334	10.80	1.115e + 10	354	7.40	1.503e + 10
315	14.50	7.543e + 09	335	10.60	1.140e + 10	355	7.50	1.521e + 10
316	14.30	7.733e + 09	336	10.40	1.158e + 10	356	7.60	1.460e + 10
317	14.20	7.923e + 09	337	10.10	1.175e + 10	357	7.70	1.400e + 10
318	14.00	8.133e + 09	338	9.88	1.192e + 10	358	7.80	1.339e + 10
319	13.90	8.303e + 09	339	9.64	1.218e + 10	359	7.90	1.279e + 10
320	13.70	8.502e + 09	340	9.40	1.236e + 10	360	8.00	1.218e + 10

Glossary

abiotic variable an environmental variable, e.g., temperature, solar radiation

algorithm a computational method for solving some mathematical problem

aliasing the inability to distinguish between the effect of two independent variables or variable combinations on the dependent variable (also called confounding)

allometric consumption function the relationship between consumption rate and animal body size

allometric length–weight relationship the relationship between the length and the weight of an organism

anabolism the process of building up of sugars and amino acids by animals through ingestion and digestion

analysis of variance (ANOVA) a statistical analysis technique that partitions the variability of a dependent variable into sources attributed to independent variables and pure error

arthropod any animal of the phylum *Arthropoda,* exemplified by spiders, insects, and crustaceans

autotroph a primary producer or self-producer of energy; a plant

basal area the area of a breast high cross-section of a tree or of all trees in a forest stand

Beer's law the physical law that states that light attenuates through some medium at a constant rate with depth

benthos biota that live on or near the bottom of a water body

bioassay a test for the effect of a toxicant on a specific test species. Often used to ascertain environmental effects of the toxicant

biomass the density of a population expressed in units of weight

branch and bound an algorithm for solving dynamic programming problems

calibration the process of changing model parameters to obtain an improved fit of model output variables to data time traces

canopy the tree foliage in a forest stand

carnivore a meat-eating animal

catabolism the process of breakdown of sugars to produce energy. This is analogous to respiration

chlorophyll *a* the photosynthetically active part of most plants

333

clearcut   a forestry practice of removing all trees from a forest stand

climax   the state achieved in theory by an ecosystem left undisturbed over a long time

cluster analysis   a statistical technique for grouping together different entities according to their similarity with regard to specific attributes

cohort   a group of animals all born at the same time and thus all of the same age. Used to denote a model where separate ages of each cohort are kept track of as opposed to a dynamic pool model

communication   the relationship between two model compartments such that flows leaving either compartment can get to the other

compartment diagram   a diagram for a simulation model where state variables are in boxes (compartments) and flows are represented by arrows

complete factorial design   an experimental design in which parameters at fixed levels are included in all possible combinations

congruential generator   an algorithm for generating pseudorandom numbers that involves starting with a seed and generating the numbers by a series of multiplications, additions, and divisions of numbers

constraint   a quantitative limit imposed on variables or combinations of variables in an optimization problem

continuous distribution   a distribution with an infinite number of states, with the states being contiguous

control   any of a number of factors that affect the rate of change of a state variable in a simulation model. These controls are usually exercised through functional relationships (equations) for process rates

control system   optimization techniques for maximizing some objective function over time by exercising continuous control over a system. The methods have been most successfully applied in physics and electrical engineering

copepod   a crustacean zooplankton common in marine and, to a lesser extent, freshwater systems

correlation coefficient   a measure of the degree of the linear relationship between two variables

corroboration   the process of checking whether a model agrees with available evidence about the ecosystem or process it represents

creel   the fisherman's sack. Creel censuses are a method of asking fishermen what they have caught and is used as a measure of catch and effort in recreational fisheries

crown closure   the time in a forest's life when all the ground surface is covered by leaf area

cumulative distribution function   a function that gives the probability that a random variable takes on any value less than or equal to its argument. Usually denoted by $F(x)$ and abbreviated c.d.f.

Daphnia   a cladoceran (zooplankton) genus that is very common in lake ecosystems and is often used as a test animal for toxicity bioassays

DBH   diameter at breast height. This is a standard measure made on a tree that has been correlated with height, leaf area, and volume for many tree species

decomposers   animals or plants that break down dead material releasing nutrients

defoliation   the removal of foliage from trees

dendrogram   a tree diagram used as a graphical display to show the similarity of different entities

determinant a scalar value involving the sums of products of elements of a square matrix, used in eigenanalysis and in matrix inversion

detritus dead organic material in the water column or on the bottom of a water body

difference equation a discrete time equation in which the state of the system at any time is expressed as a function of the state at the previous time, and the values of driving variables over the time step

differential equation an equation giving the rate of change of state variables as a function of other state variables, time and driving variables

diffusion the movement of particles or individuals out from a central location. Models of diffusion are sometimes used for animal migration

dimictic a lake that turns over twice a year. A *monomictic* lake turns over once a year. Lake turnover is when the temperature distribution with depth is constant and where waters can and do mix between all depths of the lake

discrete distribution a probability distribution that has only a finite or countably infinite number of possible values

dissolved oxygen oxygen dissolved in water and used by aquatic organisms for respiration processes

diversity index a measure of the number of species and the evenness of the distribution of population or biomass among species in an ecosystem

documentation the process of recording the model equations, rationale, data sources, variables, code, and experiments for purposes of information transmission

dominant a particular species that comprises the majority of the biota in an ecosystem

dominant eigenvalue for Leslie matrices this is the sole eigenvalue that is a positive real number. Its value relative to 1.0 determines whether the population will grow or decrease in the long-term

donor controlled a linear differential equation model in which the flows between compartments depend on the compartment from which the flows come; the donor compartment

driving variable environmental conditions that affect model behavior without being affected by the model

dynamic changing over time; in contrast to static

dynamic pool a model framework in which age, weight, or size classes are lumped together into groups of average age, weight, or size

dynamic programming an optimization problem over some planning horizon when decisions occur at discrete time stages and where each stage has a finite (usually small) number of decision options

ecosystem an ecological community, including all organisms and the abiotic environment, considered as a unit

effects matrix a matrix having coefficients that represent both the sublethal and lethal effects of toxicants on ecosystem biota

egestion the process of passing undigestible material out of the gut. Also termed defecation

eigenanalysis the finding of the eigenvalues and eigenvectors of a square matrix

eigenvalues coefficients λ associated with a matrix $\mathbf{A}$ that satisfy the equation $|\mathbf{A} - \lambda\mathbf{I}| = 0$. Used in the solution to linear compartment models and Leslie matrices

eigenvectors vectors $\mathbf{v}_i$ associated with eigenvalues λ_i of a matrix $\mathbf{A}$ satisfying the relationship $\mathbf{A}\mathbf{v}_i = \lambda_i\mathbf{v}_i$

empirical distribution   the cumulative distribution function of a sample of size n, denoted $F_n(x)$. It is $1/n \times$ {the number of values in the sample less than or equal to x}

energetics   the study of the energy transferred or expended in life processes

entanglement   the process of being tangled in net fragments or other flotsam. Used here to refer to fur seals

entrainment   the process of being captured in cooling waters of a power plant. This usually applies to small organisms such as fish larvae and zooplankton

epilimnion   the upper layer in a stratified lake

establishment   successful plant (tree) reproduction on a site

Euler technique   a numerical approximation solution for systems of differential equations in which the solution is projected forward through time in steps of size Δ in linear segments of slope $f(\mathbf{x},t)$ given by the differential equation

euphotic zone   the upper part of the water column for which there is positive net photosynthesis by the phytoplankton

eutrophication   the nutrient loading of water bodies from external nutrient sources at levels high enough to cause large plankton blooms and concern about the dissolved oxygen levels

exponential distribution   a probability distribution often used to describe the time between events in which the probability is highest for events occurring close to each other. Sometimes referred to as the negative exponential distribution

finite difference   a method in which a derivative is approximated by a difference between a function values over a small finite time difference Δ

flow   the rate of transfer of material between compartments or state variables

forbs   small grass-like plants

forecasting   methods for extrapolating the future from properties of the past. Most often applied in statistical analysis of time series

Forrester diagram   model compartment diagram including information about factors controlling flows

fractional factorial design   an experiment design in which parameters are studied at several levels with only a fraction of the possible parameter–level combinations included in the experiment

functional group   a group of species that are assumed in a model to function in a similar fashion by having the same parameter values for process equations. These are lumped together in the model as a single entity

gamma function   a function, denoted by $\Gamma(x)$, that for integers has the value $(x-1)!$. A normalized version is used here as a prototype for the effect of temperature on various processes

global variables   variables appearing in a number of simulation subroutines

half saturation level   the level of some independent variable at which some process rate is reduced to half its maximum relative to that independent variable. It is applied to such processes as the effect of nutrient availability on phytoplankton nutrient uptake or the effect of prey density on predator feeding rate

handling time   the time a predator takes to pursue, capture, and eat its prey

herbivore   a plant eating animal

heterotroph   an animal that does not produce its own energy but is dependent upon outside food sources for its sustenance

home   a place that if you have to go there they have to take you in. There is no place like it

hypolimnion   the lower layer in a stratified lake

impact assessment   the process of measuring, modeling, and monitoring a stressed ecosystem such as to predict significant change resulting from the stress or to detect such a change if it occurs

implementation   the process of coding, parameterizing, and calibrating a model

implementation document   a model document that keeps track of changes to model parameters, processes, and code involved in the implementation of the model

initial conditions   the values of the state variables at the start of a simulation

insolation   solar radiation

interaction   an interaction between two independent variables exists when the response to one of the independent variables depends on the value of the other independent variable

interaction diagram   plots used to show the form of the interaction between two variables in their effect on some quantity of interest

intermediate variable   variables that are used to compute process rates and are often of interest to the modeler, e.g., photosynthesis rate

intrinsic environment   environmental variables that are modified by the biota such as sunlight through a forest canopy

inverse matrix   for square matrices, the matrix such that when premultiplied or post-multiplied by the original matrix the result is the identity matrix I

juvenation   the observed phenomenon in harvested populations of a shift in the age structure toward a predominance of smaller animals

Latin hypercube sampling   a form of stratified sampling for sensitivity analysis in which the probability distribution of each parameter is divided into n strata of equal probability $(1/n)$ and one sample is taken from each stratum

LD_{50}   the level of some test chemical or substance at which the bioassay organisms experience 50 percent mortality

Leslie matrix   a matrix giving the survival and fecundity by age of females in a population. Used to project the future of the population from an initial state

levels   a statistical term to describe the possible conditions or states of the independent variables

Liebig's law   the assumption that the effect on a process of several factors is the minimum of the effect caused by each of the individual factors

light attenuation   the observed physical phenomenon that light levels are reduced through a water column or canopy with depth

linear programming   methods for finding the combination of variables that maximize or minimize an objective function that is a linear function of these variables, subject to constraints that are also linear functions of the variables

local variable   a variable that appears only within a simulation subroutine

lodger   a group of lions that does not control territory and that will on occasion challenge pride lions for territorial rights

logistic equation a differential equation used to describe single species growth over time, which is dominated by an intrinsic growth rate and a saturation level or carrying capacity

lognormal distribution a continuous probability distribution such that if Y is a lognormal random variable then $X = \log(Y)$ is a normal random variable

Markov chain a stochastic model having discrete states in which the probability of being in any state at any time depends only on the state at the previous time and on the probability transition matrix

memoryless the property of Markov chains of having the future depend only on the present state and not directly on any past states

Michaelis–Menton a hyperbolic curve form originally applied to the rate of nutrient uptake as a function of nutrient concentration, but also applied to a variety of other relationships

microcosm an experimental ecosystem, usually small in scale and with controlled environmental conditions

modulus function the value of the function, modulus(x,m), is the remainder when x is divided by the modulus m

monitoring program a sampling design for an ecosystem in an area subject to some impact with the objective of detecting a significant change in target biota if one occurs

monotonic function a function that never increases (or never decreases) as the independent variable increases

Monte Carlo involving the use of random numbers in a computer program or simulation model to obtain approximate solutions to some problem

normal distribution a continuous probability distribution commonly used in statistics and completely characterized by a mean μ, which it is symmetric about, and a variance σ^2, which describes the spread of the distribution. The distribution is denoted by $N(\mu,\sigma^2)$

objective function the function that is to be optimized in an optimization problem

Odum diagram a model diagram oriented toward energy transactions

optimal foraging a theory that asserts that an organism chooses its food such as to maximize its energy intake per unit time

optimization maximizing or minimizing according to some objectives. Refers to a series of techniques for choosing system controls such as to achieve the best possible performance under these objectives

order for a differential equation the highest derivative in the equation; for a numerical approximation solution to a differential equation the order denotes the accuracy in terms of the highest power of the time interval Δ included in the approximation

output variables variables that are not of major importance in running a model but that are produced as model output, e.g., producing the value of some state variable in different units than in the model for comparison with field data

parameters coefficients that control model process rates but do not change in the run of a model

parr the young of salmonid fish

partial correlation   a measure of the strength of the linear relationship between two variables after adjusting for any linear relationship the two variables have with a third (or possibly more) variable

passerine   birds belonging to the order Passeriformes, the perching birds; includes all songbirds

patchiness   the spatial process of the clumping of organisms. It has been observed in many biota; examples include fish schools and willow thickets

perturbation   any change to any ecosystem

pH   a measure of the acidity of an aqueous medium that measures the concentration of hydrogen ions. Pure water has a pH of 7, acids have lower pH and bases higher pH

phenology   the study of the seasonal changes in plant life stages

photoinhibition   reduction in photosynthesis rate at high light intensities that has been observed for both phytoplankton and terrestrial plants

phytoplankton   microscopic aquatic plants

planning horizon   some period of time over which an optimization algorithm is to be performed

Poisson distribution   a discrete probability distribution defined on the nonnegative integers having a single parameter λ. It is commonly used to describe events that occur randomly in time or space

population dynamics   the study and modeling of changes in population size over time

population projection matrix   represents the change in a population in one time step with age-specific fecundities in the first row and survival probabilities on the subdiagonal. Also called the survival–fecundity matrix or the Leslie matrix

predation   the process of eating prey thereby imposing mortality on them

pride   a mating group of lions

primary consumers   animals that consume plants (grazers)

primary producers   plants, biota that convert carbon dioxide and water into carbohydrates

probability density function   the derivative of a continuous cumulative distribution function. Denoted by $f(x)$

probability distribution   a function that describes the probability that a discrete random variable takes on any value over its range. Denoted by $p(x)$

probability transition matrix   a matrix of probabilities of transition in one time step between discrete states in a Markov chain model

pseudorandom numbers   numbers generated by an algorithm that appear to be random (when in fact they are not)

Q_{10}   a relationship for the effect of temperature on a process such that the process rate increases by the same multiple for every 10°C rise in temperature

quantile–quantile plot   a method to compare a sample distribution with a theoretical distribution by plotting the quantiles of the theoretical distribution against the sample quantiles

queueing theory   the study of the formation of lines or queues when the arrivals occur at random intervals

radionuclide   a radioactive form of a chemical element. Often used to label or trace an element through an ecosystem or individual animal or plant

random numbers   numbers generated according to some probability distribution

random sampling sampling procedures in which each point in the sample space is equally likely to be included in the sample

random variable a numerical quantity whose value is determined by the outcome of a random experiment

random walk a probability model for movement direction and angle often simulated using Monte Carlo methods

rank correlation coefficient the correlation coefficient computed from the ranks of the data rather than the data values themselves

realism the property of a model that it represents the best understanding of the composite processes available

recipient controlled a process whose rate of change depends on the compartment to which the process flows, the recipient compartment

recruit animals (usually fish) reaching maturity

recursive an algorithm that operates by repeating itself. See recursive

reflecting boundary the property of edge or end states in a model involving movement such that movement beyond the edge results in reflection of movement back to the same edge state

regression a statistical technique for fitting a function to a set of data having independent and dependent variables such as to optimize some criterion of fit between the function and the data points

replicates repetitions of the same experiment. It is applied in statistics to represent observations that have the same experimental conditions

rookery a breeding colony of birds or animals

Runge–Kutta a numerical approximation technique for differential equations described in Chap. 2

secondary consumers animals that consume other animals (predators)

seed a starting integer for a random number generating algorithm

selectivity the choice exercised by an animal in what it feeds on

self-shading the process of plants reducing light intensities with depth, thereby setting a limit on their own photosynthesis rates

sensitivity analysis methods for examining the sensitivity of model behavior to parameter change, by simulating the effects of parameter changes

sessile a nonmoving organism

shade tolerance the degree to which a tree can grow or establish under a forest canopy

simulation a representation of a series of equations on the computer that gives a dynamic picture of the system represented by those equations

simulation experiment a designed experiment to investigate some phenomenon of interest about the model, e.g., effect of perturbations, parameter sensitivity

sink a compartment to which material flowing in the model goes, but from which it does not return

source a compartment from which the material flowing in the model comes, but does not return to

spatial heterogeneity differences in density of some quantity over space

spatial homogeneity the assumption that biota are distributed uniformly in space, such that their density can be represented by a spatial average

specific dynamic action the part of animal respiration that depends on the feeding rate

stable age distribution the steady state solution to a Leslie matrix model in which the relative abundance of each age class remains constant

state variables major elements of the model whose rates of change are given by differential equations

static not changing over time, in contrast with dynamic

stationary the property of a stochastic process of not changing in distribution over time or space

stationary state the set of state probabilities of a Markov chain model that do not change over time

steady state the solution to a differential or difference equation after a long time when the short-term part of the solution has dropped out. This solution usually has some properties that do not change over time

steepest ascent an algorithm for solving nonlinear optimization problems with stepwise search proceeding in the direction of steepest increase in the objective function

stochastic having a state characterized by a probability. Contrasted with deterministic where the state is known with certainty

stochastic process a collection of random variables that have some common structure or relationship, usually over time or space

stock the reproductive females of a population; may be expressed as numbers or biomass

stoichiometric ratio The average ratio for some biota of one element to another; for example, the nitrogen to phosphorous ratio for phytoplankton

stratified lake a lake having no mixing between layers

stratified sampling a sampling procedure in which the sample space is partitioned into groups or strata and each stratum is sampled randomly. This helps assure that the sample has values over the full range of the sample space. One example used here is Latin hypercube sampling

subadult an animal that has not reached sexual maturity, yet displays many other adult behaviors

subprocess one of a number of subunits comprising a process controlling material flow in an ecosystem model, e.g., the effect of temperature on plant photosynthesis

succession the change over time in species composition in ecosystems after major disturbance, leading to their "native" state

survival–fecundity matrix see population projection matrix

systematic sampling sampling procedures in which the sample space is sampled at a priori determined points

Taylor series expansion an infinite series representation of a function $f(x)$ in the neighborhood of a point a involving powers of $(x - a)$ and the derivatives of the function evaluated at a

temporary variables simulation model variables that are used in the code or documentation to group together a number of variables for notational or computational convenience

tertiary consumers animals high on the food chain that consume predators, e.g., hawks

thermal loading the increased temperature in a water body caused by power plant cooling systems

thermocline the water depth in a stratified lake at which the temperature changes very rapidly between the epilimnion and hypolimnion. See the definition of dimictic for more information

threshold the level of some independent variable at which some process changes from one distinct state to another; for example, a temperature threshold above which fish feeding ceases

time scale   the run time of a model

time step   the unit of time used in a numerical approximation to a differential equation

time traces   graphs of model behavior or of data from the ecosystem over time

toxicant   any foreign addition to an ecosystem that can adversely affect ecosystem biota

traffic offered   the ratio of the arrival rate to the service rate in a queueing model

transition matrix   a matrix of probabilities of one time step transfers between states of the system

translocation   the active transfer of nutrients or carbohydrates between parts of a plant

uniform distribution   a probability distribution in which the random variable has equal probability of taking on any value over its range. The notation $U(0,1)$ means that the range of values is between 0 and 1

update   the simulation process of changing the values of the state variables at each time step to reflect process rates at that time

uptake   the process of phytoplankton of actively absorbing nutrients from the surrounding water column

upwelling   the physical process in nearshore ocean systems of rising of nutrients and colder bottom waters to the surface because of constant wind patterns along the shoreline. Upwelling areas are highly productive of plankton and fish

validation   a term common in the literature referring to the process of evaluating model behavior by comparing it with data. It is not preferred because of its connotations of truth

verification   the process of checking that the model code in fact represents the model equations formulated

water column   the water in some water body treated as spatially homogeneous having light reducing with depth from the surface

work gate   an active energy processing unit that combines energy from two or more sources; used in Odum diagrams

wrap-around boundaries   the property of edge states that movement out of one edge of a grid results in movement into the edge states at the opposite end of the grid

yearlings   animals in their first year of life; also called young-of-the-year

zooplankton   small aquatic animals

Notation

A,B,C,D,E	parameter effects for sensitivity analysis by factorial design
A	a matrix of flow rate coefficients for a compartment model
a	arbitrary parameter used to explain the second order Runge–Kutta technique
	the multiplier in a congruential random number generator
	ratio of encounter to digestion rate of prey in a queueing theory based larval fish feeding model
B	a matrix of coefficients of the transfers between compartments in one time step in a difference equation model
b	arbitrary parameter used to explain the second order Runge–Kutta technique
c	the increment in a congruential random number generator
c_j	boundary condition coefficient associated with the jth eigenvalue in the solution to linear differential equations
DBH	tree diameter at breast height (cm)
$F(x)$	cumulative probability distribution for a random variable X, which gives the probability that $X < x$ for all possible x
f_i	fecundity for age class i (eggs/female)
f_{ij}	the flow rate from compartment j to i in a linear model
$f(\mathbf{g})$	a function of intermediate variables
$f(x)$	the probability density function for a continuous random variable X
$f(\mathbf{x},t)$	a function of the state variables and time
$f(\mathbf{x},\mathbf{z},t)$	a function of state variables, driving variables and time
g	generic intermediate variable
g	an arbitrary vector of intermediate variables
$g(\cdot)$	generic function for an intermediate variable
	a density-dependent function for fur seal pup survival on land
$gareabs$	basal area for all trees on a 10×10 m plot in forest succession model (cm^2)
$garealf$	sum of the leaf area in a 10×10 m plot for all trees above a particular height (cm^2)

gbirth	bird natality (birth) rate (no./wk)
gcapeff	medusa capture efficiency feeding on larval herring
gcapeff$_{ij}$	capture efficiency of prey type i of length j
gchl$_{rt}$	lion challenge ratio that determines the probability of success of a territorial challenge by lodger males
gchl$_{th}$	lion challenge threshold, an age-dependent variable that determines the likelihood of success of a territorial challenge for equal numbers of lodgers and pride males
gdisnit	nitrogen discharge into the river segment being modeled (gN/m^3/day)
geattim$_{ij}$	average time taken to eat a prey item of type ij (day)
gemig	bird fall outmigration rate (no./wk)
genctr	encounter rate of medusae with their prey (/sec)
	prey arrival rate to larval fish (/day)
genctr$_{ij}$	encounter rate with size class j of prey species i by predator
generg$_{ij}$	energy content per size class j prey individual of species i
gexcnit	zooplankton nitrogen excretion (μgC eq./L/day)
gextlt	light extinction coefficient by phytoplankton in the water (m^2/g)
gfd	zooplankton feeding rate (/day)
	fish ingestion rate in a stochastic feeding model (g/day)
gfdpy	the effect of prey density on fish feeding rate
	the effect of prey density on zooplankton grazing rate
gfdtp	the effect of temperature on fish feeding or zooplankton grazing rate
gfdwt	the effect of fish weight on feeding rate
gflownit	nitrogen flow rate into the river segment being modeled (gN/m^3/day)
gflowp	phytoplankton flow rate into the river segment being modeled (mgC/L/day)
gflowz	zooplankton flow rate into the river segment being modeled (mgC/L/day)
ggrow	tree growth increment (/yr)
ggrowlt	effect of solar radiation on tree growth
ggrowtp	effect of temperature on tree growth
ggut	fish gut contents (g)
ggut$_{mx}$	maximum gut content in stochastic feeding model (g)
ghantim$_{ij}$	handling time per individual prey item of size class j and species i (day)
gimmig	bird spring immigration rate (no./wk)
gintak	fish energy intake rate (/unit time)
glayegg	bird egg laying rate (eggs/breeding female/wk)
glenprp$_{ij}$	proportion of prey species i that are in length class j
glt	annual solar radiation level at any height in the forest canopy (normalized)
gmixnit	nitrogen mixing rate (μgC eq./L/day)
gmixp	phytoplankton mixing rate (μgC/L/day)

gmois	moisture stress on tree growth
gmrt	bird mortality rate (no./wk)
	zooplankton mortality rate (/day)
	tree slow growth mortality rate (/yr)
gmrtage	tree mortality rate due to aging (/yr)
gpht	phytoplankton or plant photosynthesis rate (/day)
gphtlt	the instantaneous effect of light on plant photosynthesis
gphtnit	effect of nitrogen concentration on photosynthesis rate
gphttp	the effect of temperature on photosynthesis rate
gpred	medusa predation rate feeding on herring larvae (/sec)
	predation mortality rate on zooplankton (/day)
gprefwt	feeding selectivity by fish of weight *xwt* on prey of weight *xpreywt*.
gprey	total prey density available to foraging fish (g/m^3)
$gpurtim_{ij}$	average pursuit time for prey type *ij* (day)
grsp	bird respiration rate (g/bird/wk)
	phytoplankton respiration rate (/day)
grsptp	the effect of temperature on fish respiration rate
grspwt	the effect of fish weight on respiration rate
grspz	zooplankton respiration rate (/day)
gspace	effect of space competition on tree growth
gsrch	search rate for prey in a queueing theory fish feeding model (/day)
gstren	lion challenge strength—the ratio of numbers of lodger to pride males in a territorial challenge
$gsuc_{pr}$	probability of challenge success in a lion territorial challenge
gswim	medusa swimming rate (mm/sec)
guptnit	nitrogen uptake rate by phytoplankton (/day)
$h(\mathbf{x},\mathbf{g},\mathbf{z},t)$	a functional relationship between intermediate variables and state variables, driving variables, intermediate variables, and time
i	generic index denoting compartment number, age class, and a number of other countable groups
I	the identity matrix
j	generic index denoting compartment number and other countable groups
k	another generic index
k_{ij}	flow rate in a linear compartment model from compartment *j* to *i* (/time)
k_{jj}	flow rate out of compartment *j* (/time)
$kage_{mx}$	maximum possible tree species age (yr)
	maximum possible zooplankton age (day)
karealf	parameter in the DBH-leaf area empirical relationship
kassim	bird and zooplankton assimilation efficiency
$kareabs_{mx}$	maximum possible basal area on a 10 × 10 m plot in forest growth model (cm^2)

$kbiop_{th}$	phytoplankton threshold below which there is no zooplankton grazing
$kbirth$	fecundity (eggs/female)
$kCChl_{rt}$	phytoplankton carbon to chlorophyll ratio (mgC/mgChl)
kd	depth of the euphotic zone (zone in the water column above which photosynthesis rate exceeds respiration rate) (m)
$kdbh_{mx}$	maximum DBH for a tree species (cm)
$kdbh_{th}$	diameter increment below which slow growth mortality occurs (cm/yr)
$kdbhht_{1,2,3}$	parameters used in calculating tree height as a function of tree DBH
$kdegdy_{mn}$	minimum number of annual degree days above a threshold for tree growth (°C)
$kdegdy_{mx}$	maximum number of annual degree days above a threshold for tree growth (°C)
$kdisnit$	direct nitrogen discharge rate into water segment modeled (mgN/day)
$kdig$	digestion rate for feeding fish
$kdryprp$	proportion of growing season a tree species can withstand soil moisture below the wilting point
$keggeff$	egg conversion efficiency
$kenctr$	encounter rate of feeding fish with their prey
$kent$	rate of fur seal entanglement in fishing net debris (/yr)
$kent_{rt}$	ratio of total seal to entangled seal population
$kevap_{mn}$	minimum soil moisture requirement for tree establishment (mm/yr)
$kevap_{mx}$	maximum soil moisture for tree establishment (mm/yr)
$kexcnit$	zooplankton nitrogen excretion per unit respiration (µgC eq./µgC)
$kextlt$	light extinction coefficient (/m)
kfd	bird consumption rate (g/wk)
kfd_{mx}	maximum zooplankton grazing rate (/day)
$kfdpy_{hs}$	phytoplankton density level at which zooplankton grazing is reduced to half maximum rate (mgC/L)
	prey density at which fish feeding rate is reduced to half its maximum rate (g/m³)
$kfdq_{10}$	a parameter derived from the ratio of the fish feeding rate at 20°C to that at 10°C
$kfdtp_{1,2}$	parameters in the effect of temperature on zooplankton grazing rate
$kfdtp_{mx}$	temperature above which fish feeding ceases (°C)
$kfdtp_{op}$	temperature at which fish feeding rate is optimal (°C)
$kfdwt$	exponent in the effect of weight on zooplankton feeding rate
$kfdwt_{1,2}$	parameters in the effect of weight on fish feeding rate
$kflownit$	nitrogen concentration flowing into the segment being modeled (mgN/L)

$kflowp$	phytoplankton density flowing into the segment being modeled (mgC/L)
$kflowz$	zooplankton density flowing into segment being modeled (mgC/L)
$kgrow_{mx}$	maximum annual tree growth rate (cm/yr)
$kgrowdy$	number of days in the growing season
$kgrowlt_{1,2,3}$	parameters in light effect on tree growth
$kgut_{mx}$	maximum gut contents of feeding fish
$khatch$	time to hatch bird eggs (days)
kht_{mx}	maximum tree height for forest succession model (cm)
$klenwt$	coefficient in fish length-weight allometric relationship (g/cm^3)
$kmix$	mixing rate between epilimnion and hypolimnion (/day)
$kmrt$	zooplankton mortality rate (/day)
	slow growth mortality parameter for trees
	natural mortality rate for fur seals (/yr)
$kmrtage$	age mortality parameter for trees
$kmrtlg$	mortality rate for seals entangled in large net fragments (/yr)
$kmrtent$	mortality rate for entangled fur seals (/yr)
$knit$	initial nitrogen concentration (μgC eq./L)
$knitC_{rt}$	phytoplankton nitrogen to carbon ratio (mgN/mgC)
$kphtnit_{hs}$	nitrogen concentration at which photosynthesis is reduced to half its maximum rate (mgN/L or μgC eq./L)
$kpht_{mx}$	maximum photosynthesis rate (/day)
$kphtlt_{hs}$	light intensity at which photosynthesis rate is half maximum (ly/day)
$kphtlt_{mx}$	light intensity at which photosynthesis rate is maximum (ly/day)
$kphttp$	parameter for temperature effect on photosynthesis (/°C)
$kpred_{mx}$	maximum zooplankton predation mortality (/day)
$kpredn_{th}$	numbers threshold below which there is no zooplankton predation mortality (no./L)
$kpredwt_{th}$	weight threshold below which there is no zooplankton predation mortality (μgC)
$kpredpy_{hs}$	zooplankton biomass at which predation on them is half its maximum rate (μgC/L)
$kprefpy$	logarithm of the most preferred ratio of fish to prey weight in the size selective fish feeding function
$kpref_{vr}$	a measure of the spread in size selectivity for fish
$krsp$	maximum ingestion independent zooplankton respiration rate (μgC$^{-krspwt}$/day)
$krspsda$	maximum ingestion dependent zooplankton respiration rate (μgC$^{-krspwt}$/day)
$krsptp$	parameter for temperature effect on phytoplankton respiration (/°C)
$krsptpz$	parameter for temperature effect on zooplankton respiration (/°C)
$krspwt$	exponent in the effect of zooplankton weight on respiration rate

$krspwt_{1,2}$	parameters in the effect of fish weight on daily respiration rate
$ksex_{rt}$	bird sex ratio (female fraction of population)
	fur seal pup sex ratio
$ksrch_{mx}$	maximum search rate for prey
$ksurv_i$	survival from age class i to $i + 1$
$ksurvegg$	bird egg survival
$kuptnit_{mx}$	maximum nitrogen uptake rate (μgC eq./μgC/day)
$kvol$	volume of the water segment being modeled (L)
kwt_{mn}	minimum zooplankton weight (μgC)
kwt_{mx}	maximum zooplankton weight (μgC)
M_i	four slopes used to produce the slope Φ used in the fourth order Runge–Kutta
$\mathbf{M}_i$	a vector of slopes M_i
m	the modulus for a congruential random number generator
n	number of compartments in a linear compartment model
	number of parameters in a sensitivity analysis
$n_i(t)$	number of individuals (females) in age class i at time t
$\mathbf{n}(t)$	population vector for the Leslie matrix model
P_d, P_w	probability of a dry or wet day in a Markov chain model example
p_i	percentage of biomass in species i in Benares grazing land model
$p_{k0}(t)$	probability the fish gut is empty at time t given there were k items in it at time 0 in a feeding experiment
$\mathbf{p}(t)$	vector of probabilities of being in different states at time t
$p(x)$	the probability function of a discrete random variable
q_i	probability for the ith range in a histogram
R	period of driving variable sinusoid (day)
r	age of fur seal maturity, delay from birth to reproduction (yr)
$r(\lambda_j)$	the number of repeats of the eigenvalue λ_j
s	the stationary (equilibrium) probability vector for a Markov chain
s	index denoting the number of distinct eigenvalues
s_i	fraction of population at age i surviving in one time step to age $i + 1$
T	probability transition matrix for one step transitions between states in a Markov chain
t	time
t_0	time shift (from zero) for the average value in a sinusoid driving variable curve
t_1, t_2	temporary variables used in the effect of temperature on fish ration
U_n	the nth generated uniformly distributed pseudorandom number
$\mathbf{v}_i$	the ith eigenvector used in solving linear differential and difference equations

V_i	the value of various biotic responses to grazing management in a Benares grazing land model
v_{ij}	the jth element of the ith distinct eigenvector
X, Y, Z	random variables
X_n	the nth generated random integer. The original number, called the seed is denoted by X_0
x	generic state variable. Used to denote compartments for linear models, as in x_i being the ith compartment
$\mathbf{x}$	a vector of state variables
$\mathbf{x}(t)$	a vector of state variables at time t
$\dot{\mathbf{x}}(t)$	the derivative of the vector $\mathbf{x}$ with respect to time at t
$\mathbf{x}_{eq}$	vector of equilibrium values of state variables in a linear model
$\mathbf{x}_j^0$	the contribution of the jth eigenvector to the initial or boundary condition in the solution of a linear differential equation
$xage$	tree age (yr)
	zooplankton age (day)
$xarealf$	tree leaf area (cm^2)
$xbio$	bird biomass density (g/100ha)
$xbioegg$	zooplankton egg biomass (μgC/L)
$xbiop$	phytoplankton biomass (mgC/L) or (μgC/L)
$xbioz$	zooplankton biomass (mgC/L) or (μgC/L)
$xdbh$	tree diameter at breast height (DBH) (cm^2)
$xent$	number of fur seals entangled in fishing net debris
$xgut$	fish gut contents
xht	tree height (cm)
$xlen$	fish or herring larvae length (mm)
$xlodger$	numbers of lodger male lions in a particular group
xn	bird population (no./100ha)
	herring larvae density when prey for medusae (no./L)
	tree numbers (no./100m^2)
	zooplankton numbers (no./L)
xn_{tt}	total unentangled fur seal population
$xnit$	nitrogen concentration (mgN/L or μgC eq./L)
$xprey$	prey density in queueing theory fish feeding model (no./L)
$xpride$	number of male lions in a particular pride
$xpup_{eq}$	equilibrium number of fur seal pups leaving the Pribilof Islands
$xpreywt$	fish prey weight (g)
$xradenctr$	encounter radius for medusae
$xrecrt$	number of recruits into a fishery
$xstock$	fish stock (numbers of eggs)
	fur seal fecund stock
$xumbrad$	medusa umbrella radius (mm)
$xvol$	tree volume (cm^3)
xwt	average bird weight (g)
	average fish weight (g)
	average zooplankton weight (μgC)

z	generic driving variable
	depth into the water column (m)
$\mathbf{z}$	an arbitrary vector of driving variables
z_{mn}	generic minimum for a driving variable sinusoid curve
z_{mx}	generic maximum for a driving variable sinusoid curve
$zdry$	number of drought (low moisture days in growing season (day)
$zevap$	soil moisture tension measured by evapotranspiration (mm/yr)
$zflow$	flow rate into and out of a water segment being modeled (L/day)
$zinsol$	solar radiation incident to the top water column (ly/day)
	total annual insolation in forest growth model (normalized)
zlt	solar radiation at any level in the water column (ly/day)
ztp	environmental temperature (air or water) (°C)
	average number of degree days above a threshold temperature in the forest growth model
ztp_{av}	average environmental temperature used in Monte Carlo generation of temperature driving variables (°C)
ztp_{mn}	minimum annual average daily environmental temperature (°C)
ztp_{mx}	maximum annual average daily environmental temperature (°C)
α	arbitrary fraction of Δ moved in second order Runge–Kutta algorithm
	parameter in the Beverton–Holt stock–recruit relationship
α_i	sum of all loss rates including entanglement for fur seals (/yr)
β	parameter in the Beverton–Holt stock–recruit relationship
β_i	sum of all mortality rates for age class i entangled seals (/yr)
γ	entrainment rate in striped bass model (/yr)
Δ	an arbitrary increment in time used in numerical solutions to differential equations
ϵ	a measure of error in the Euler approximation
Λ	a population projection or Leslie matrix
λ	the rate parameter for a Poisson or exponential probability distribution
λ_j	the jth eigenvalue of a matrix, using in solving linear differential and difference equations
μ	the mean of a normal probability distribution
σ^2	the variance of a normal probability distribution
$\Phi(\mathbf{x},t,\Delta)$	the vector of slopes used to go a distance Δ in the Runge–Kutta numerical solution to a system of differential equations

Literature Cited

Abramsky, Z., and G. M. Van Dyne: Field studies and a simulation model of small mammals inhabiting a patchy environment. *Oikos,* 35:80–92, 1980.

Ackoff, R. L., and M. W. Sasieni: *Fundamentals of Operations Research,* Wiley, New York, 1968.

Andersen, K. P., and E. Ursin: A multispecies extension to the Beverton and Holt theory of fishing, with accounts of phosphorus circulation and primary production. *Medd. Dan. Fisk. Havunders,* 7:319–435, 1977.

Anthony, T. F., and B. W. Taylor, III: Analyzing the predictive capabilities of Markovian analysis for air pollution level variations. *J. Environ. Manage.,* 5:139–149, 1977.

Arya, J. C., and R. W. Lardner: *Mathematics for the Biological Sciences,* Prentice Hall, Englewood Cliffs, N.J., 1979.

Bailey, K. M., and R. S. Batty: A laboratory study of predation by *Aurelia aurita* on larval herring (*Clupea harengus*) experimental observations compared with model predictions. *Mar. Biol. (Berl.),* 72:295–301, 1983.

Bailey, N. T. J.: *The Elements of Stochastic Processes with Applications to the Natural Sciences,* Wiley, New York, 1964.

Beanlands, G. E., and P. N. Duinker: *An Ecological Framework for Environmental Impact Assessment in Canada,* Federal Environmental Assessment Review Office, Hull, Quebec, 1983.

Beddington, J. R., and R. M. May: Harvesting natural populations in a randomly fluctuating environment. *Science,* 197:463–465, 1977.

Beverton, R. J. H., and S. J. Holt: On the dynamics of exploited fish populations. *Fish. Invest. Ser. II Mar. Fish. G. B. Minist. Agric. Fish. Food,* 19, 1957.

Beyer, J. E.: *Survival of Fish Larvae—A Single Server Queue Approach,* Institute of Mathematical Statistics and Operation Research, Technical University of Denmark, Copenhagen, 1976.

Beyer, J. E.: *Aquatic Ecosystems: An Operational Research Approach,* University of Washington Press, Seattle, 1981.

Bierman, V. J. Jr.: Mathematical model of the selective enhancement of blue-green algae by nutrient enrichment, in *Modeling Biochemical Processes in Aquatic Ecosystems,* ed. R. P. Canale, Ann Arbor Science, Ann Arbor, Mich., 1976, pp. 1–31.

Bierman, V. J. Jr., F. H. Verhoff, T. L. Poulson, and M. W. Tenney: Multi-nutrient dynamic models of algal growth and species competition in eutrophic lakes, in *Mod-*

eling the Eutrophication Process, ed. E. J. Middlebrooks, D. H. Falkenborg, and T. E. Maloney, Ann Arbor Science, Ann Arbor, Mich., 1974, pp. 89–108.

Bierman, V. J. Jr., D. M. Dolan, E. F. Stoermer, J. E. Gannon, and V. E. Smith: *The Development and Calibration of a Spatially Simplified Multi-Class Phytoplankton Model for Saginaw Bay, Lake Huron*, Great Lakes Environmental Planning Study, Contribution No. 33, 1980, p. 126.

Bledsoe, L. J., and G. M. Van Dyne: A compartment model simulation of secondary succession, in *Systems Analysis and Simulation in Ecology, vol. 1*, ed. B. C. Patten, Academic Press, New York, 1971, pp. 479–511.

Botkin, D. B., J. F. Janak, and J. R. Wallis: Some ecological consequences of a computer model of forest growth. *J. Ecol.*, 60:849–872, 1972.

Bowie, G. L., W. B. Mills, D. B. Porcella, C. L. Campbell, J. R. Pagenkopf, G. L. Rupp, K. M. Johnson, P. W. H. Chan, S. A. Gherini, and C. E. Chamberlin: *Rates, Constants, and Kinetics Formulations in Surface Water Quality Modeling, 2nd ed.*, EPA Environmental Research Laboratory Report EPA/600/3-85/040, Athens, Georgia, 1985, p. 455.

Box, G. E. P., W. G. Hunter, and J. S. Hunter: *Statistics for Experimenters: An Introduction to Design, Data Analysis and Model Building*, Wiley, New York, 1978.

Brett, J. R.: Satiation time, appetite and maximum food intake of sockeye salmon (*Oncorhychus nerka*). *J. Fish. Res. Board Can.*, 28:409–415, 1971.

Broecker, W. S., T. Takahashi, H. J. Simpson, and T. H. Peng: Fate of fossil fuel carbon dioxide and the global carbon budget. *Science*, 206:409–418, 1979.

Brown, M. P.: A mathematical model of PCB bioaccumulation in plankton. *Ecol Model.*, 15:29–47, 1982.

Canale, R. P. ed.: *Modeling Biochemical Processes in Aquatic Ecosystems*, Ann Arbor Science, Ann Arbor, Mich., 1976.

Caperon, J.: Population growth in micro-organisms limited by food supply. *Ecology*, 48:715–722, 1967.

Caswell, H.: The validation problem, in *Systems Analysis and Simulation in Ecology, vol. 4*, ed. B. C. Patten, Academic Press, New York, 1976.

Chambers, J. M., W. S. Cleveland, B. Kleiner, and P. A. Tukey: *Graphical Methods for Data Analysis*, Wadsworth, Belmont, Calif., 1983.

Charnov, E. L.: Optimal foraging, the marginal value theorem. *Theor. Popul. Biol.*, 9:129–136, 1976.

Chen, C. W.: Concepts and utilities of ecologic modeling. *J. Sanit. Eng. Div. Proc. Am. Soc. Civ. Eng.*, 96:1085–1097, 1970.

Chen, C. W., J. D. Dean, S. A. Gherini, and R. A. Goldstein: Acid rain model— Hydrologic module. *J. Environ. Eng.*, 108:455–472, 1982.

Child, G. I., and H. H. Shugart, Jr.: Frequency response analysis of magnesium cycling in a tropical forest ecosystem, in *Systems Analysis and Simulation in Ecology, vo. 2*, ed. B. C. Patten, Academic Press, New York, 1972, pp. 103–135.

Clark, C. W.: *Mathematical Bioeconomics: The Optimal Management of Renewable Resources*, Wiley, New York, 1976.

Cleveland, W. S., and R. McGill: Graphical perception and graphical methods for analyzing scientific data. *Science*, 229:828–833, 1985.

Clow, D. J., and N. S. Urquhart: *Mathematics in Biology: Calculus and Related Topics*, W. W. Norton & Co., New York, 1974.

Cochran, W. G., and G. M. Cox: *Experimental Designs, 2nd ed.*, Wiley, New York, 1957.

Conte, S. D.: *Elementary Numerical Analysis: An Algorithmic Approach,* McGraw-Hill, New York, 1965.

Cosby, B. J., G. M. Hornberger, and M. G. Kelly: Identification of photosynthesis–light models for aquatic systems. II. Application to a macrophyte dominated stream. *Ecol. Model.,* 23:25–51, 1984.

Costanza, R.: Articulation, accuracy and effectiveness of mathematical models: A review of freshwater wetland applications. *Ecol. Model.,* 27:45–68, 1985.

Coughenour, M. B., W. J. Parton, W. K. Lauenroth, J. L. Dodd, and R. G. Woodmansee: Simulation of a grassland sulfur-cycle. *Ecol. Model.,* 9:179–213, 1980.

Cox, D. R., and W. L. Smith: *Queues,* Methuen, London, 1961.

Dale, V. H., and G. L. Swartzman: Simulating the effects of increased temperature in a plankton ecosystem: A case study, in *Algae as Ecological Indicators,* ed. L. E. Shubert, Academic Press, London, 1984, pp. 395–427.

Dale, V. H., T. W. Doyle, and H. H. Shugart: A comparison of tree growth models. *Ecol. Model.,* 29:145–170, 1985.

DeAngelis, D. L., and G. T. Yeh: An Introduction to modeling migratory behavior of fishes, in *Mechanisms of Migration in Fishes,* ed. J. D. McCleave, G. P. Arnold, J. J. Dodson and W. H. Neill, Plenum, 1984, pp. 445–469.

DeAngelis, D. L., R. A. Goldstein, and R. V. O'Neill: A model for trophic interaction. *Ecology,* 56:881–892, 1975.

DeWit, C. T., and J. Goudriaan: *Simulation of Ecological Processes,* Centre for Agricultural Publishing and Documentation, Wageningen, Netherlands, 1974.

Di Toro, D., D. J. O'Connor, and R. V. Thomann: A dynamic model of the phytoplankton populations in the Sacramento-San Joaquin Delta. *Adv. Chem. Ser.,* 106:131–150, 1971.

Di Toro, D., D. J. O'Connor, R. V. Thomann, and J. L. Mancini: Phytoplankton–zooplankton–nutrient interaction model for western Lake Erie, in *Systems Analysis and Simulation in Ecology, vol. 3,* ed. B. C. Patten, Academic Press, New York, 1975, pp. 423–474.

Draper, N. R., and H. Smith: *Applied Regression Analysis, 2nd ed.,* Wiley, New York, 1981.

Eggers, D. M.: The nature of prey selection by planktivorous fish. *Ecology,* 58:46–59, 1977.

Elliott, J. M.: The energetics of feeding, metabolism and growth of brown trout (*Salmo trutta*) in relation to body weight, water temperature and ration size. *J. Anim. Ecol.,* 45:923–948, 1976.

Eppley, R. W.: Temperature and phytoplankton growth in the sea. *U. S. Nat. Mar. Fish. Serv. Fish., Bull.,* 70:1063–1085, 1972.

Forrester, J. W.: *Industrial Dynamics,* MIT Press, Cambridge, Mass., 1961.

Fowler, C. W.: Interactions of northern fur seals and commercial fisheries, *Transactions of the 47th North American Wildlife and Natural Resources Conference,* Wildlife Management Institute, Washington, D.C., 1982.

Frankiel, F. M., and D. W. Goodall, eds.: *Simulation Modeling of Environmental Problems,* Wiley, Chichester, 1978.

Gardner, R. H., D. D. Huff, R. V. O'Neill, J. B. Mankin, J. B. Carney, and J. Jones: Application of error analysis to a marsh hydrology model. *Water Resour. Res.,* 16:659–664, 1980a.

Gardner, R. H., R. V. O'Neill, J. B. Mankin, and D. Kumar: Comparative error analysis of six predator-prey models. *Ecology,* 61:323–332, 1980b.

Gardner, R. H., R. V. O'Neill, J. B. Mankin, and J. H. Carney: A comparison of sensitivity analysis and error analysis based on a stream ecosystem model. *Ecol. Model.*, 12:173–190, 1981.

Garratt, M.: *Statistical Techniques for Validating Computer Simulation Models*, U. S. IBP Grassland Biome Technical Report 286, Colorado State University, Fort Collins, Colo., 1975, p. 68.

Getz, W. M.: The ultimate-sustainable-yield problem in nonlinear age structured populations. *Math. Biosci.*, 48:279–292, 1980.

Getz, W. M.: Production models for nonlinear stochastic age-structured fisheries. *Math. Biosci.*, 69:1–20, 1984.

Getz, W. M.: Optimal and feedback strategies for managing multicohort populations. *J. Optim. Theory Appl.*, 46:505–514, 1985.

Goh, B. S.: *Management and Analysis of Biological Populations*, Elsevier, New York, 1980.

Gold, H. J.: *Mathematical Modeling of Biological Systems—An Introductory Guidebook*, Wiley, New York, 1977.

Gordon, A. D.: *Classification, Methods for the Exploratory Analysis of Multivariate Data*, Chapman and Hall, New York, 1981.

Grant, W. E.: *Systems Analysis and Simulation in Wildlife and Fisheries Science*, Wiley, New York, 1986.

Griffiths, P., and J. D. Hill: *Applied Statistics Algorithms*, Ellis Horwood Ltd., Chichester, 1985.

Hadley, G.: *Linear Programming*, Addison-Wesley, Reading, Mass., 1962.

Hall, C. A. S., and J. W. Day, Jr.: Systems and models: Terms and basic principles, in *Ecosystem Modeling in Theory and Practice: An Introduction with Case Studies*, ed. C. A. S. Hall and J. W. Day, Jr., Wiley, New York, 1977a, pp. 5–36.

Hall, C. A. S., and J. W. Day, Jr., eds.: *Ecosystem Modeling in Theory and Practice: An Introduction with Case Studies*, Wiley, New York, 1977b.

Hedgpeth, J. W.: Models and muddles some philosophical observations, *Helgol. Wiss. Meeresuntersh*, vol. 30, 1977, pp. 92–104.

Hett, J. M.: *Land-Use Changes in East Tennessee and a Simulation Model Which Describes These Changes for Three Counties*, U.S. IBP Eastern Deciduous Forest Biome Report ORNL-IBP-71-8, Oak Ridge National Laboratory, Oak Ridge, Tenn., 1971, p. 56.

Hett, J. M., and R. V. O'Neill: Systems analysis of the Aleut ecosystem, *Arctic Anthrop.*, 11:31–40, 1974.

Hilborn, R., C. J. Walters, R. M. Peterman, and M. J. Stanley: Models and fisheries: A case study in implementation. *N. Am. J. Fish. Manage.*, 4:9–13, 1984.

Hoel, P. G., S. C. Port, and C. J. Stone: *Introduction to Probability Theory*, Houghton Mifflin, Boston, 1971.

Hogg, R. V., and A. T. Craig: *Introduction to Mathematical Statistics, 4th ed.*, Macmillan, New York, 1978.

Holling, C. S.: The functional response of predators to prey density and its role in mimicry and population regulation. *Can. Entomol.*, 41:385–398, 1965.

Holling, C. S.: *Adaptive Environmental Assessment and Management*, Wiley, New York, 1978.

Holm, J., and S. Kellomaki: Comparison of growth strategies of two competing plant species in forest ground cover. *Ecol. Model.*, 23:135–150, 1984.

Horn, H. S.: *The Adaptive Geometry of Trees*, Princeton University Press, Princeton, N.J., 1971.

Hornberger, G. M., and R. C. Spear: An approach to the preliminary analysis of environment systems. *J. Environ. Manage.*, 12:7–18, 1981.

Hutchinson, G. E.: *A Treatise on Limnology, I. Geography, Physics and Chemistry*, Wiley, New York, 1957.

Iman, R. L., and W. J. Conover: Small sample sensitivity analysis techniques for computer models, with an application to risk assessment. *Commun. Statist. Theor. Meth.*, A9:1749–1842, 1980.

Innis, G. S. ed.: *New Directions in the Analysis of Ecological Systems, Parts 1 and 2*, Simulation Councils Proceedings, 5, Society for Computer Simulation, La Jolla, Calif., 1977.

Innis, G. S. ed.: *Grassland Simulation Model*, Ecological Studies, 26, Springer-Verlag, New York, 1978.

Innis, G. S.: A spiral approach to ecosystem simulation, I., in *Systems Analysis of Ecosystems*, ed. G. S. Innis and R. V. O'Neill, Statistical Ecology, 9, International Co-operative Publishing House, Fairland, Maryland, 1979, pp. 211–386.

Innis, G. S., and R. V. O'Neill, eds.: *Systems Analysis of Ecosystems*, Statistical Ecology, 9, International Co-operative Publishing House, Fairland, Maryland, 1979.

Ivlev, V. S.: On the utilization of food by planktophage fishes. *Bull. Math. Biophys.*, 22:371–389, 1960.

Ivlev, V. S.: On the quantitative relationship between survival rate of larvae and their food supply. *Bull. Math. Biophys.*, 27:215–222, 1965.

Jeffers, J. N. R.: *An Introduction to Systems Analysis: With Ecological Applications*, University Park Press, Baltimore, 1978.

Jorgensen, S. E.: A model of fish growth. *Ecol. Model.*, 2:303–313, 1976.

Jorgensen, S. E.: *Handbook of Environmental Data and Ecological Parameters*, International Society of Ecological Modelling, Copenhagen, Denmark, 1979.

Jorgensen, S. E. ed.: Modeling Primary Productivity. *Ecol. Model.*, 23:1–184, 1984.

Jorgensen, S. E. ed.: *Ecological Modelling*.

Kaluzny, S. P., K. A. Rose, P. J. Sullivan, and G. L. Swartzman: *Evaluation of Ecosystem Models in Power Plant Impact Assessment: A Case Study Using Lake Ontario*, U. S. Nuclear Regulatory Commission Report NUREG/CR-3308, 1983, p. 57.

Kennedy, W. J. Jr., and J. E. Gentle: *Statistical Computing*, Marcel Dekker, New York, 1980.

Kernighan, B. W., and P. J. Plauger: *The Elements of Programming Style, 2nd ed.*, McGraw-Hill, New York, 1978.

Kessell, S. R.: *Gradient Modeling, Resource and Fire Management*, Springer-Verlag, New York, 1979.

Ketchum, B. H.: The absorption of phosphate and nitrate by illuminated cultures of Nitzschia closterium. *Am. J. Bot.*, 26:399–407, 1939.

Kickert, R. N.: Names of published computer models in the environmental biological sciences: A partial list and new potential risks. *Simulation*, 43:22–39, 1984.

Kitchell, J. E., J. F. Koonce, R. V. O'Neill, H. H. Shugart, J. J. Magnuson, and R. S. Booth: Model of fish biomass dynamics. *Trans. Am. Fish. Soc.*, 103:786–798, 1974.

Kitchell, J. F., D. J. Stewart, and D. Weininger: Applications of a bioenergetics model to yellow perch (*Perca flavescens*) and walleye (*Stizostedion vitrem vitrem*). *J. Fish. Res. Board Can.*, 34:1922–1935, 1977.

Knuth, D. E.: *The Art of Computer Programming, vol. 2, 2nd ed.*, Addison-Wesley Publishing Co., Reading, Mass., 1981.

Kremer, J. N., and S. W. Nixon: *A Coastal Marine Ecosystem,* Ecological Studies, 24, Springer-Verlag, New York, 1978.

Krishnamurthy, L.: Herbage dynamics of some semi-arid Indian grazing lands. *Ecol. Model.,* 5:77–84, 1978.

Laevastu, T., and H. A. Larkins: *Marine Fisheries Ecosystem: Its Quantitative Evaluation and Management,* Fish News Books Ltd., Farnham, England, 1981.

Landry, M. R.: The structure of marine ecosystems: An alternative. *Mar. Biol. (Berl.),* 35:1–7, 1976.

Larsen, D. P., H. T. Mercier, and K. W. Malueg: Modeling algal growth dynamics in Shagawa Lake, Minnesota with comments concerning projected restoration, in *Modeling the Eutrophication Process,* ed. E. J. Middlebrooks, D. H. Falkenborg, and T. E. Maloney, Ann Arbor Science, Ann Arbor, Mich., 1974, pp. 15–31.

Lassiter, R. R.: *Modeling Dynamics of Biological and Chemical Components of Aquatic Ecosystems,* EPA Ecological Research Series, EPA-66013-75-012, Corvallis, Oregon, 1975.

Lauenroth, W. K., G. V. Skogerboe, and M. Flug, eds.: *Analysis of Ecological Systems: State-of-the-Art in Ecological Modelling,* Developments in Environmental Modelling, 5, Elsevier, New York, 1983.

Lawler, J. P.: *The Effect of Entrainment at Indian Point on the Population of the Hudson River Striped Bass,* Testimony Before the Atomic Safety and Licensing Board in the Matter of Indian Point Unit No. 2, USAEC Docket No. 50-247, 1972.

Lehman, J. T., D. B. Botkin, and G. E. Likens: The assumptions and rationales of a computer model of phytoplankton population dynamics. *Limnol. and Oceanogr.,* 20:343–364, 1975.

Leidy, G. R., and R. M. Jenkins: *The Development of Fishery Compartments and Population Rate Coefficients for Use in Reservoir Ecosystem Models,* Final Contract Report Y-77-1, Environmental Effects Laboratory, U.S. Army Engineer Waterways Experiment Station, Vicksburg, Miss., 1977, p. 134.

Leidy, G. R., and G. R. Ploskey: *Simulation Modeling of Zooplankton and Benthos in Reservoirs: Documentation and Development of Model Constructs,* Technical Report E-80-4, Environmental Effects Laboratory, U.S. Army Engineer Waterways Experiment Station, Vicksburg, Miss., 1980, p. 221.

Lemmon, H.: Comax: An expert system for cotton crop management. *Science,* 233:29–33, 1986.

Leslie, P. H.: On the use of matrices in certain population mathematics. *Biometrika,* 33:183–212, 1945.

Levin, S. A., and K. D. Kimball: *New Perspectives in Ecotoxicology, Ecosystem Research Center Report No. 14a,* Cornell University, Ithaca, New York, 1983, p. 158.

Levins, R.: The strategy of model building in population biology. *Am. Sci.,* 54:421–431, 1966.

Lewis, E. G.: On the generation and growth of a population. *Sankhya,* 6:93–96, 1942.

MacCormick, A. J. A., O. L. Loucks, J. F. Koonce, J. F. Kitchell, and P. R. Weiler: *An Ecosystem Model for the Pelagic Zone of Lake Wingra,* U.S. IBP Eastern Deciduous Forest Biome Memo Report 72-122, University of Wisconsin, Madison, 1972, p. 103.

McKellar, H. N. Jr.: Metabolism and model of an estuarine bay ecosystem affected by a coastal power plant. *Ecol. Model.,* 3:85–118, 1977.

Middlebrooks, E. J., D. H. Falkenborg, and T. E. Maloney, eds.: *Modeling the Eutrophication Process,* Ann Arbor Science, Ann Arbor, Mich., 1974.

Miller, P. C., B. D. Collier, and F. L. Bunnell: Development of ecosystem modeling in the tundra biome, in *Systems Analysis and Simulation in Ecology, vol. 3,* ed. B. C. Patten, Academic Press, New York, 1975, pp. 95–115.

Mitsch, W. J., R. W. Bosserman, and J. M. Klopatek, eds.: *Energy and Ecological Modelling,* Developments in Environmental Modelling, 1, Elsevier, New York, 1981.

Morgan, B. J. T.: *Elements of Simulation,* Chapman and Hall, London, 1984.

Morrison, K. A., N. Therien, and B. Coupal: Simulating fish redistribution in the LG-2 reservoir after flooding. *Ecol. Model.,* 47, 1985.

O'Brien, J. J., N. A. Slade, and G. L. Vinyard: Apparent size as a determinant of prey selection by bluegill sunfish (*Lepomis macrochirus*). *Ecology,* 57:1304–1310, 1976.

O'Neill, R. V.: *Examples of Ecological Transfer Matrices,* U. S. IBP Eastern Deciduous Forest Biome Report ORNL-IBP-71-3, Oak Ridge National Laboratory, Oak Ridge, Tenn., 1971.

O'Neill, R. V., and O. W. Burke: *A Simple System Model for DDT and DDE Movement in the Human Food-Chain,* U. S. IBP Eastern Deciduous Forest Biome Report ORNL-IBP-71-9, Oak Ridge National Laboratory, Oak Ridge, Tenn., 1971, p. 18.

O'Neill, R. V., R. A. Goldstein, H. H. Shugart, and J. B. Mankin: *Terrestrial Ecosystem Energy Model,* U. S. IBP Eastern Deciduous Forest Biome Memo Report 72-19, 1972.

O'Neill, R. V., R. H. Gardner, and J. B. Mankin: Analysis of parameter error in a non-linear model. *Ecol. Model.,* 8:297–311, 1980.

O'Neill, R. V., R. H. Gardner, L. W. Barnthouse, and G. W. Suter: Ecosystem risk analysis: A new methodology. *Environ. Toxicol. Chem.,* 1:167–177, 1982.

Odum, H. T.: Trophic structure and productivity of Silver Springs, Florida. *Ecol. Monogr.,* 27:55–112, 1957.

Odum, H. T.: *Environment, Power and Society,* Wiley, New York, 1971.

Odum, H. T.: An energy circuit language for ecological and social systems: Its physical basis, in *Systems Analysis and Simulation in Ecology, vol. 2,* ed. B. C. Patten, Academic Press, New York, 1972, pp. 140–211.

Odum, H. T.: *Systems Ecology: An Introduction,* Wiley, New York, 1983.

Park, R. A., and C. D. Collins: Realism and ecosystem models. *Perspect. Comput.,* 2:18–27, 1982.

Park, R. A., R. V. O'Neill, J. A. Bloomfield, H. H. Shugart, Jr., R. S. Booth, R. A. Goldstein, J. B. Mankin, J. F. Koonce, D. Scavia, M. S. Adams, L. S. Clesceri, E. M. Colon, E. H. Dettmann, J. A. Hoopes, D. A. Huff, S. Katz, J. F. Kitchell, R. C. Kohlberger, E. J. LaRow, D. C. McNaught, J. L. Peterson, J. T. Titus, P. R. Weiler, J. W. Wilkinson, and C. S. Zahorcak: A generalized model for simulating lake ecosystems. *Simulation,* 23:33–50, 1974.

Parrish, J. D.: Marine trophic interactions by dynamic simulation of fish species. *U. S. Nat. Mar. Fish. Serv. Fish. Bull.,* 73:695–716, 1975.

Parzen, E.: *Stochastic Processes,* Holden-Day, San Francisco, 1962.

Patrick, R.: Some effects of temperature on freshwater algae, in *Biological Aspects of Thermal Pollution,* ed. P. A. Krenkel and F. L. Peter, Vanderbilt University Press., Nashville, Tenn., 1969, pp. 161–185.

Patten, B. C.: A primer for ecological modeling and simulation with analog and digital computers, in *Systems Analysis and Simulation in Ecology, vol. 1,* ed. Bernard C. Patten, Academic Press, New York, 1971a, pp. 3–121.

Patten, B. C. ed.: *Systems Analysis and Simulation in Ecology, vol. 1,* Academic Press, New York, 1971b.

Patten, B. C.: A simulation of the shortgrass prairie ecosystem. *Simulation,* 19:117–186, 1972.

Patten, B. C., and M. Witkamp: Systems analysis of cesium 134 kinetics. *Ecology*, 48:813–824, 1967.

Patten, B. C., D. A. Egloff, and T. H. Richardson: Total ecosystem model for a cove in Lake Texoma, in *Systems Analysis and Simulation in Ecology, vol. 3*, ed. B. C. Patten, 1975, pp. 205–421.

Pielou, E. C.: *An Introduction to Mathematical Ecology*, Wiley, New York, 1969.

Putter, A.: Studien uber physiologische Ahnichkeit. *Pflugers Arch. Ges. Physiol.*, 180:298–340, 1920. (Sea. Dep. Agr. Fish. Scotl., Mar. Res., vol. 5, 1968)

Ralston, A.: *A First Course in Numerical Analysis*, McGraw-Hill, New York, 1965.

Reckhow, K. H., and S. C. Chapra: Confirmation of water quality models. *Ecol. Model.*, 20:113–133, 1983a.

Reckhow, K. H., and S. C. Chapra: *Engineering Approaches for Lake Management, Vol. 1 Data Analysis and Empirical Modeling*, Ann Arbor Science, Ann Arbor, Mich., 1983b.

Reed, K. L., and S. G. Clark: *SUCcession SIMulator: A Coniferous Forest Simulator Model Documentation*, U. S. IBP Coniferous Forest Biome Bulletin No. 11, University of Washington, Seattle, 1979.

Reed, K. L., E. Hamerly, B. E. Dinger, and P. G. Jarvis: An analytical model for field studies of photosynthesis. *J. Appl. Ecol.*, 13:925–942, 1976.

Reed, W. J.: The steady state of a stochastic harvesting model. *Math. Biosci.*, 41:273–307, 1978.

Reed, W. J.: Optimum age-specific harvesting in a nonlinear population model. *Biometrics*, 36:579–593, 1980.

Riley, G. A.: Factors controlling phytoplankton populations in George's Bank. *J. Mar. Res.*, 6:54–73, 1946.

Riley, G. A.: A theoretical analysis of the zooplankton population of Georges Bank. *J. Mar. Res.*, 6(2):104–13, 1947.

Riley, G. A.: Quantitative ecology of the plankton of the western North Atlantic. *Bull. Bingham Oceanogr. Collect. Yale Univ.*, 12:1–169, 1949.

Riley, G. A.: Oceanography of Long Island Sound, 1952–1954, II: Physical oceanography. *Bull. Bingham Oceanogr. Collect. Yale Univ.*, 15:15–46, 1956.

Ripley, B. D.: Computer generation of random variables: a tutorial. *Int. Stat. Rev.*, 51:301–319, 1983.

Rose, K. A.: *A Review and Comparison of Parameter Sensitivity Methods Applicable to Large Simulation Models*, Masters Thesis, University of Washington, Seattle, 1981, p. 50.

Rose, K. A.: *Evaluation of Nutrient–Phytoplankton–Zooplankton Models and Simulation of the Ecological Effects of Toxicants Using Laboratory Microcosm Ecosystems*, PhD Dissertation, University of Washington, Seattle, 1985, p. 283.

Russell, C. S. ed.: *Ecological Modeling in a Resource Management Framework*, Resources for the Future, Inc, Washington, D.C., 1975.

Ryther, J. H.: Photosynthesis in the ocean as a function of light intensity. *Limnol. and Oceanogr.*, 1:61–70, 1956.

Ryther, J. H., and D. W. Menzel: Light adaption by marine phytoplankton. *Limnol. and Oceanogr.*, 4:492–497, 1959.

Sauer, R. H.: A simulation model for grassland primary producer phenology and biomass dynamics, in *Grassland Simulation Model*, ed. G. S. Innis, Ecological Studies, 26, Springer-Verlag, New York, 1978, pp. 55–87.

Scavia, D., and A. Robertson, eds.: *Perspectives in Lake Ecosystem Modeling*, Ann Arbor Science, Ann Arbor, Mich., 1979.

Scavia, D., B. J. Eadie, and A. Robertson: *An Ecological Model for Lake Ontario; Model Formulation, Calibration and Preliminary Evaluation,* NOAA Technical Report ERL 371-GLERL 12, Great Lakes Environmental Research Laboratory, Ann Arbor, Mich., 1976, p. 63.

Schneider, H., and G. P. Barker: *Matrices and Linear Algebra,* Holt, Rinehart and Winston, New York, 1968.

Schrage, L.: A more portable Fortran random number generator. *ACM Trans. Math. Software,* 5:132–138, 1979.

Searle, S. R.: *Matrix Algebra for the Biological Sciences,* Wiley, New York, 1966.

Sellers, W. D.: *Physical Climatology,* University of Chicago Press, Chicago, 1965.

Shields, P. C.: *Elementary Linear Algebra, 3rd ed.,* Worth Publishers, Inc., New York, 1980.

Shoemaker, C. A.: Mathematical construction of ecological models, in *Ecosystem Modeling in Theory and Practice: An Introduction with Case Studies,* ed. C. A. S. Hall and J. W. Day Jr., Wiley, New York, 1977, pp. 75–114.

Shugart, H. H.: *A Theory of Forest Dynamics: An Investigation of the Ecological Implications of Several Computer Models of Forest Succession,* Springer-Verlag, New York, 1984.

Shugart, H. H., and R. V. O'Neill, eds.: *Systems Ecology,* Benchmark Papers in Ecology, 9, Dowden, Hutchinson and Ross, Stroudsburg, Penn., 1979.

Shugart, H. H., and D. C. West: Forest succession models. *Bioscience,* 30:307–314, 1977.

Shugart, H. H., M. S. Hopkins, I. P. Burgess, and A. T. Mortlock: The development of succession model for subtropical rain forest and its application to assess the effects of timber harvest at Wianagree state forest, New South Wales. *J. Environ. Manage.,* 11:243–265, 1980.

Simons, T. J.: Analysis and simulation of spatial variations of physical and biochemical processes in Lake Ontario. *J. Great Lakes Res.,* 2:215–233, 1976.

Sjoberg, S.: Zooplankton feeding and queuing theory. *Ecol. Model.,* 10:215–225, 1980.

Snedecor, G. W., and W. G. Cochran: *Statistical Methods, 7th ed.,* Iowa State University Press, Ames, Iowa, 1980.

Sokal, R. R., and F. J. Rohlf: *Biometry, 2nd ed.,* W. H. Freeman and Co., San Francisco, 1981.

Solomon, D. S., and H. H. Shugart: Integrating forest-stand simulations with paleoecological records to examine long-term forest dynamics, in *Proceedings Europ. Sci. Fund. Workshop on Forest Dynamics, Uppsala Sweden,* Elsevier, Holland, 1983.

Starfield, A. M., P. R. Furniss, and G. L. Smuts: A model of lion population dynamics as a function of social behavior, in *Dynamics of Large Mammal Populations,* ed. C. W. Fowler and T. D. Smith, Wiley, New York, 1981, pp. 121–134.

Starfield, A. M., and A. L. Bleloch: *Building Models for Conservation and Wildlife Management,* Macmillan, New York, 1986.

Steele, J. H.: Environmental control of photosynthesis in the sea. *Limnol. and Oceanogr.,* 7:137–150, 1962.

Steele, J. H.: *The Structure of Marine Ecosystems,* Harvard University Press, Cambridge, Mass., 1974.

Steinhorst, R. K.: Parameter identifiability, validation and sensitivity analysis of large system models, in *Systems Analysis of Ecosystems,* ed. G. S. Innis and R. V. O'Neill, Statistical Ecology, 9, International Co-operative Publishing House, Fairland, Maryland, 1979, pp. 33–58.

Steinhorst, R. K., H. W. Hunt, G. S. Innis, and K. P. Haydock: Sensitivity analyses of the ELM model, in *Grassland Simulation Model,* ed. G. S. Innis, Ecological Studies, 26, Springer-Verlag, New York, 1978, pp. 231–255.

Stewart, D. J., D. Weininger, D. V. Rottiers, and T. A. Edsall: An energetics model for lake trout. *Can. J. Fish. Aquat. Sci.,* 40:681–698, 1983.

Straskraba, M., and A. H. Gnauck: *Freshwater Ecosystems Modelling and Simulation,* Developments in Environmental Modelling, 8, Elsevier, New York, 1985.

Stross, R. G., P. A. Nobbs, and S. W. Chisholm: SUNDAY, a simulation model of an arctic *Daphnia* population. *Oikos,* 32:349–362, 1979.

Sullivan, P., G. Swartzman, and A. Bindman: *Process Notebook for Aquatic Ecosystem Simulations, 2nd ed.,* U. S. Nuclear Regulatory Commission Report NUREG/CR-3392, 1983, p. 182.

Swartzman, G. L.: *A Preliminary Bird Population Dynamics and Biomass Model,* U. S. IBP Grassland Biome Technical Report No. 3, Colorado State University, Fort Collins, Colo., 1969.

Swartzman, G. L.: Simulation modeling of material and energy flow through an ecosystem: Methods and documentation. *Ecol. Model.,* 7:55–81, 1979a.

Swartzman, G. L.: A comparison of plankton simulation models emphasizing their applicability to impact assessment. *J. Environ. Manage.,* 9:145–163, 1979b.

Swartzman, G. L.: *Factors Bearing on the Present Status and Future of the Eastern Bering Sea Fur Seal Population with Special Emphasis on the Effect of the Subadult Male Harvest,* National Technical Information Service PB84-172329, 1984, p. 77.

Swartzman, G. L., and R. Bentley: A review and comparison of plankton simulation models. *ISEM Journal,* 1:30–78, 1979.

Swartzman, G. L., and K. A. Rose: Simulating the biological effects of toxicants in aquatic microcosm systems. *Ecol. Model.,* 22:123–134, 1984.

Swartzman, G. L., and J. S. Singh: A dynamic programming approach to optimal grazing strategies using a succession model for a tropical grassland. *J. Appl. Ecol.,* 11:537–548, 1974.

Swartzman, G. L., and G. M. VanDyne: An ecologically based simulation–optimization approach to natural resource planning. *Annu. Rev. Ecol. Syst.,* 3:347–395, 1972.

Swartzman, G. L., and T. M. Zaret: Modeling fish species introduction and prey extermination: The invasion of *Cichla ocellaris* to Gatun Lake, Panama, in *Analysis of Ecological Systems: State-of-the-Art in Ecological Modelling,* ed. W. K. Lauenroth, G. V. Skogerboe and M. Flug, Developments in Environmental Modelling, 5, Elsevier, New York, 1983, pp. 361–371.

Taub, F. B., and M. E. Crow: Synthesizing aquatic microcosms, in *Microcosms in Ecological Research,* ed. J. P. Geisy, vol. 52, U.S. Dept. Energy, 1980, pp. 69–104.

Teal, J. M.: Energy flow in the salt marsh ecosystem of Georgia. *Ecology,* 43:614–624, 1962.

Tiwari, J. L., and J. E. Hobbie: Random differential equations as models of ecosystems II initial condition and parameter specifications in terms of maximum entropy distributions. *Math. Biosci.,* 31:37–53, 1976.

Tomovic, R.: *Sensitivity Analysis of Dynamic Systems,* McGraw-Hill, New York, 1963.

Ursin, E.: A mathematical model of some aspects of fish growth. *J. Fish. Res. Board Can.,* 24:2355–2453, 1967.

Usher, M. B.: Extensions to models, used in renewable resource management, which incorporate an arbitrary structure. *J. Environ. Manage.,* 4:123–140, 1976.

Vandergraft, J. S.: *Introduction to Numerical Computations,* Academic Press, New York, 1978.

Vaughan, B.: Problems in evaluating radiation dose via terrestrial and aquatic pathways. *Environ. Health Perspect.,* 42:149–161, 1981.

Weisberg, S.: *Applied Linear Regression, 2nd ed.,* Wiley, New York, 1985.

Werner, E. E., and D. J. Hall: Optimal foraging and the size selection of prey by the bluegill sunfish (*Lepomis macrochirus*). *Ecology,* 55:1042–1052, 1974.

West, D. C., S. B. McLaughlin, and H. H. Shugart: Simulated forest response to chronic air pollution stress. *J. Environ. Quality,* 9:43–49, 1980.

Wichmann, B. A., and I. D. Hill: An efficient and portable pseudo-random number generator. *Applied Stat.,* 33:188–190, 1982.

Wiegert, R. G.: Simulation models of ecosystems. *Annu. Rev. Ecol. Syst.,* 6:311–338, 1975.

Winberg, G. G.: *Intensivnost Obmena i Pishchevye Potrebnosti Ryb,* Minsk, 1956. (Rate of Metabolism and Food Requirements of Fishes, Fish. Res. Board Can. Transl. Ser. no 194., 1960)

Yezzi, D. J. Jr., and A. P. Uzzo, Jr.: Dynamic model of nutrification in Huntington Bay, New York. *Ecol. Model.,* 6:59–75, 1979.

Author Index

362

Subject Index